The State of the Earth

Contemporary Social Sciences

Series Editor: *Ali Kazancigil*

Volume I *Sociology* Neil J. Smelser
Volume II *The State of the Earth* Akin Mabogunje

The State of the Earth

Contemporary Geographic Perspectives

Akin Mabogunje

First published 1997

2 4 6 8 10 9 7 5 3 1

Blackwell Publishers Ltd
108 Cowley Road
Oxford OX4 1JF
UK

Blackwell Publishers Inc.
350 Main Street
Malden MA 02148
USA

British Library Cataloguing in Publication Data

A CIP catalogue record for this book is available from the British Library.

Library of Congress Cataloging-in-Publication Data

The state of the earth : contemporary geographic perspectives /
[edited by] Akin Mabogunje.
p. cm. — (Contemporary social sciences ; v. 2)
Includes bibliographical references and index.
ISBN 0-631-20243-9. — ISBN 0-631-20244-7 (pbk)
1. Geography. I. Mabogunje, Akin L. II. Series.
G116.S73 1997
910—dc20 96-36188
CIP

Typeset by Grahame & Grahame Editorial, Brighton

Printed in Great Britain by T. J. International, Padstow, Cornwall

This book is printed on acid-free paper

Contents

Figures

Tables

Foreword

This volume is the second in the UNESCO and Blackwell Series, Contemporary Social Sciences. The series offers international surveys of the social science disciplines, each accurately representing the current state of the field and written for university and college professors, researchers, and graduate students. Subsequent volumes will cover other disciplines, such as anthropology, psychology, political science, economics, demography, linguistics, history and international relations.

Volumes within Contemporary Social Sciences will be *authoritative*, involving world-class scholars as principal authors, and *comprehensive*, representing the widest range of perspectives in each discipline. Moreover, the emphasis is *international* in two senses: first, information and analyses are provided by contributing specialists throughout the world; and second, each volume reflects the ongoing transformations of the social sciences' basic units of analysis – including the concepts of the nation-state and society – that are brought about by accelerating forces of internationalization and globalization.

These international accounts of the disciplines are also intended as a contribution to the international circulation of social-scientific information. The internationalization of the social sciences has made notable advances, particularly since the end of the Second World War. At the same time, communication between social science communities across countries and regions – an essential ingredient of scientific progress – is still hampered by parochialism. The wide diversity of epistemological, theoretical and methodological perspectives of the social sciences themselves, differing cultural and institutional traditions among nations and cultures, and the realities of contemporary cultural and political competition among nations and regions, may impede free, unbiased scientific communication.

UNESCO has a long-standing commitment to and sustained experience in international scientific communication. It is the designated UN agency for promoting social science teaching and research, as well as fostering international cooperation within and among its disciplines. More particularly, the Organization has a comparative advantage in developing international state-of-the-art surveys, because of its global and regional perspectives and its involvement with the professional associations and networks it has helped to create and continues to support.

Some of UNESCO's past activities have been dedicated to purposes similar to those of this series. For example, between 1951 and 1968, the Organization produced a series of volumes surveying some 15 disciplinary and subject fields, under the title of The University Teaching of the Social Sciences. UNESCO's professional periodical, the *International Social Science Journal*, now in its 47th year, is another such activity. Each issue of the *ISSJ*, published in six language editions (English, French, Spanish, Arabic, Chinese, and Russian) and distributed worldwide, is an international state-of-the-art assessment of a social science discipline or subject.

The most ambitious UNESCO survey, however, was the *Main Trends of Research in the Social and Human Sciences*, carried out between 1965 and 1978. The project produced three ambitious volumes, written by outstanding scholars, for example Jean Piaget, Roman Jakobson, Paul Lazarsfeld, and Stein Rokkan. Several hundred specialists, national and international bodies, councils, foundations, universities, research centres, and professional associations from all parts of the world participated in the projects. The volumes themselves proved a landmark and the subsequent paperback editions of the individual chapters became large sellers in several languages.

Drawing on the collaborative approach of the *Main Trends* project, we designed a new method, combining speed and cost-efficiency in production with breadth of coverage, reliability, and quality of coverage for the current series. The major challenge was to combine two ingredients: the coherence of logic and style provided by a single author, and the wealth of information and breadth of coverage offered by an international team of specialists. We addressed that challenge by selecting a reputed scholar to act as designer and author of each volume, supported by contributing specialists selected from different countries and parts of the discipline. The author of each volume will use contributors' texts as resources from which to fashion and write a book. The contributions of the specialist scholars will be appropriately acknowledged in the volume, and their papers published in the *International Social Science Journal*.

The present volume conforms to this model. All those who have been involved in the enterprise consider it a success. With few exceptions, the contributors invited to participate understood and accepted their assignments, lived by the rules of the game entailed by those assignments, and delivered manuscripts of high quality. We thank them for their share in the enterprise.

The key person in the successful completion of this volume is Professor Akin L. Mabogunje, as its intellectual leader and author. An internationally respected geographer, he has an excellent record of scholarship, many publications and an extensive international experience. He is open to all significant trends, theories, and schools of thought in the discipline. On each of these counts he met our expectations. We are grateful to Professor Mabogunje for his outstanding collaboration.

Ali Kazancigil
Editor of the Series

Preface

In March 1994, I received an invitation from Mr Ali Kazancigil, the Director of the UNESCO Division for the International Development of Social and Human Sciences. It was to the effect that I should serve as author/editor for a special volume on geography which UNESCO was sponsoring in collaboration with Blackwell Publishers. The volume was to be the second in the series on contemporary social sciences, coming after that on sociology which was published in 1994. The purpose of the book is to provide readers with an international perspective on the current state of research in geography, as a discipline in the social sciences. It is also to contribute to the circulation of ideas and information among social scientists and promote cross-fertilization among the research and teaching communities generally.

I must confess that at that point in time I had no clear idea of what the assignment entailed. The invitation, however, requested that I stop-over at the UNESCO Headquarters in Paris any time I was in Europe to plan the volume. This visit did not take place until September 1994. Discussions at the planning meeting quickly left me in no doubt that this was an unusual and more challenging type of assignment than any I had undertaken hitherto. It required that I should author/edit a volume based in part on contributions by an international panel of distinguished geographers. The idea was to combine the coherence of logic and style provided by a single author with the breadth of coverage offered by an international team of highly competent contributors. I also had the responsibility of both planning the volume and identifying members of this international team. More than this, my brief stressed that the volume should not be just a series of essays representing distinctive intellectual or institutional schools of thought or national orientation in the discipline, nor should it be a handbook or textbook covering the various subfields of the discipline. Rather, it should try to focus on specific topics and issues that constitute areas of current geographical interest and address some of the fundamental challenges facing the human race at the end of the twentieth century. In short, the assignment required not only that I know what developments have been taking place in the geographical discipline in recent times, but also the individuals who are best placed to capture and discuss these developments effectively.

It was something of a tall order. Fortunately, I was guest of the Royal Scottish Geographical Society soon after my Paris meeting. This gave me the opportunity to discuss the project at length with Emeritus Professor Terry Coppock, my former teacher, friend and mentor from our University College London days. Our discussions and conclusions concerning the content of the volume and a first list of potential contributors proved immensely valuable as to leave me with an overwhelming sense of gratitude for the time and interest he showed in the project.

However, since the assignment required that the list of contributors should show a distinctive international spread, I had to consult other colleagues and friends in the United States, Brazil, Hong Kong and, of course, Nigeria, not just with regard to possible contributors but also with regard to the topics and issues I had identified as constituting areas of current geographical research interest. Among those consulted were Emeritus Professor William Robert Kates, who was then President of the Association of American Geographers, Professora Bertha Becker of the Federal University of Rio de Janeiro in Brazil, and Professor Yue-man Yeung of the Chinese University of Hong Kong. I want to thank all of them for responding so willingly and promptly to my enquiries. I found their advice and suggestions remarkably fruitful with regard to ideas concerning what the volume should seek to emphasize and who the additional contributors should be from their particular region of the world.

The task of contacting the contributors was undertaken by UNESCO, specifically by Ms Nadia Auriat, Programme Specialist within the Division of Social Science, Research and Policy who worked directly under Mr Ali Kazancigil. It has not been the easiest of tasks, especially where identified contributors could not, for various reasons, accept the invitation. The constant fax transmissions to invite other scholars must have been something of a hassle. The strain of having to wait to hear whether they accepted the invitation or not and, even when they had accepted, whether they would honour their commitment, must have been enormous. I owe Ms Auriat a debt of gratitude since throughout all of this she maintained a stoical and most encouraging posture. I must not forget Ms Maria Gutiérrez, her senior secretary, who also had to take on some of this responsibility in the final stages of producing the volume.

In the event, some fourteen colleagues from seven countries in four continents (United States, Russia, United Kingdom, France, Brazil, Argentina and Hong Kong) contributed to the volume. I believe the range of topics and issues covered by the volume has an internal coherence to impress anyone of the distinctive strengths and contributions of geography to the understanding and resolution of some of the fundamental challenges of our age.

What has been most gratifying to me in all this has been the expeditiousness with which all the contributors have responded to their assumed obligation. I wish to acknowledge with sincere gratitude the collaboration and cooperation I received from all of them in the production of this unique volume. Indeed, in the case of one of them it had been at some cost to the present state of his health. The support has been truly heart-warming. Of course, according to the terms of the invitation, all contributions will be published in full and under the names of their authors in subsequent issues of the *International Social Science Journal*, specifically nos 150 and 151 of December 1996 and March 1997, respectively. In the meantime, the agreement gave me full liberty to acknowledge, use and adapt their contributions as I think fit so as to give some coherence of logic and style to the present volume. Nonetheless, I hope that in spite of the adaptations and transformations to their contributions, the authors can still find a lot in the text which they can identify as theirs.

I wish to thank Professor Peter Gould of Pennsylvania State University, who on his own volition offered to help with providing me with a second editorial opinion on the various chapters in the volume. In spite of the difficulties of communication between Nigeria and the United States, the interaction has gone a long way towards improving the overall quality of the volume. I realize also that I had to impose the tight time schedule of producing the final version of the manuscript on him in spite of his own other commitments. For his forbearance and patience with me on this score, I will always be in his debt.

My Nigerian colleagues both at the Department of Geography, University of Ibadan and at the Development Policy Centre, also at Ibadan, have simply been wonderful. Professors Christopher Ikporukpo and Stanley Okafor gave me freely of their time in reading over and correcting the manuscript, and raising questions concerning some of the issues raised in the chapters. Professor Bola Ayeni was particularly forthcoming in respect of the chapter on geographical methodology, while Professors Sylvester Abumere and Abosede Iyun, Vice-President of the International Geographical Union, also looked over other sections of the volume and gave me the benefit of their opinion. Professor Isaac Adalemo, himself a geographer, but currently the managing director of the Development Policy Centre, put the facilities of the centre at my disposal and, in particular, allowed me free access to the services of Mr 'Lekan Bello, the computer specialist at the centre. The latter was particularly generous with his time in helping me produce the final draft of the volume.

Finally, my sincere and loving gratitude goes to members of my family, particularly my wife, Titilola. Their forbearance of my neglect over the last few months of getting the final version of the manuscript ready has been very touching. I want to thank them all, including my grandchildren, for their patience, support and understanding of the vagaries of life for a 'retired' academic.

Akin L. Mabogunje
Ibadan, Nigeria

Introduction

Geography has been defined variously as the study of the earth as the home of human beings, as the study of human–environment interaction, or as the study of spatial processes and regional development. There are, in fact, many other definitions of the field. What they all have in common is interest in the curiosity human beings have always shown in and about the nature and characteristics of the planet they are occupying and whose resources they are using and transforming with consequences they do not always fully grasp. Such curiosity, in terms of organized knowledge, goes back to ancient times when it was then no more than creative speculation. It has continued right up to the present as information-gathering and analytical techniques have advanced with the increased technological sophistication of human societies. This has made it possible to start to better understand the intricate relationships between the enhanced human capacity to exploit the environment. It is also helping to ensure that such relationships are kept more positive than negative and can be sustained over the long run.

This volume is thus an attempt to provide readers with a general picture of the present state of the geographical discipline. In doing this, the book has derived considerable advantage from the format of an earlier volume on sociology. As a result, it does not attempt to treat the volume just as a series of essays representing distinctive intellectual or institutional schools of thought or national orientation in the discipline, nor does it attempt to provide a handbook or textbook covering the various subfields of the discipline. Moreover, it does not seek to engage in some of the theoretical and methodological controversies that are indicative of the healthy growth of the discipline. Rather, the thrust of the volume is to focus on specific topics and issues that constitute areas of current geographical research interest and address some of the fundamental challenges facing the human race at the end of the twentieth century.

This approach has proved extremely valuable in many ways. In the first place, it has made it unnecessary to reflect the heuristic division of the discipline into human and physical geography and made it possible to concentrate instead on the interrelations between the two subfields. In the second place, notwithstanding the social sciences orientation of the volume, the present format has facilitated the discussion of the efficacy of social processes in the context of current and challenging issues of environmental sustainability. Finally, the approach has made it possible to give due weight to the acceleration in the rate of geographic change following on the significant and portentous technological and organizational developments that have taken place particularly since the end of the Second World War.

The result is a volume divided into four broad sections. Part I, comprising three chapters, considers the fundamentals of Geography. It begins in chapter 1 by focusing on how geography has gone about organizing knowledge about the world and its fascinating but ever-changing perspectives on the problems of understanding the patterns resulting from the human occupance of the earth's surface. It emphasizes that, from the very beginning, geography found itself confronted by the artificiality of partitioning human knowledge and inquiry among discrete scientific disciplines. This problem was less compelling as long as knowledge about the world was considered purely within natural philosophy. But as human knowledge progressed, especially from the seventeenth century onwards, and various sciences were hiving off from natural philosophy, the problem of such partitioning became increasingly serious. By the time geography emerged as a university discipline in the nineteenth century it resulted in the subject recognizing a physical and a human world, both of which interact on each other in a complex and intricate manner.

Understanding the nature of these interactions has been the challenge of geography over the years. Gould, in his chapter 1 contribution to the book, suggests that the metaphor that best captures what geography has been trying to do with regard to this artificial partitioning of knowledge is that of 'the braided stream', with its waters constantly flowing from one channel to another. The aptness of this metaphor is underscored by the fact that many of the most important and difficult problems of today demand the skills and perspectives of a number of disciplines connected by such

Cars used as a barrier against flood erosion.
Source: F. Le Diascorn/Rapho.

'cross-braiding' of individual channels of thought. Against this background, geography can be seen to have been influenced by and is influencing other areas of scientific knowledge and inquiry. Such influence is particularly noticeable in the changed perspective from the initial dalliance with environmental determinism in the late nineteenth century to the present-day, postmodernist reaffirmation of real world complexity which argues that the geography of any area cannot be totally understood through some grand theoretical constructs. This is because such grand theories tend to ignore analysing the human conditions of groups presently marginalized such as women, the disabled, the aged and the generally vulnerable members of society.

Chapter 2 treats of geographical theories. It points out the relative recency of theoretical preoccupations in the discipline and the initial concern with location theories. In the initial phase, most location theories borrowed heavily from spatial economics, especially from the works of German spatial economists such as von Thunen for agricultural land uses and Alfred Weber for industrial locations. The subject then began to adapt the process of theorizing to other areas of geographical interest such as settlement patterns, spatial diffusion, spatial interaction, spatial spread of diseases, and to urban and regional development in general. The development made it possible to group geographical theories into four categories, namely those concerned with location, those examining spatial interaction, those investigating problems of growth and development, and those exploring issues of decision-making and choice. This grouping, while not claiming to be exhaustive, has been extremely valuable in capturing a significant part of theoretical work going on in the discipline. Some of the theories, especially those of location and spatial interaction, lend themselves to modelling and quantitative representation using mathematical equations and inferential and multivariate statistics. Thus, the initial period of theoretical development in the subject had the impact of being something of a 'quantitative revolution'.

Chapter 3 reviews the methodology of the subject, especially in terms of the tremendous development that has taken place in this area in an age of supercomputers and electronic remote-sensing. Maps and various cartographical techniques provided the initial tools for the analysis of the varied characteristics of the Earth's surface and for the location of various features on it. Although statistical information was often collected and used to illustrate the spatial differentiation of the Earth's surface, very little analysis was done to the data themselves. They were simply represented on maps using a variety of visually effective techniques. These were expected to show the viewers important aspects of regional distribution of a particular variable of interest. The quantitative and theoretical revolution changed all that. It encouraged geography to be more circumspect about the reliability of the data it collects or seeks to analyse, and what the data itself says about the phenomenon that is being investigated. The arrival of the computer brought about a tremendous change in geography's capabilities to deal with a large volume of data. Over time, these data have given the discipline enhanced capacity to integrate into its analysis features of geographical locational specificity. All of this heralded the advent of the Information Revolution, the growing international importance of remote sensing, the versatility of the geographical information system (GIS), and the remarkable possibilities of geographical visualization and multimedia representation for adding a dynamic dimension to data analysis and presentation.

Part II, also in three chapters, considers issues relating to the mutual relationship between the Earth's environment and human society. Chapter 4 reviews the tremendous transformation that human beings have wrought on the Earth's surface. It emphasizes the role that technological creativity and social organization have played in equipping human beings with the capacity to completely change the face of the Earth. In the long history of human habitation of the Earth's surface, such a capacity has developed only in very recent times, especially since the invention of agriculture. This has enabled human society to produce adequate amounts of food to sustain its occupation of specific areas of the Earth's surface over a sufficiently long time as to affect it for good or ill. Over such a period, the vegetation cover has been removed to make way for domesticated plants and animals, the soil cover itself has periodically been worked over. In areas where such soil cover has a rather fragile structure, erosion destroyed extensive areas and made them no longer productive, leading in some cases to the end of human occupancy of such areas. Swamps have been drained and dry lands have been irrigated and turned into luscious farmlands. Trade has developed between distant lands and brought about the rise of cities. But it is the emergence of power-driven manufacturing in the cities of the world since the mid-eighteenth century that has proven decisive in terms of the human impact on environmental resources. This has magnified and intensified human capacity to pollute the environment and to impact negatively on the global ecosystems. The situation has brought human society dangerously close to damaging its 'nest' irreversibly and affecting some elements of the environment with consequences that are still far from being fully appreciated.

One measure of such dramatic impact could be through significant fluctuations in global climatic phenomena. This issue is considered in chapter 5. Evidence is provided to show that climate change is a constitutive part of environmental history. But the character of the change in recent times may not be unconnected with the manner in which human beings have been interfering with the ecosystems in different parts of the world since the nineteenth century. Increasingly, it is being appreciated that we do not know enough about changes in the manifestation of the forces that drive the great air masses round the earth and that influence the insulation the earth receives from the sun. The implications of the fluctuations in the incidence of these forces are now of considerable interest for the global sustainability of human development. Considerable controversies have thus been generated on the whole issue of climate change and the future of the human environment. The available evidence is evaluated in this chapter and provides some indication of the present state of knowledge as to what the present fluxes in global climate portend.

Chapter 6 considers present knowledge about societal responses to environmental hazards and natural disasters. Periods of excessively cold weather, of floods, hurricanes and typhoons, of droughts, famine, and the encroaching desert, as well as landslides, earthquakes, and volcanic eruptions, all indicate the fragility of human occupancy in many parts of the world. These extreme deviations from natural norms can be a catastrophe. Settlements and valuable economic assets can be wiped out in a matter of hours and numerous lives can also be lost. In recent times there has also been the hazard resulting from human pollution of the environment. The over-exploitation of natural resources and the emission of relatively large quantities of toxic and radioactive substances into the environment have become causes of serious concern for human health and well-being. Yet, societal responses to these hazards do not always appear enlightened or far-sighted. This apparent contrariness of human societies can be best understood in terms of social perception and the extent to which technical and organizational capabilities tend to blunt societal sensitivity to the hazardousness of a particular environment.

Part III, which comprises four chapters, looks more specifically at societal processes and their implications for spatial structures. Chapter 7 examines the overriding role of economic and cultural factors in determining the manner in which different human communities occupy and use space. It emphasizes how market forces in particular influence locational decisions, but argues that the situation is a lot more complex than simple concern with profitability or cost minimization. Human locational behaviour springs from many sources, not least of which are cultural factors. Indeed, it is possible to conceive of the market economy itself – in the sense of the capitalist free-market economy – as culture, and to note that not all societies are as yet acculturated to this way of life. The resulting diversity of cultural forms and the implication for locational processes provide a fascinating richness to studies in cultural geography.

Chapter 8 on modes of production and spatial differentiation calls attention to the fact that since the late eighteenth century the capitalist mode of production has sought to integrate older modes of production in different parts of the world into one global system with varying degrees of success or effectiveness. The result is that for most of the world today the spatial differentiation is to a large extent one of degree rather than of kind. A basic notion is thus to see the whole world as operating a single capitalist mode of production, differentiated as between the core area comprising largely western Europe, North America and Japan and an extensive, highly variegated periphery comprising the rest of the world. Changes continue to take place in both areas as a result of the continuing struggle between capital and organized labour. As a result, favourable conditions are created for some countries in the periphery to emerge to challenge the dominance of others in the core. Beyond this trend, however, there is also the growing emphasis in global geography not on regional differentiation or even differences among nation-states but on network relations, especially among cities. This late twentieth-century development stands out as the axes of change to watch as we move into the future.

Chapter 9, on transportation, trade, tourism and the world-economy system, takes the argument further. These various processes define the spatial structures that help to emphasize the growing importance of network relations in the world-economy system. They have been instrumental in expanding the spatial span of the capitalist mode of production throughout the world and promoting current globalization trends. They have also fostered the increased mobility that now characterizes virtually all factors of production except land. Increasing technological innovations had made it possible to free labour and to reduce its importance in the production process. They have also served to reduce the raw material components in products. The result is the rapidly rising relative importance of the tertiary sector of transportation, trade, and services. In this regard, technological innovations in maritime transportation have been of singular significance. Tourism, or the leisure-seeking mobility of people, has blossomed in the aftermath of rising personal incomes and continuing technological innovation in air transportation. All of these developments go to reinforce those forces operating to make the whole world one huge, global village.

At the centre of this whole process is of course the extreme volatility of financial flows on a global scale. Chapter 10 considers spatial financial flows and future global geography. It notes the recency of geography's interest in the world of finance, of monetary exchanges, credit and debt. It attempts to explore the relationship between monetary forms and territorial structures with particular emphasis on the relationship between urban centres, states and the financial systems. It also recognizes the fact that money and credit have to be produced and distributed and the special position of the three major world cities of New York, Tokyo and London in this process. The increasing innovativeness in the development of different forms of virtual money is affecting the importance of national borders in the regulations of all kinds of financial institutions. This development is taking place hand-in-hand with a contradictory process of 'financial exclusion' arising from the debt crisis of many developing countries and some developed ones. This latter process is encouraging the rise of 'alternative' financial institutions which mount a challenge to conventional financial systems and are more likely to echo social and cultural concerns in future global geography.

Part IV looks at spatial organization and the increasing process of organizing the world into one, huge global entity. This process can be said to date back to the rise of the modern nation state in the late fifteenth century. These states emerged in Europe from the fragmented principalities of the medieval period, consolidating and modernizing their power using information acquired through exploration and mapping with considerable intellectual and political dexterity. The rise of the absolutist nation states thus succeeded in homogenizing large areas of geographical space. Their rivalries and competition for political and economic ascendancy were projected to other parts of the world. By the nineteenth century, most of these had become colonial or former colonial territories of these absolutist European states. Constitutional and political changes within the nation states heralded the age of representative government and foreshadowed the political independence of colonial territories in the nineteenth and twentieth centuries. The result is a world more predisposed to the next round of major spatial reorganization.

Of considerable importance for this next round of spatial organization is the role of transnational corporations. These are giant enterprises with assets/sales larger than the gross national product of many nation states. Their activities are multi-locational and are spread over many countries. Their considerable size allows them to switch and re-switch their operations from one country to another in response to changing conditions and, therefore, to constrain the power of national governments to implement their own economic policies. Their access to satellite electronics also enables them to propagate particular life-styles right across large regions of the world. Clearly, transnational corporations represent a new force in the equation of spatial organization with which nation states have to come to terms. Indeed, there have been arguments as to whether their emergence sounds the death-knell of nation states. Whatever the validity of such arguments what is clear is that their emergence has provoked nation states to seek greater cooperation amongst themselves and to conceive of new forms of spatial organization in the form of supranational regional blocs of states. The European Union represents the most developed form of this new type of spatial organization. But others are being established in different parts of the world. Their growth and development over the next few decades are bound to give a very dramatic turn to the geography of the world in the twenty-first century.

It is, however, not only the transnational corporations that appear to have the capacity to determine the structure of that geography. All over the world, there are emerging mega-cities or cities with a population of over eight million, representing a significant concentration of productive and consumption forces on both national and international scales. At present, most of these mega-cities are found in developing countries and their levels of development are still far from reflecting their potential for growth and societal transformation. Those of them in the developed countries and a few in the newly industrializing countries have already become world or global cities in the sense that, although located within specific nation states, their interactions are increasingly global in reach and above the constraints of national borders. These interactions also tend to be more intense with other global or world cities. The implications of these developments for geography in the twenty-first century are clearly open-ended but they represent global changes the full import of which will take time to unravel.

In all, this volume on the state of the art in geography depicts a subject in considerable ferment. The complexity of the reality it is attempting to understand is being matched by the growing sophistication of its own philosophical, theoretical and analytical capabilities. The major transformation of the subject in the period of the quantitative and theoretical revolution has ensured that henceforth its disciplinary interests will be closely related to the mainstream concerns of the social sciences. This has meant that its particular spatial perspective has come to be better appreciated, its techniques and methodologies, especially the geographical information systems, paid increasing attention, and its insights into social problems and issues given due recognition. The interaction has not, however, been one way. Geography too has been placed in a position that it is infinitely more sensitive to the perspectives, methodologies and insights of other social science disciplines than it used to be. The expectation is thus of an even more exciting future in which the imagery of the 'braided stream' will also imply a well-watered field giving a bountiful and varied harvest of ideas and conceptions for improving the human condition.

Part I

Fundamentals

1 Space, Time and the Human Being

In spite of centuries of human curiosity about the earth's surface and the numerous and different characteristics of its inhabitants, a moment in the history of the development of geography came when, in the second half of the nineteenth century, it was being actively canvassed as a university discipline. The great age of explorations and expeditions had been in full flood and so much knowledge had been accumulated about different parts of the world. Professional associations for the promotion of geographical knowledge had been established in France, Germany, Britain and Russia in the first half of the century. The stage was set for a deepening of the social awareness of space and place, of change over time and of the human ability to reflect carefully upon the natural and the human worlds. It was becoming very patent that there was considerable difference between the practical and enterprising endeavours of exploration and expedition and the intellectual justification of geography within the educational curricula of the emerging new generation of universities.

As a formal field of inquiry in the education of the young, geography, like history and the social sciences, is thus a relatively recent discipline. Many of the questions it poses used to be framed and answered under the rather spacious intellectual umbrellas of philosophy and theology. But in the same way that individual physical sciences hived off from natural philosophy at the end of the seventeenth century, with the biological sciences following in the eighteenth, so the human sciences, including geography and history, began to define themselves as distinct areas of inquiry in the nineteenth century. It is then that we find separate departments forming in universities (Taylor, 1985), inevitably in some tension

This chapter is based on a contribution by Peter Gould.

with established fields like philosophy, theology, medicine and jurisprudence.

The names of two Germans – Alexander von Humboldt and Carl Ritter – were closely associated with this development. Both personalities, however, died in 1859 – well before the first university department was established in 1874. Other departments followed rapidly in France, the United Kingdom, the United States and elsewhere in the world. All were active in defining the intellectual claims of geography and the special perspective it provides to human knowledge.

The Human-Environment Perspective and the Regional Concept

Perhaps the earliest perspective Geography sought to provide to human knowledge was the close relation between human beings and the environmental characteristics of the part of the world where they live. Halford Mackinder (1887), one of the earliest pioneering British geographers, defined geography as the science that sought to examine the influence of the environment on human societies. This influence was conceived as operating in one direction only, with the environment not only setting the scene but also determining the direction and scope of human progress. The publication in 1882 of the volume, *Anthropogeographie oder Grundzuge der Anwendung der Erdkunde auf die Geschichte*, by the German geographer, Friedrich Ratzel, set the tone and direction for the development of the subject at this time. In the English-speaking world, the publications of two other geographers were to dominate this period in the intellectual growth of the discipline. The first set was by Ellen Semple. Her most classic text, *Influences of Geographic Environment* (1911), not only sought to build on the work of Ratzel but also drew somewhat extreme conclusions about the influence of the environment on the development of human societies on different parts of the earth's surface. An example of the underlying tenet of this perspective states as follows:

> *Geographical environment, through the persistence of its influence, acquires particular significance. Its effect is not restricted to a given historical event or epoch, but, except when temporarily met by some counteracting force, tends to make itself felt under varying guises in*

The Geographer: painting by David Teniers the younger. *Source:* Frankfurt Museum/Roger-Viollet.

its succeeding history. It is the permanent element in the shifting fate of the races.
(Semple, 1911:6)

In much the same way, Ellsworth Huntington investigated the influence of climate on the development of human societies. In the volume *Civilization and Climate* (1915) he tried to show that the civilization of the world varies almost precisely as we should expect if human energy were one of the essential conditions, and if energy were in large measure dependent upon climate (1915:314).

In spite of the somewhat simplistic generalizations of these publications and much of the work of geographers in this early phase in the intellectual development of the subject, it would be unfair not to acknowledge that they did recognize that the relationship between environment and human society was a lot more complex and required considerable and continued investigation. Besides, other geographers, even at this early phase, were already very critical of the extreme determinism in much of contemporary geography and sought to counter with an alternative perspective, namely that of possibilism. This other perspective was strenuously canvassed by the French school of geography, dominated at the time by the work of Paul Vidal de la Blache. This school recognized not only the complex interrelationship between the environment and human societies but also the fact that in this relationship human beings were not passive pawns but really active agents evaluating and exploiting the various possibilities presented by the environment.

Thus, from the very beginning of its academic career, geography demanded that we do not lose sight of the connections between the physical and the human world. Indeed, scholars in other traditions, when viewing contemporary geography from the outside, are often puzzled that physical geography and human geography can exist side by side. Their often unspoken assumption is that a science dealing strictly with the physical-biological world of things, where strong regularities may well be labelled laws leading to predictions, cannot be compatible with a human world where regularities, if they can be found at all, never lead to predictions which cannot be obviated by those about whom the predictions are made. So clear has such a separation seemed at times that even geography departments have divided into two curricula, or have sloughed off the physical component to adjacent earth sciences like geology and meteorology. Nonetheless, such a drastic separation appears today increasingly unrealistic.

For one thing there are the innumerable events reinforcing the already heightened awareness that the human and natural worlds are intimately interconnected. That the latter can affect the former is obvious every time a tectonic plate slips, a hurricane lashes a coast, rains and spring meltwaters overflow riverbanks, or warm water slops from the west Pacific to cover the cold upwelling waters off Peru – the well-known El Niño effect. Increasing evidence indicates that the Sahelian drought is in part caused by global teleconnections originating in the hot box of the Pacific. But the causal arrow is not just one way, and may involve very complex chains of events. Physical effects on the human world may be exacerbated by human intervention, not the least of which are those interventions displaying a certain amount of *hubris*. Huge levees are built to contain all but the 200-year flood, but the Mississippi River takes no notice of such computed probabilities, and inundates vast built-up areas constructed on its flood plain. The Volta dam reduces sediment flow so much that the sea now eats into some of Ghana's most fertile areas of food production. The Aswan Dam of Egypt halts fertility-renewing floods, wipes out the protein-rich fisheries of the eastern Mediterranean, and uses much of the electric energy to produce artificial fertilizers. Moreover, the sheer pressure of the human presence impacts with ever greater causal strength upon the physical world, whether it be local groundwater pollution, the cutting and burning of tropical rainforests with their extraordinarily delicate ecosystems, or the increase in gaseous emissions leading to global warming and ozone holes. The human being and the natural world appear increasingly inseparable, and therefore should be studied together.

Whether belonging to the determinist or possibilist school, the curriculum of geography at its origin also emphasized the importance of regional variation of the earth's surface and the critical associational or correlational significance in the juxtaposition of physical and human factors within each region. The regional method thus came to dominate the approach to the study of the earth's surface. There were, however, many different varieties of regional geography even at this time. The more popular, such as that stressed in the work of A. J. Herbertson, was concerned with identifying what were termed natural regions. These were defined as definite areas of the surface of the earth considered as a whole, not the configuration alone but the complex of land, water, air, plant, animal and man, regarded in their special relationship as together constituting a definite characteristic portion of the earth's surface (Herbertson, 1905:301). A natural region, according to this earliest perspective, has four dimensions – configuration (relief, altitude and structure), climate, vegetation and population density. The last was a feature dependent on the other three and to that extent was the least important. In terms of the world as a whole, it identified six major or systematic types of natural regions – polar, cool temperate, warm temperate, tropical, subtropical and equatorial. Each of these types was usually described in terms of the four dimensions, a method that spilled over even when the unit of interest became the nation state.

Other types of regional geography were provoked as geographers attempted to grapple with the changing characteristics of areas within the same broad geographical zones. This often led to the appreciation of the fact that the environment, at best, only sets limits to human endeavours within a given region and that the cultural factor in those endeavours could not be ignored as it often exerted a considerable impact on the way portions of the same region were occupied by different groups. Vidal de la Blache (1926) clearly illustrates this situation through the regional geography of *pays* in France. This encouraged the rise of the perspective that each area or region had its own personality, which could be experienced and identified in the field. Assiduous data collection, an astute determination of the boundaries of the various regions and a creative presentation of all these through the use of various cartographic techniques were regarded as the hallmark of good geographical practice.

A later generation of geographers was to argue that geography's concern with environmentalism and possibilism in the early years of its establishment as a university discipline was not unconnected with the prevailing intellectual ideology, or legitimizing principle, deriving from Darwinism. Peet (1985:310), for instance, suggests that environmental determinism was geography's contribution to social Darwinist ideology, providing a naturalistic explanation of which societies were the fittest in the imperial struggle for world domination. In this manner, the subject, consciously or unconsciously, sustained European countries in their colonial enterprise and their pursuit of global supremacy.

However, by the beginning of the Second World War, considerable doubts had been expressed about the one-directional nature of environmental determinism, the perspective of the subject had begun to shift in other directions. This shift did not seek to dissociate the subject from the acknowledgement of the close interrelationship between the physical and the human world. Indeed, if there is one major contribution of geography over the decades, it is in emphasizing the intimate interconnectedness between the two worlds. Thus, although a shift in the perspective of the subject became noticeable most markedly after the Second World War, it still sought to provide a rationale for the juxtaposition of the physical and human worlds. The new element in the shift is a greater concern for the role of time (as process) in the evolution of features of the earth's surface.

The Temporal Perspective in Geography – Structure, Process and Stage

One feature of the approach to the Second World War, was the growing industrialization of most parts of Europe. This meant, increasingly, that the correlation between the physical world and the essentially agricultural societies of the world was becoming less tenable in explaining the varying characteristics of the earth's surface. It also provoked an interest in understanding how any part of the world comes to vary in its characteristics over time. This interest in the changes that occur in a region over time became of special interest both in geomorphology and cultural (or historical) geography.

Geomorphology came to be concerned not just with physiography but with how a particular physical landscape evolved over time. The work of the American geographer, W. M. Davis (1915), was seminal in this approach. His emphasis was to try to articulate the genesis or evolutionary basis for any given land-form bearing in mind the geological structure on which it was developed. This approach attempted to identify stages in the growth of particular land-forms noting that it was possible to identify youthful, mature and old stages in this evolution. The idea that one can apply the concept of evolutionary development to features of the earth's surface was extended to other areas of physical geography including climatology, hydrology and biogeography within an analytical framework reflecting a concern with structure, process and stage.

In the field of human geography, the pioneering work of another American geographer, Carl Sauer, opened a new vista in the perspective of geography. He emphasized the long-term historical changes that are noticeable in the material culture of a people and called attention to the need to study each stage in the evolution of the landscape occupied by a particular culture. In Britain, H. C. Darby was to advance the interest in historical geography further by attempting to reconstruct the geographies of different periods in the past. His most notable achievement in this regard was his geography of Domesday Britain or Britain in 1088, the year in which William the Conqueror insisted that a full census be taken of all of Britain, the population, their assets, their amenities and the taxes they paid. Working on the records of this unique census, Darby (1952–77) was able to take the reader through what Britain must have looked like in the eleventh century. Again, there was interest in looking at the landscape of different periods and identifying stages in their evolution.

The introduction of what Carl Sauer called temporal relations in geography led to greater interest in more systematic observation of the landscape, using data from the newly emerging field of aerial photography and remote sensing as well as from archival historical materials. Detailed local field-data collection and measurement also started to emerge as an interest of the field. In the geomorphological area, this encouraged an initial attempt at inductive theorizing, about the possible origins of particular land-forms. This development further provided increasing opportunities for collaboration of geographers with scholars in neighbouring disciplines of geology and history. The impact of including the temporal dimension in geographical investigation did not, of course, preclude the continuing interest in the regional perspective. Rather, it made it possible within each region to seek to understand how the present landscape evolved and what factors or forces had been predominant in bringing about changes over a given time period.

The period after the Second World War was catalytic in many respects. Although geographers had played various roles during the war period, the end of the war was a period for serious reflection on the contribution of the discipline to the war effort, and on its future prospects in the brave new world unfolding. One of the major criticisms at this time was that the subject concentrated too narrowly on regional geography to the neglect of its more systematic branches. Edward Ackerman, a one-time President of the Association of American Geographers, was perhaps the most articulate in giving expression to the nature of this criticism. According to him:

> *The regional method of research, once wartime geographic compilation and investigation was started, proved to have no more value than the past literature. Where anything more than superficial analysis was required in government work, the only possible course was one of systematic specialization. Dependable accumulations of data, and reliable interpretations of those data, would not be had otherwise . . . If our literature is to be composed of anything more than a series of pleasant cultural essays, and if our graduates are to hope for anything more than teaching professions, we shall do well to consider a more specialized, or less diffuse approach.*
> (1945:128–9)

The appeal then was for greater specialization in particular topics or themes, linking together both the systematic and the regional approaches of the subject. Equally important was the concern for the subject to strengthen its analytical capabilities through the adoption of more rigorous quantitative techniques to complement its traditional use of descriptive statistics and cartographic methods.

The period from the 1950s onwards saw the emergence within geography of a whole series of new fields or specializations formed by the splitting off of subdisciplines, which sometimes strengthened into new disciplines in their own right. But, as Peter Gould observed in his contribution to this chapter, if modern intellectual life sometimes seems like a tree constantly sending out finer and more specialized twigs, it is also possible to view it through another metaphor – that of the braided stream, with its water constantly flowing from one channel to another. Many of the important and difficult problems in modern times demand the skills and perspectives of a number of disciplines connected by such cross-braiding of individual channels. And, if this is true of the larger intellectual realm today, it is also true within the traditionally eclectic field of geography. In the five decades following on the astute comments of the likes of Ackerman, two developments have been characteristic of geography. The first is the heightened awareness of the intimate and intricate interconnectedness between the human and the natural world but with an educational purpose not to treat the relationship in the rather mechanistic or deterministic manner of the past. The second is the continued acquisition of increasingly sophisticated quantitative capabilities, which have allowed refreshing insights into earlier approaches of the subject in the light of newer global perspectives, supplemented by vastly increased analytical possibilities through remote-sensing, geographical information systems, and supercomputing power. These changes have led to a more vigorous contestation of ideas within the field and witnessed a period of remarkable redefinition of the educational purpose and intellectual value of the subject.

Quantitative Geography and the Spatial Scientific Perspective

The new development began with a heightened disgruntlement with existing approaches. This was almost equally matched by the attraction to the more scientific orientation of other social sciences, such as economics, sociology and psychology, an orientation which would make it easier for the two broad areas of the subject – physical and human geography – to be able to communicate more easily. Besides, science was academically and socially respectable and so was social science. Thus, it was reckoned that if geography became more scientific, it could become more useful, thereby advancing the esteem of both the discipline and its professionals.

The first step in the process of providing geography with a new perspective was, thus, to identify conceptual frameworks that provided a basis for some analytical understanding of the distribution of objects and events in space (Harvey, 1969:191). In this renewed interest in the where and why of various phenomena, the network characteristics of their distribution and the distance element became predominant factors of interest. It was greatly emphasized in this new conception that distances represented a hindrance, an impedence, in the possible interactions between two places on the earth's surface. Any such interactions, therefore, entail a cost of overcoming the friction created by this impedence. That cost can be represented in different ways in terms of time or money (such as transport costs). Human society, however, tries in different ways to minimize such costs and, in consequence, organizes itself in space in a rational manner.

The discovery of the logic behind such spatial organization of various human objects and events, notably human settlements and human productive activities, became a major focus of the new perspective of spatial science. One of the earliest proponents of this new trend was William Bunge. In the introduction to the second edition of his book, *Theoretical Geography*, Bunge (1962:xv–xvi) had argued that:

> *The earth is not randomly arranged. Locations of cities, rivers, mountains, political units are not scattered around helter-skelter in whimsical disarray. There exists a great deal of spatial order, of sense, on our maps and globes.*

The task of geography, therefore, was to strive assiduously to understand the logic behind these locational distributions, to become, indeed, the science of locations, seeking to predict locations, to establish clear relationships in space, to display the correlations between spatial distributions and to account for the characteristics of particular places in the light of these correlations. Geography was therefore expected no longer to be content with simply describing, but like a strict science, should be able to formulate predictive laws of spatial behaviour as well as uncover laws or rules governing observable spatial regularities.

Some of the more fruitful concepts for studying the human world in space and time were drawn from analogies from the physical world, especially the formal field of classical physics. As the theoretical and quantitative perspectives became prominent in geography, particularly with the extraordinary advances in computing, many fruitful analogical perspectives emerged into greater prominence. Concepts generally considered under the rubric of gravity models underpin many formal approaches to spatial interaction, diffusion, the daily journey-to-work, and even global transmission of diseases. Their principles may be derived directly from analogous theories of entropy maximization (Wilson and Bennett, 1985). Extremum principles underlie all maximizing and minimizing approaches. In the case of goal programming they may well incorporate human choices about the priorities given to various goals when perhaps not all of them can be immediately met. Many formal mathematical models employed in the physical sciences, such as Markov chain analysis, appear to have direct and illuminating applicability to the human realm. The ethical dangers of reifying the human world, of treating human beings *en masse* as things, have been noted (Olsson, 1980), although any theoretical or collective view must inevitably consider individual idiosyncracy.

For physical geography, the quantitative and theoretical perspective provided a refreshing interlude. Since the formal framework for theorizing about physical processes, whether terrestrial, atmospheric or oceanic, is still Newtonian physics, the development in the subject was most welcome. The framework, however, appears to serve its practitioners well at what might be called the meso-scales at which earth processes are examined. Such studies require neither the microworld of quantum mechanics nor the relativistic conceptions appropriate to astronomical scales. For example, all global circulation models employ differential equations that would be perfectly familiar to late eighteenth-century mathematicians, the only difference being that the consequences of their interrelationships on a global grid are now computable to levels of finite observability.

At the same time, there is a growing awareness that traditional continuum mathematics, while sometimes useful as a language for very general theorizing, may bear little relation to finite computing abilities (often with hard-won data). This is particularly true when attempting to compute dynamic models over long stretches of time, stretches that may range from a few years in an epidemiological model, to centuries or millennia in atmospheric, oceanic or terrestrial models, some of which are linked together. Such realizations have opened up new possibilities for thinking in physical geography, but have inevitably produced some intellectual fads. For example, catastrophe theory has made us much more sensitive to the fact that small changes in critical values can flip-flop a large system from one stable mode to another. Moreover, we have known since Lorenz's computing experiments with atmospheric models that our always limited abilities to measure, when combined with digital computing, will always accumulate error terms which eventually reduce purported predictions to random chaos. Similarly, there are other earth scientists, whose grasp on the classical scientific paradigm of prediction is difficult to dislodge. These scientists spend large amounts of time and money looking for strange attractors, regions of a state space for which a dynamic system may have a certain predilection. Even if such an attractive region can be delimited, it still serves no useful, that is, predictive purpose, since, by definition, the system is still computably chaotic.

That the criterion of utility should be invoked for research in physical geography and adjacent fields should come as no surprise in these days of global warming, in which the role of the human use of fossil fuels appears to play a part. The issue of global warming has become highly politicized, both nationally and internationally, and is now the focus of a large academic industry, with large sums of money and much personal prestige at stake. Its conflicts could well become the focus of a penetrating study of science as a socially negotiated enterprise rather than an ideal endeavour seeking the truth in all its purity. Prudence dictates that the topic be taken seriously, even though evidence marshalled by one side or another may be highly selective. For example, to compute a half-degree centigrade rise in global temperatures over the past 100 years from thousands of observation points around the world, one must have a great deal of faith in the accuracy and reliability of those original observations. As for the atmospheric circulation models, which come in half-a-dozen varieties, they are all so mechanistically simple and crude in the specifications that doubling the carbon dioxide content, an act of simulation that has almost achieved the status of the sacred, is bound to produce a heating effect. That we appear to be at the top of a Milankovitch cycle, one that we can record repeatedly over the past 400,000 years from oxygen isotope ratios (Imbrie, 1985), seems to be discounted, as is the fact that basic components are either wished away or given great precision when only order of magnitude estimates are really available. For example, the global carbon cycle, the revolving door both generating and fixing carbon in its many forms, has to be able to balance the global account book made up of atmospheric, oceanic and terrestrial components. The role of the latter has only been estimated with great uncertainty. Such error terms inevitably impinge in still unknown ways on estimates made for the atmospheric and oceanic portions of the global budget.

As for the possible human consequences, such research, which must inevitably be highly speculative to the point of being no more than futuristic scenario-writing, has no foreseeable utility in the sense that anything can conceivably be done with the results. Any effects over the next 100 years are irreversible, and this minimum impact time already represents four human generations. Such speculations and scenarios have no meaning for the vast majority of farmers in the world, whether the huge agri-businesses in the United States or the peasant farmers in Chad or Pakistan. This is because any possible consequences will be so slow that they will be beyond most finite perceptions and so be meaningless.

Even in an inevitably selective review such as the present, to give the impression that all quantitative effort in physical geography focuses on global warming would be quite misleading. The causal arrow in human–nature relations appears more and more prominent to point from the former to the latter, and the impact time may extend far into the future. We recall that the half life of plutonium-239, of which 7.5 kilograms could kill every human being on earth if effectively distributed, is 24,000 years, or roughly five times the portion of human history since the pyramids of Egypt were constructed. The Chinese Academy of Sciences, in an exemplary publication (1979), has documented the geographic distribution of numerous human diseases. Many of these raise very strong hypotheses of environmental cause, especially those resulting from industrial pollution. In North America, it becomes more and more difficult, which means more and more expensive, to produce public water supplies that are not contaminated by chemicals or bacteria in the aquifers of other water sources. Many European rivers can no longer sustain aquatic life. Mammals, such as the Baltic seal and the otter, are being wiped out by PCBs and other organo-chlorides. Physical geographers have enormously important roles to play in environmental education at all levels and at all geographic scales. Apart from long and costly court actions, normally beyond most citizens, education is the only road to a world with environmental balance and health.

With regard to human geography, details of the theoretical formulations that came with the quantitative and spatial science perspective are the subject of the next chapter and need not concern us here. But a major corollary of interest in a scientific approach to the subject was the preoccupation with issues of measurement, data collection and the statistical testing of hypotheses and relationships. This process of quantification became very sophisticated within a relatively short space of time. It went from interest in inferential statistics to the use of various multivariate and non-linear analytical models and to dalliance with topological geometry. Indeed, the very active interest in quantitative techniques led to the discipline having to grapple with other issues of scientific methodology and to engaging in a rigorous pursuit of intellectual inquiry within a scientific framework. The publication of *Locational Analysis in Human Geography* (Haggett, 1965) and *Models in Geography* (Chorley and Haggett, 1967) recorded quite impressively the remarkable achievements which the subject had made within less than two decades not only in quantitative sophistication but also in modelling and predicting human spatial behaviour.

The Cartographic Dimension: Between the Scientific and the Humanistic

Few areas in geography today better demonstrate the remarkable development in technical capabilities that has taken place in the subject as a result of the quantitative-theoretical revolution than cartography. Indeed, few people outside of the discipline are fully aware of the technical explosion, ranging from computer cartography, the growing capabilities of geographic information systems (GIS), to analytical abilities in the spatial and temporal domains simply unthinkable only 20 years ago. Few students today ever touch a pen or pencil to a sheet of paper since, in developed countries at least, even beginning classes in cartography, start with computer graphics and laser printing to produce images of the highest publishable quality. Whereas geography students in the past would patiently compute the coordinates of a particular map projection and then slowly fill in outlines and features from atlases and topographic sheets, today students at the same level click on five basic projection menus, each containing a dozen or more variations. These can be drawn, scaled and rotated instantly to find by experiment the most effective view, the computer quickly redrawing the basic outlines required. These can then be overlaid with text and symbols, or even filled in by natural colour satellite images.

Many areas of cartographic-based analysis require huge amounts of data to be handled quickly. So, increasingly machines with large random access memories and high speed are needed. For many smaller research programmes, the size and speed of microcomputers have kept pace but for projects relying on remote sensing, super-computers are generally required. It is difficult to make a meaningful distinction today between GIS and analytical programmes since both often are part of the same routines made up of linked software packages. In terms of analytical perspectives, however, they have opened up a world of spatial analysis on the theoretical or wish-list horizons of a few years ago. Many analyses involve the interrogation of spatial series or what are known more simply as map distribution, particularly to answer questions of significant clustering. To give but one example: many statisticians denied that childhood leukemia clustered significantly close to an atomic reprocessing plant in northern England until the geographer, Openshaw (1987) used GIS capabilities to test nine million separate hypotheses at different scales, plotting the significant ($p = 0.002$) ones on the map. At the end, a huge black blob was centred over Sellafield, indicating significant clustering at nearly all scales of analysis.

Nor has the dynamic perspective been ignored. Variations on the expansion method (Jones and Casetti, 1992) employ spatial and temporal series simultaneously, using these approaches to predict not simply values down the time horizon but *where* these values are to be expected. Here spatial adaptive filtering (Kabel, 1992; Gould et al., 1991) allows the geographer to use all the spatial information in the (x,y,t) cube (or, in less jargonistic terms, a pile of maps showing the changes in a single variable over a succession of time periods) in order to predict further maps. Predicting the geographic outcome of a process, treated traditionally in a simplistic temporal way, has employed neural nets as transformations, or mappings in the strict mathematical sense, to predict often hard to measure variables from those more easily observed (Hewitson and Crane, 1994).

Few geographers today view cartography as a slow march towards the ideally truthful map but recognize that what the map maker and the map reader bring to the map controls the interpretation and the meaning. This goes far beyond the traditional use of the map for propaganda purposes, for we realize today that *every* map constitutes a text of power and that the silences of things left out may be as important as the things shown (Harley, 1988). In brief, a map, any map, is not a neutral, mute text with the same meaning for any reasonable person, but a human construction to be interpreted from many different perspectives, and so one with possibly many different meanings.

It is little wonder then that there has been intense and continuing interest in visualization. Part of such interest arises from the simple fact that no human being can comprehend the literally billions of bits of information making up a multi-channelled stream from satellite remote sensing. Somehow one has to compress, filter, enhance and simplify everything into a visual image to ensure an intellectual grasping of complexity to give it meaning. But part of this comprehension also arises from the deeper understanding of dynamic processes when animation of successive images is possible. For example, there is much, and quite understandable, interest in the question of the global transmission of diseases, especially since few airports in the world are more than 36 hours apart. In principle, one could simulate the movement of a virus from any one of the world's 4,028 airports as a succession of probabilistic smears that intensify over time (Gould, 1995). If a series of 1,000 of such global images were linked together by animation, one could produce a dramatic and dynamic sequence disclosing major routes and possible bottle-necks. The serendipity effects of animation may be startling as spatial patterns, previously thought to display no temporal change or any regularity of interest, suddenly take on animated movement leading to speculations or formal hypotheses not even thinkable before.

Within the spatial scientific perspective, cartography helps to emphasize in graphical terms the spatiality of human existence. However, while recognizing that no discipline can make an exclusive claim on this aspect of human life and society, it acknowledges that there is something peculiarly geographic about the geographer's emphasis upon space. The concept of geographic space can either conjure up the image of the traditional map lying open on an atlas page or the homogeneous and smooth space of the theoretical social physicist. Neither is totally irrelevant although both today are inadequate. The reason is simple: the smooth spaces of theory are only simplifying starting points while the topographic surfaces of the atlas display a type of structured space (mountain barriers, river channels, etc.) that has become less and less important over the past two centuries of technological advance. In an electronic age of internet and e-mail, there may appear to be no space, but a dimensionless point, and in that sense no geography.

It is, however, not the geographical space of the map that is important, except perhaps for traditional display purposes but how that space has been structured by the human presence and its technology. It is these structured, often multi-dimensional spaces that control much of the dynamics of human society. Indeed, one of the reasons for often taking examples from medical geography is simply that diseases are forms of data often recorded and so can be used as tracers. In the same way that minute quantities of short-lived radioactive materials may be used in medicine to investigate the pathways and blockages of a human body so disease records may disclose the complex spatial structures of compelling interest to geographers. For example, in the New York metropolitan region, containing about 18 million people, it is possible to transform the traditional geographic map of boroughs and counties into a commuter space whose distances are measured by the intense flows of the daily journey-to-work (Wallace and Wallace, 1995). If a virus, in this case the HIV virus, is injected at any point into this highly structured space, it is drawn quickly into Manhattan, which sits like a spider at the centre of its geographic and commuter webs. Here the virus quickly proliferates in marginalized populations and is carried by commuters to the furthermost suburbs. Indices measuring the intensity with which each county is connected to all other parts of the daily commuting system predict AIDS rates over time and at different places to a high degree ($r = 0.93$). It will be of interest to other social and behavioural scientists that difficult, sometimes impossible to measure parameters, such as transmission rates from one social group to another, are simply irrelevant in such analyses. Once the *structure* of the appropriate space has been discovered, other information follows to allow close predictions in both space and time.

In this and many other examples, human interaction itself structures the geographic space of the map. It is therefore hardly surprising to find gravity model ideas at the heart of the many spatio-temporal processes, including the diffusion of news, ideas, innovations and diseases. But such a structural perspective also recognizes that movements of all kinds may be acutely affected by cultural, economic, religious and even ethnic barriers. In other words, human movements and interactions may not be some relatively simple mathematical function of size, distance or cost, but shaped by truly human factors difficult to measure in any precise quantitative form. Indeed, it was in reaction to the first wave of perhaps over-enthusiastic quantifying and reifying of human spaces that we see the first reactions and objections that led to the revival of social and cultural geography. Such a counter-reaction reflected the dismay of many geographers that human aspects of importance, delicacy and nuance were being ignored or being overrun by spatial theoretical juggernauts and their quantifying acolytes.

Not unexpectedly, a critical reaction set in. First, there was concern with the positivist philosophical stance that the discipline was now extolling. Much of the theoretical and quantitative research being undertaken rested on a series of implicit assumptions, most notable of which was a hidden order in creation and in spatial occurrence. The distribution of various geographical phenomena was assumed to be governed by natural or unswerving laws whose discovery would reconnect geography with the various philosophical traditions of positivism. Such a position was challenged (Cloke, Philo and Sadler, 1991:14) as creating a false sense of objectivity by artificially separating the observer from the observed and denying the existence of strong correspondence links in the sense of the former seeing only what he wanted to see. Secondly, this assumed objectivity gives the impression that the results of research investigations were value-neutral. The emphasis on quantification, whilst not a demerit in itself, had the tendency of filtering out concern with social or ethical questions and, therefore, allowed no room for consideration of values or consideration of how society could or should be organized. Finally, spatial science, in its positivist frame, was said to suffer from an inability to see beyond the map for it gave little recognition to two crucial aspects of spatial patterns and processes, namely the working out of deep-seated economic, social and political structures that direct and constrain the paths of human existence as well as the perceptions, intentions and actions of human beings as conscious agents operating on the earth's surface. In short, as Harvey (1973) argued, it left geography being unable to say anything really meaningful about events as they unfold around us.

Many of these criticisms have subsequently been recognized within the discipline and were to lead to a new period of active intellectual ferment involving the revisiting and restructuring of problems with which the subject had been engaged almost from its beginning. The perspectives this time, however, relate to a whole series of analytical constructs (that came later to be referred to as grand theories and meta-narratives) drawn eclectically from the domain of the social sciences. Those which were concerned with the spatial impact of the deeper issues of economic, social and political processes took off from a Marxist analytical frame; others concerned with how far appreciation of spatial structures and organizations were reflections of human perceptions, intentions and praxis sought an alternative humanist approach to geographical understanding. The flowering of perspectives from these two approaches has provided some of the rich excitement in the field since the late 1970s.

The Grand Theoretical Perspectives – Marxist and Humanist

As with the spatial science perspective, Marxist geography placed the economy at the centre of its discourse. It sought to analyse landscapes in terms of the prevailing mode of production and the social processes of capitalist accumulation that produced them, placing great attention on the impact of local and national government policies, and the activities of financial institutions. The type of equilibrium of interest to spatial science was now seen as problematic, if not illusory. A dialectical process entailing crises and contradictions came to be associated with the production and reproduction of space. Regional transformations were no longer merely topological operations but the complicated outcomes of the mode of production and the social struggles by human actors, basically of capitalists and labour. In short, the Marxist perspective, as it were, made it possible to reconnect the spatial structures of spatial science with the bundles of social relations that were embedded within them and to think of social practices in a number of critical ways. Geographical research thus came to be grounded in the material conditions of existence and within a historical materialist mode of analysis.

At first, the emphasis was a crude analytical construct which saw the unequal development of regions and nations as fundamental to capitalism, as is the direct exploitation of labour by capital (Soja, 1980:219). Development and underdevelopment came to be represented spatially as involving a dialectical relationship between a core region and a periphery. Always, the core was where the action and development were to be found; the periphery being an impoverished, usually extensive region, dependent on the core. This core–periphery relation was seen as extending also to the global level and came to be important in the spatial analysis of imperialism and neo-colonialism.

Further development of the Marxist perspective on geography led to the search for mature reflections and came to recognize that whilst it is difficult to deny that regions are unevenly developed and that capital had a hand in such unevenness, there was need to go beyond these insights. The response to this problem produced a rich and varied literature in the discipline. These contend, for instance, that different levels of production require different qualities of labour power. Capital searches for the appropriate quality of labour to satisfy its needs amongst available regional alternatives and locates production accordingly, giving rise in the aggregate to different rounds of capital accumulation. These changing needs of capital result in some region's labour forces being discarded, others adapted, and still others opened up to new forms of exploitation through industrial restructuring leading on to regional restructuring. At the end of the process of struggle between capital and labour, the state of play can be summarized as a spatial division of labour (Massey, 1984a). In time, as successive cycles of capital accumulation come and go, so the successive spatial divisions of labour they engender combine with one another to form a rich and complex palimpsest of human occupance. With time, therefore, the Marxist perspective came to draw more heavily on other concerns and versions of social theory, notably realism and critical theory, which strive to link structural influences analytically to the workings of the world. In the process, this perspective broadened its range of concerns and came to engage in a healthy and lively dialogue with alternative analytical constructs, especially those of a humanist orientation.

Although an important strand of the critique of spatial science in the 1970s, the humanist tradition had a much longer history in geography, going back at least to the treatment of the human agency in possibilist versions of the environmental tradition. In part, its resurgence was a result of impulses from within the spatial scientific field, arising from the application of stochastic principles and an interest in human behaviour. However, the evident failure of spatial science to take seriously the complexity of human beings as creative individuals led to a growing oppositional critique founded on more strictly humanist principles. Whilst Marxist critics of spatial science reacted against social inequality and the uneven distribution of power in society, humanist geographers focused initially on the role of the human agency, and in particular on the way in which the researcher's subjectivities mediate the research process.

The humanist perspective thus emphasized attempts by geographers to reflect carefully on the intellectual tradition and development of their own field, including their own personal experiences and development (Buttimer, 1983; Haggett, 1990). It sought to direct attention not only to the deeper phenomenological and existential connections people have with places but also with sensitizing geographers to the everyday and yet intimate attachments all sorts of people have to the places where they live. It thus insists on taking seriously the intersubjectively constituted 'lifeworlds' – the shared meanings and common-sense knowledges – associated with groups of people who lead similar lives under similar circumstances in similar places. Place, group and lifeworld are hence seen as three closely linked entities that should stand at the heart of humanist geography.

In recent years there has been a remarkable sea-change in such studies from the previous hagiographic and too often defensive writings trying to delimit intellectual boundaries. Part of the change is undoubtedly due to the greater maturity and sophistication gained by a thorough grounding in adjacent areas of intellectual history, particularly of science and philosophy. It is difficult now to consider the development of geographic thought as something occurring in isolation, and as a smooth progression along a path leading to geographic truth from which the young must be taught not to digress. Rather, the humanist perspective has grown out of a more fragmented process than the evolution of Marxist thought in geography and covers a diversity of different and sometimes incompatible intellectual positions. It has been in part a product of the openness of geographical writing to other fields. Whether drawing upon, or contributing to, such diverse fields as literary theory (Barnes and Duncan, 1992), language (Pred, 1990), aesthetic stances (Tuan, 1993), intellectual history (Glacken, 1967) or cultural studies (Watts, 1991; Western, 1992), geographic writing is open in its willingness to draw upon other traditions, and confident that it can provide perspectives from its own spatial heritage to contribute to a larger enterprise of understanding.

If one is prepared to distance oneself somewhat from all the intellectual swirls of the moment, perhaps one of the most remarkable developments in the humanist tradition has been the renewal or renaissance of social and cultural geography. This is an extremely complex development in which the braided stream analogy is worth recalling to emphasize the many informing and influential connections internal to the discipline's development. Here any attempt to provide an overview must inevitably oversimplify. In social geography, for instance, there has been what Jackson and Smith (1984) referred to as a hermeneutic revival (hermeneutics being the interpretation of meanings). This revival has encouraged interest in the exploration of the relationship between people and place, going beyond attempts to offer an interpretation of human experience in its social and spatial setting. Such a social geography stresses the ways in which communities of people intersubjectively build up an understanding of how their local worlds work and how this understanding gets translated, in a sense, into the practices that determine their use of space. These intersubjective understandings are seen as critical dimensions of the culture sustained by particular people in particular places. And it is this perspective that connects up social geography to cultural geography, ensuring that the latter gives less emphasis to the material artefacts and technologies of peoples, as in the tradition of Carl Sauer and Fred Kniffen, and more to the immaterial modes of thinking and living (Cosgrove and Jackson, 1987). Thus, culture comes to be perceived as the glue of locally established meaning and knowledge through which particular peoples in particular places make and remake their lives.

What is apparent from this new perspective is the way it depicts the role of the outsider, and more particularly researchers, in interpreting the signs and symbols of particular cultures and peoples in particular places. Like ethnography, such a perspective of geography requires time-consuming research involving full immersion in the life of the particular people studied, observing and interviewing them and participating as fully as possible in their lives and rituals. And then they must seek to understand and represent their worlds in terms as faithful as possible to their own interpretations. Humanist geographers of this persuasion thus have to confront the same problem of mediation that ethnographers have had to contend with as they bring their frame of reference into contact with the people being researched.

Nonetheless, whether Marxist or humanist, the underlying assumption is of a world with a fundamental *order* to it, which can be teased out by the ordered and rational processes of intellectual inquiry. This impact of the natural sciences upon the social sciences – including human geography – has led the latter to embark upon quest after quest for the true order of the human world. That is, to attempt to discover the way in which human agents interact with one another, with their institutions and with other processes, forces, mechanisms. This concern with order in the universe leads to an approach to knowledge which sees it only in terms of science, whether it is dealing with natural or human phenomena, and in terms of scientific enterprises that, according to Lyotard (1984), anchor themselves in particular sets of claims about how the world operates. This led Lyotard to use the term modern to designate any science that legitimates itself with reference to a meta-discourse, making an explicit appeal to some grand narrative, such as the dialectics of spirit, the hermeneutics of meaning, the emancipation of the rational or working subject, or the creation of wealth (ibid. p. xxiii).

Postmodernism Perspective – Retreat from Grand Theories and Meta-narratives

Not unexpectedly, therefore, serious criticism of the all-embracing pretension of the Marxist or humanist perspective in geography came to be seen as part of the postmodern movement in intellectual life. Both perspectives had conceived a human world ordered coherently around some clearly defined conceptual core, be this a mode of production (as in the Marxist perspective) or the subjectivity of human beings (as in the humanist perspective). They had encouraged intellectual thought to emphasize the commonality, the essential sameness, among phenomena of interest and to view these through coherent societal organizing principles. This resulted in a tendency to pay little or inadequate attention to everything else that could not be captured within these broad theoretical constructs. In contrast, it is argued that reality confronts us with a chaotic multiplicity of phenomena, which are remarkable by their disorder and incoherence. The chaos is such that it is impossible to understand the totality; consequently no conceptual core can be fashioned.

Postmodernism urges a greater sensitivity and alertness to the differences that exist between phenomena, events and processes; its concern is to understand the basis of these differences instead of ignoring or sweeping them under the carpet of some grand theory or meta-narrative.

Thus, the postmodern perspective in geography emphasizes a concern for the innumerable variations that exist among the many types of human beings studied within the subject. Such variations include those between men and women, between social classes, between ethnic groups and between various other categorizations of human groups, and the varied implications of the different inputs and experiences these diverse groups bring to socio-spatial processes. Sensitivity to these differences demands an alertness to all other human groupings such as children, women, the aged, workers, gays, disabled people, nomads and so on who depart from the modern norms of a middle-class, middle-aged, male, straight society that frame much of present-day intellectual and practical activity. Such sensitivity to different peoples, in turn, should bring the discipline down from its high horse of spatial science to confront again its long-standing geographical sensitivity to different places, neighbourhoods, districts, regions and countries.

Consequently, instead of grand theories providing totalizing accounts of social processes and necessarily steamrolling over the delicate details of how things happen differently in different places, the thrust of postmodern perspective is for less pretentious, more eclectic and more empirically grounded attempts at understanding spatial phenomena in their details, their differences and their geography. Thus, postmodern geography exhibits considerable interest not just in conventional social processes but also in various social pathologies (madness, sickness, criminality, deviance and abnormal sexuality). It seeks to understand the factors shaping the location of the institutions invented by society to deal with them (asylum, hospitals, prisons, workhouses and confessionals); their spatial distribution and environmental associations, their nearness and farness from one another and from other phenomena such as human settlements; the spatial arrangements present in their plans and architecture; and the geographies of the discourses that identify these pathologies and propose institutional solutions to them (Cloke, Philo and Sadler, 1991: 196). Furthermore, by taking differences seriously, postmodernism seeks not just to produce sufficiently detailed local knowledge of particular peoples in particular places purely for its own sake but also as part of a practical strategy of empowering those groupings – labour, women, ethnic minorities, the unemployed, the sick and so on – who repeatedly lose out in the face of the more material restructurings endemic to postmodern conditions.

Another postmodern perspective in geography highlights Marx's own view of the fragmentation and transcience of capitalist societies: the notion that all that is solid melts into air. Harvey (1989) represents this alternative postmodern perspective, calling attention to the replacement of so-called Fordist organization of capitalist production and consumption by a supposedly more flexible organization. He suggests that a Fordist era of mass production and mass consumption – an era that coincided with class politics and state interventionism – is now being replaced (in some sectors and localities of economic activity) by a rather different assemblage of practices. On the ground, this shift from the rigidities of Fordism to more flexible labour processes, labour markets, products and patterns of consumption, is characterized by the emergence of entirely new sectors of production, new ways of providing financial services, new markets and, above all, greatly intensified rates of commercial, technological and organizational innovations (Harvey, 1989:147). And beneath these dramatic restructurings lies an incredibly rich and detailed set of changes reshaping technologies, work practices, labour relations, management strategies and intra- and inter-firm links.

These two different views of the postmodern perspective, it has been argued, reflect a distinction between those who consider postmodernism to be an attitude of mind and those who consider it to be an object of study. The former comprise those who reject as modernist an obsessive quest for order in phenomena and who show concern for the role of deeper economic and social forces in the making of that order controlled by human beings. For them, no such order exists in real life. On the other hand, the latter considers postmodernism as a condition of the contemporary world that involves a distinctive shift and fragmentation in the temporal and, more pertinently, the spatial organization of economic, social, political and cultural processes. This perspective then is seen as an attempt to provide a modernist explanation of the postmodern object.

Nonetheless, irrespective of either of its variants, the postmodern perspective has produced within geography a highly sensitized and greatly simplified social concern, usually expressed as a deep dissatisfaction with the way things are. Such dissatisfaction inevitably leads to questions about the reasons for why things came to be this way, questions asked at geographic scales from the local city neighbourhood, through the region and nation, to the global level (Corbridge, 1986; Smith, 1984). The underlying catalyst appears to be an ethical concern for obviously grave disparities between peoples and nations, heightened by a sense of injustice that inevitably challenges the status quo. At its best, research motivated by such concern illuminates, often in ways unique to geographic inquiry (Harvey, 1985), the marginalization of groups of people on the basis of economic, religious, political, ethnic, or other grounds.

The sensitivity of the postmodern perspective has, within geography, fuelled in particular the increasingly strong feminist movement based on a sense of injustice concerning half the human race, most of whom live in strongly dominant patriarchical societies. This movement provides an awareness that gender shapes perceptions, and so geographies. That there might literally be a woman's world, or better still worlds, was barely thinkable 25 years ago. Today, there are standard undergraduate courses on the geography of gender, emphasizing that space and place may have radically different meanings for men and women. At its best, the writings of feminist geographers have greatly heightened sensitivity to such issues, either in thoughtful, general statements (Hanson, 1992), or by exemplary research that demonstrates by concrete example the intellectual richness of a feminist perspective (Mårtensson, 1979; Massey, 1994; Jarosz, 1994). At its less than ideal, it has become a shrill protest from which even strongly feminist geographers demur (Pile and Rose, 1992). Feminist traditions in geography, to the degree that they inform geographic inquiry, vary in sophistication and maturity. The Anglo-American variety still appears to be maturing, especially when compared to the more intellectually secure forms in Scandinavia and German-speaking countries.

Another aspect of the postmodernist perspective may be labelled post-colonialist, with strong ties to literary criticism (Said, 1993) and subaltern studies (Spivak, 1988; O'Hanlon, 1988), from which major influences can be traced. Research in this tradition is intent on opening up and re-examining the human and geographic consequences and experiences of a once-colonized people. In one sense, this is not necessarily a difficult task; most former colonial regimes provide ample material from which to fashion evidence of paternalism, cultural arrogance, economic dominance, and all the other ills which appear when one people take over another, whether by force of arms or not. What is difficult is to provide a fair, balanced and objective picture when the prior assumptions are that all colonial regimes are inherently unfair to the point of evil, and that objectivity is impossible in light of both the writer's moral certainty and his new sensitivity to language and power. Like a person of convinced ideological perspective and persuasion always finding evidence for his or her position, post-colonial geographers display on occasion a tendency to find only what they are looking for. The added irony is that virtually all geographic research and writing is conducted by a generation that has never known colonialism, either as colonizer or colonized. The result is often a genre of geographic writing that might fairly be labelled cathartic geography, to the degree that no apparent distinction is made between guilt (for events for which one can be held directly responsible) and shame (which can be felt for any horror of the human condition).

Like any field, geography generates its intellectual fads and enthusiasms, but it would be unjust and ignorant to dismiss these as intellectually unimportant, for these often constitute new and enriching perspectives from which to view the human geographic world (Gregory, 1994). As successive bandwagons roll by, and every social science experiences these, they come under intense scrutiny and criticism by those who are not caught up in the enthusiasms and latest revelations of truth – even the revelation that there is no truth! The result is that most perspectives-of-the-moment do not dry up entirely and blow away, but leave valuable residues of insight and provide the possibility for thinking and undertaking new and more perceptive geographic research in the future. They often also point to strong and informing links to other areas of the humanities and human sciences.

Conclusion: A Spatial Century?

Contemporary geography is thus an enormously exciting discipline both intellectually and practically. From a somewhat inward looking and defensive field in the 1950s, it has exploded during the past two generations of geographers to form one of the most vital components of university education. Its bases are varied and have complex histories, but include a series of healthy reactions and counter-reactions, starting with challenges to the Establishment's views in the late 1950s and early 1960s. There is little question that what came to be called the quantitative revolution provided a strong challenge to entrenched methods of geographic research and writing (Billinge, Gregory and Martin, 1984), some of which consisted of making maps from census materials and then writing about the maps. Few reflected that there might be important variables not captured by the official census, a contretemps analogous to a quantitative generation's chagrin upon finding that many urban factorial ecologies probably told you more about the structure of the census than the structure of the city.

There is also little question that both quantitative and theoretical developments were greatly catalyzed by the rapid growth of computing power which provided the practical means of handling large quantities of data, and the means of exploring methodological approaches previously beyond finite human capacities. These not only included many areas of multivariate analysis, often contemporary versions of an older tradition of synthesis, but entirely new areas of statistical analysis that were still unthought of by statisticians (Wrigley, 1995). Reactions to such approaches were not long in coming and these generated in their turn perspectives on geographic analysis grounded by strong ideological positions. Yet these have been challenged in their turn by postmodern views which, while highly varied in themselves, have at their core a clear sense that one perspective or approach to geographic inquiry can never capture all that is essential. Such postmodern viewpoints have been perceived as challenges to ideological orthodoxy for they have obviated any claim to essentialist positions (Graham, 1990). Fortunately, for the future health of geography, postmodernists themselves must face the paradox that most of their own strong claims against essentialist viewpoints come close to essentialism of an often highly dogmatic kind.

Such intellectual ferment is to be prized not only internally to the discipline itself but in terms of its interdisciplinary contact and influence. In the last ten years, more and more adjacent areas of the human and physical sciences, as well as the humanities, have exhibited a much deeper awareness of the crucial importance of space and place. People, both scholars and practitioners, are slowly recalling that almost every physical process and human endeavour has a geography as well as a history. At one level, to say that everything exists in space and time may be little more than a rather banal truism. At a more reflective and intense level it develops into an acute awareness of spatio-temporality that can open up and illuminate a topic in extraordinarily effective ways. Indeed, it is not impossible to think of the twenty-first century as the spatial century, a time when a sense of the geographic emerges prominently once again into human thinking. Perhaps, in a refurbished form unthinkable to the Chinese, Arab and European explorers opening up their worlds from the twelfth to the eighteenth centuries, geographic awareness will permeate intellectual and practical life once again. Intellectually, more and more scholars are sensing that place is always defined in a larger, often multi-dimensional space. In other words, and to use the title of a book written by a strong advocate of such a view, *Geography Matters*! (Massey, 1984b). In practical terms, it means that every planning decision involves not simply a *when* but also a *where*, always positioned in a space structured in complex ways by prior decisions.

In such a resurgence into general thinking, geographic education has an enormously important role to play, whether at the level of the school, university or in the public realm. Events in today's world are too interconnected and immediate not to have meaning, that is, not to make sense to an educated citizen of the world. An accident at a nuclear plant, the emergence of a new virus, the emission of atmosphere-destroying gases, an aggressive local political decision – all impact around a world now drawn in so many ways into tighter contact. In a very deep sense, therefore, nothing is unconnected in today's world, which is perhaps why it is important not to channel braided intellectual streams but to let them mix their currents and find their own respected ways.

References

Ackerman, E.A. (1945) Geographic training, wartime research and immediate professional objectives, *Annals of the Association of American Geographers*, 35:121–43.

Barnes, T. and Duncan, J. (eds) (1992) *Writing Worlds: Discourse, Text and Metaphor in the Representation of Landscape* (London: Routledge).

Billinge, M., Gregory, D., and Martin, R. (eds) (1984) *Recollections of a Revolution: Geography as Spatial Science* (London: Macmillan).

Bunge, W. (1962) *Theoretical Geography* (Lund Studies in Geography, Series C: General and Mathematical Geography, Paper no. 1) (Lund:Gleerup).

Buttimer, A. (1983) *The Practice of Geography* (London: Longman).

Chinese Academy of Sciences (1979) *Atlas of Cancer Mortality in the People's Republic of China* (Shanghai: China Map Press).

Chorley, R.J. and Haggett, P. (eds) (1967) *Models in Geography* (London: Methuen).

Cloke, P., Philo, C. and Sadler, D. (1991) *Approaching Human Geography: An Introduction to Contemporary Theoretical Debates* (London: Paul Chapman).

Corbridge, S. (1986) *Capitalist World Development: A Critique of Radical Development Geography* (Totowa, NJ: Towman and Littlefield).

Cosgrove, D. and Jackson, P. (1987) New directions in Cultural Geography, *Area*, 19:95–101.

Darby, H.C. (1952–77) *The Domesday Geography of England*, 7 vols (Cambridge: Cambridge University Press).

Davis, W.M. (1915) The principles of geographical description, *Annals of the Association of American Geographers*, 5:61–105.

Glacken, C. (1967) *Traces on the Rhodian Shore* (Berkeley: University of California Press).

Gould, P. (1995) Spatiotemporal cartography and global diffusion, *Sistema Terra* (in press).

Gould, P. et al. (1991) AIDS, predicting the next map, *Interfaces*, 21:80–92.

Graham, J. (1990) Theory and essentialism in Marxist geography, *Antipode*, 22:53–66.

Gregory, D. (1994) *Geographical Imaginations* (Oxford: Blackwell).

Haggett, P. (1965) *Locational Analysis in Human Geography* (London: Edward Arnold).

Haggett, P. (1990) *The Geographer's Art* (Oxford: Blackwell Publishers).

Hanson, S. (1992) Geography and feminism: worlds in collision?, *Annals of the Association of American Geographers*, 82:569–86.

Harley, B. (1988) Secrecy and silences: the hidden agenda of cartography in early modern Europe, *Imagio Mundi*, 40:111–30.

Harvey, D. (1969) *Explanation in Geography* (London: Edward Arnold).

Harvey, D. (1973) *Social Justice and the City* (London: Edward Arnold; reissued 1988 by Blackwell Publishers).

Harvey, D. (1985) *Consciousness and the Urban Experience* (Baltimore: Johns Hopkins University Press; Oxford: Blackwell).

Harvey, D. (1989) *The Condition of Postmodernity: An Enquiry into the Origins of Cultural Change* (Oxford: Blackwell).

Herbertson, A.J. (1905) The major natural regions: an essay in systematic geography, *Geographical Journal*, 25:300–12.

Hewitson, B. and Crane, R. (1994) *Neural Nets: Applications in Geography* (Dordrecht: Kluwer Academic).

Huntingon, E. (1915) *Civilization and Climate* (New Haven, CT: Yale University Press).

Imbrie, J. (1985) A theoretical framework for the Pleistocene ice ages, *Journal of the Geological Society of London*, 142:417–32.

Jackson, P and Smith, S.J. (1984) *Exploring Social Geography* (London: Allen & Unwin).

Jarosz, L. (1994) Agents of power, landscapes of fear: the vampires and heart thieves of Madagascar, *Society and Space*, 12:421–36.

Jones, J. and Casetti, E. (1992) *Applications of the Expansion Method* (London: Routledge).

Kabel, J. (1992) *A Geographic Perspective on AIDS in the United States* (University Park: Ph.D. dissertation, Pennsylvania State University).

Lyotard, J.F. (1984) *The Postmodern Condition: A Report on Knowledge* (Manchester: Manchester University Press).

Mackinder, H.J. (1887) On the scope and methods of Geography, *Proceedings of the Royal Geographical Society*, 9:141–60.

Mårtensson, S. (1979) *On the Formation of Biographies in Space–Time Environments* (Lund: Lund Universitet).

Massey, D. (1984a) *Spatial Divisions of Labour* (London: Macmillan).

Massey, D. (1984b) *Geography Matters* (Cambridge: Cambridge University Press).

Massey, D. (1994) *Space, Place and Gender* (Cambridge: Polity Press).

O'Hanlon, R. (1988) Recovering the subject: subaltern studies and the histories of resistance in colonial South Asia, *Modern Asian Studies*, 22:189–224.

Olsson, G. (1980) *Birds in Eggs/Eggs in Bird* (London: Pion).

Openshaw, S. (1987) *Building a Mark I Geographical Analytical Machine* (Newcastle-upon-Tyne: Northern Regional Laboratory).

Peet, R. (1995) The social origins of environmental determinism, *Annals of the Association of American Geographers*, 75:309–33.

Pile, S. and Rose, G. (1992) All or nothing? Politics and critique in the modernism–postmodernism debate, *Society and Space*, 10:123–12.

Pred, A. (1990) *Lost Words and Lost Worlds: Modernity and the Language of Everyday Life in Late Nineteenth-Century Stockholm* (Cambridge: Cambridge University Press).

Said, E. (1993) *Culture and Imperialism.*

Semple, E.C. (1911) *Influences of Geographic Environment: On the Basis of Ratzel's System of Anthropo-Geography* (London: Constable).

Smith, N. (1984) *Uneven Development: Nature, Capital and the Production of Space* (Oxford: Blackwell).

Soja, E.W. (1980) The socio-spatial dialectic, *Annals of the Association of American Geographers*, 70:207–25.

Spivak, G. (1988) *In Other Worlds: Essays in Cultural Politics* (New York: Routledge).

Taylor, P. (1985) The value of a geographical perspective, pp. 92–110 in R. Johnston (ed.) *The Future of Geography* (London: Methuen).

Tuan, Y.-F. (1993) *Passing Strange and Wonderful* (Washington DC: Island Press).

Vidal de la Blache, P. (1926) *Principles of Human Geography* (London: Constable).

Wallace, R. and Wallace D. (1995) US apartheid and the spread of AIDS to the suburbs: a multi-city analysis of the political economy of spatial epidemic spread. *Social Science and Medicine* (in press).

Watts, M. (1991) Mapping meaning, denoting difference, imagining identity: dialectal images and postmodern geography, *Geografiska Annater*, 738:7–16.

Western, J. (1992) A *Passage to England: Barbadian Londoners Speak of Home* (Minneapolis: University of Minnesota Press).

Wilson, A. and Bennett, R. (1985) *Mathematical Models in Human Geography and Planning* (London: Wiley).

Wrigley, N. (1995) Revisiting the modifiable areal unit problem and the ecological fallacy, in A. Clith et al. (eds) *Diffusing Geography* (Oxford: Blackwell Publishers).

2 Geographical Theories

Although a long-established tradition within human geography is concerned with the *location* of activities and their associated infrastructures and with the *interaction* between locations, it was only after the Second World War that a decisive interest emerged in systematizing geographical knowledge and using it to explore new areas of the human condition. Today, the usefulness of theory and predictive models in geography is a matter of record but this was not always the case. Considerable dispute and contestation surrounded this development.

Most geographers refer to the disciplinary upheaval of the late 1950s and 1960s as the quantitative revolution. To provide a detailed and faithful account of the development during this period, which was, quite literally, an explosion of theoretical concern in the subject, would be quite impossible. That it was bound up with quantification is hardly surprising since, with many analogical roots in the physical sciences and with a concern for distance and spatial transformation (Tobler, 1961), most of the empirical research informed by theoretical constructs required counting and relating (McCarty, 1954, 1959; McCarty, Hook and Knos, 1956). Such counting and relating had been used more extensively in physical geography where the rigours of scientific experimentation, careful experimental design, accurate measurement and explanatory model building had been more widespread. Such was not the case in human geography. Much of human geography was descriptive rather than analytic; interpretative rather than logical; and dependent on the judgemental expertise of persons claiming specific knowledge of systematic areas or of regional systems. Consequently, in the eyes of many geographers, introducing theory and introducing measurement, mathematical modelling and

This chapter is based on a contribution by R. G. Golledge.

statistical analysis into the subject were more or less the same thing. They entailed and brought about a wholesale philosophical reorientation into the discipline. Initially, positivist philosophical method was the primary mechanism by which theory was introduced and spread within the discipline.

The beginnings of the theoretical revolution in geography can be traced to a remarkable constellation of young geographers at the University of Washington in the late 1950s. Encouraged by William Garrison (1956), whose own reviews in the geographic literature (Garrison, 1959, 1960) opened up possibilities for a new 'geographic way of looking', many of the pioneering works in theoretical geography were published under the names of these 'space cadets' (Berry and Garrison, 1958a, 1958b; Bunge, 1962; Dacey, 1960; Garrison and Marble, 1957, 1961; Getis, 1963; Morrill, 1962, 1963; Nystuen, 1963; Nystuen and Dacey, 1961; Tobler, 1963). Simultaneously, theoretical innovations started in departments at Northwestern University (Taaffe, 1956, 1959, 1962; Thomas, 1960), Iowa University (McCarty and Salisbury, 1961) and at the American Geographical Society (Warntz, 1957, 1959, 1964; Stewart and Warntz, 1958). It was from these centres that a concern for the theoretical, for seeing the general in the particular, gradually spread to other departments and came to influence the subject as a whole throughout the world.

To do justice to these developments in the subject, however, two further names must be given prominence. The spatial-regional economist, Walter Isard, was a major moving force in these early developments. Dissatisfied with the virtually a-spatial orientation of traditional neoclassical economics, and mostly ignored by his own discipline, he founded the Regional Science Association as a focus for both the theoretical development and rigorous empirical testing of spatial patterns, essentially subsidizing the publication of his first book *Location and Space-Economy* (1956) because no established publisher would take the risk without considerable subvention. Matters were, however, different four years later with his *Methods of Regional Analysis* (1960), which went through many editions. The second name is that of Peter Haggett, who realized the importance of both the Swedish and North American developments, and started a few years after his return to Cambridge in 1957 to deliver a series of then quite unorthodox lectures (Chorley, 1995; Thrift, 1995). This series was later to become his *Locational Analysis in Human*

Computer projection of a proposed new international airport onto a photograph of Macao.
Source: Luc Olivier/ Gamma.

Geography (1965). This work, drawing together the often scattered developments of the previous decade, gave a coherence and direction to theoretical developments few had seen before, and became a source of inspiration to whole generation of students.

As with any innovation, however, there were innovators who went beyond the bounds of the discipline to search for training that was required to encourage other new ways of thinking. Such excursions were documented by Gould (1985), Johnston (1979), Gale and Olsson (1979), Amedeo and Golledge (1975) and others. It is not the purpose here to trace the history of theoretical pursuit in geography. Rather, it is to try to understand how this happened and what has been the impact on the development of the discipline itself.

The word 'theory' is Greek in its roots. In a fine act of what he called 'retrieval', the philosopher, Martin Heidegger (1977) sought the original meaning in an etymological reflection on the verb *theōrein* and its nominal derivative *theōria. Thea* refers to the outward appearance of a thing, while *horaō* means to look closely. Thus, one original meaning of the word that may still stand a chance of appealing to modern sensibility is 'to examine carefully that which is outwardly before us'. A second etymological possibility is that *theōria* derives from *Thea*, a goddess, and *ōra*, to honour or esteem. 'To theorize' will thus mean to honour the goddess of truth. Accordingly, Heidegger notes that as the Greek *theōria* is taken over into the Latin *contemplari*, the very essence of its meaning undergoes a radical reversal. This is because *contemplari* means to cut something off, and this meaning is heard still today in the word *template*. But a template is something pre-formed, which is then imposed on something else to make it conform to what we want it to be. Theory, in this sense, can thus be thought of not just as a gentle and respectful standing open to examine that which is before us, but as an authoritarian framework which we bring to something and make it conform to our preconceptions. Against this latter definition, one wonders how often we approach a study or a research project knowing what we are going to find before we have looked? How often do our 'theoretical templates' shape our seeing, making us oblivious to some things, while forcing our attention on others?

Notwithstanding the dual possibilities of its original meaning, theory represents systems of reasoning concerned with explaining real-world situations. Providing an exact explanation or an exact prediction for an event or events that happened in the real world, however, requires an extremely complex and complete system of reasoning. Rarely are such exact explanations or predictions possible. Instead approximations become the best solutions. Such approximations, however, command different degrees of confidence. Using them, we may be able to achieve some accurate predictions and, indeed, some acceptable explanations. In the geographic domain, little success has been achieved in formulating well-defined theories or models of spatial structure that allow exact predictions and explanations of real-world phenomena to take place. Both deterministic and stochastic inferential processes have been used in the attempt to develop geographic theories and models to provide appropriate levels of satisfaction.

Theories and the models that represent them are usually presented in a logical form. Throughout the early period of the development of geographic theories, little was known about the general influence of geographic space on human activity. Little information was formally structured concerning the patterns and distributions of phenomena over the earth's surface and the interactions between them. As a result, the idea of producing reliable and valid geographic theory with exact predictive or explanatory capabilities was regarded as a hopeless dream. However, geographers began the long and arduous task of moving in the direction of this distant goal. Like many other disciplines, they initially embraced the idea that one can simplify the real world with an appropriately selected set of assumptions about the environment, the people and the human–environment relations. Similarly, since relatively little was known about the stochastic nature of geographic events or about human behaviour in space, the initial step taken by many geographers was to embrace deterministic rather than stochastic inferential processes.

In this way, geographers began developing theories and models that described conditions as they *might be* or *ought to be* rather than as they were. Thus, while some geographers were content with descriptive models of change (such as the detailed analysis of the nature of growth and change within spatial systems), others adopted a more rigorous but, to many, less satisfying normative mode. The descriptive modellers made little attempt to abstract from reality and were content to analyze and interpret empirical data. As Amedeo and Golledge (1975:270) put it, they were in fact concretely tied to actual events in the real world.

On the other hand, those choosing to work in normative worlds sought to describe in a more general way the variations in spatial structure that could conceivably exist (or perhaps even should exist) if specific constraints were adhered to. The value of this latter approach lay in the fact that it was possible to work with a well-structured logical system, use a deductive method of inference, and produce successive waves of information through unravelling the effects of changing assumptions or linking heretofore unlinked axioms. In a real sense, any new output from this approach represented only the logical outcome of individual and joint consideration of the basic axioms and propositions that formed the base for the theory or its model. Given this type of approach, the worth of the consequent theory and model could not be estimated by attempting to compare them with real-world situations, unless of course the theory or model is so complete that the real-world situation closely approximates the conditions under which the theory or model was constructed, or the intention is to use the normative theory to pass judgement on the real-world situation. Rather, the efficacy of the theory or the model built in a normative framework lay in the accuracy of its logical structure and the legitimacy and validity of the inferences made from that structure. These are then used as yardsticks to assess real-world situations in terms of concepts like spatial efficiency or some other criteria on which the theories or models are based. The bulk of these normative theories were developed in the context of economic geography. It is not surprising, therefore, that in their initial phase, the focus of geographical theories was on the spatial manifestations of sets of economic premises such as least cost, maximum profit, level of competition, economic rationality, scale economies, and so on.

Notwithstanding, it is possible to categorize theories that have evolved in geography under four broad categories. First, there are theories of location which seek to explain the laws of spatial distribution; second, there are theories of interaction which are concerned with explaining the laws of movement and spatial behaviour; third, there are theories of growth and development which seek to explain the nature of past, present and future states of being and finally, there are the theories of decision making and choice which are attempts at explaining observable regularities and repeatable trends in individual, group, institutional and governmental behaviours.

Theories of Spatial Location

In general, theories of location focus on certain aspects of geographical space, notably the role of space as an impedence factor. To cross space involves time, various costs and other inconveniences. These costs are to be avoided as much as possible. Consequently, location theories seek to explain and predict how the organization of society is spatially structured to keep down and minimize the so-called frictions of distance. Thus, to many, human geography became the study of spatial science and attempts were made to develop theories reflecting this position (Nystuen, 1963). Emphasis on the new position evolved from previous concern about the subject's characteristics of places and was a positivist approach to the problem of areal differentiation. If areal differentiation can be explained by the articulation of relevant laws, then these laws can in turn be articulated to form theories from which models and hypotheses can be derived and tested.

In its initial phase, the greatest impact in this area came from attempts to articulate the spatial implications of various economic theories. Perhaps the earliest influence in this regard derived from the work of German spatial economists. Certainly Johann Heinrich von Thunen (translated 1966) and Alfred Weber (1909) have become household names to virtually every geography student who takes an introductory course in human geography or an advanced course in economic geography.

The pervasive influence of space had, for the most part, been ignored in classical economic theory. For example, in formulating a theory of land rent, Ricardo (1817) argued only that as soil fertility changed, productivity levels would also change, thus ensuring that land of the highest fertility or quality would provide more output per unit area than land of inferior quality. It was only this difference in fertility that for him determined 'rent' which he interpreted as a return to the specific characteristics of land as a factor of production. But von Thunen wondered if there were other deep-seated and not so obvious factors influencing not just 'rent' but also the way land uses developed and were distributed in space. He experimented on his family holdings at Tellow near Rostock in Germany. He found that some land uses required intensive rather than extensive cultivation, that some products were required at market in a fresh and unspoiled manner, that other products could be transported longer distances without any deterioration in their quality or usefulness, and that the longer distance a product had to be transported, the greater transportation costs played in the total cost picture. Von Thunen thus established the importance of the tyranny of distance in the process of production and movement of agricultural and pastoral land products to market. He argued that even if the land was uniformly homogeneous with respect to climate, soil fertility, and slope, distance from market would still mitigate the types of land use that could economically be undertaken.

It was on the basis of this idea that land-use types would change as their marginal rent changed that von Thunen built his model. Given a number of assumptions, von Thunen poses the question: how will agricultural production arrange itself spatially? The answer, as demonstrated by him, is that concentric land-use zones will emerge and the types and quantities of products demanded in the city and the consequent prices that these commodities will command in the market will influence the total volume of production of each type of land use. Land uses would thus enter into competition with one another to determine which use will occupy a given location. The land use that can command the greatest price will consequently bid the highest for the right to occupy a land unit. Following on this, land uses will generate families of bid-rent curves for different sites based on their value and need. It can then be deduced that as one use achieves a capacity to outbid others it will then, therefore, occupy the land area. Thus, using the notion of bid rents, a simple scheme of land-use gradients can be developed (figure 2.1).

Geographers gradually became aware of this model, regarding it as a norm for interpreting land-use patterns (Chisholm, 1962) at both a national scale (particularly when a country or a region was dominated by a single large market) and, on the other hand, a micro-scale, for looking at the gradient of land uses around individual cities. The theory has established without doubt the importance of geographic space even when many of the normal environmental features such as slope, soil fertility, climate, temperature and rivers were assumed unimportant. The von Thunen model as it was first introduced and used in geography was largely descriptive. It was not until the quantitative revolution that researchers with the requisite mathematical and model-building background were able to examine the full range of geographic implications.

This descriptive model was, in 1954, operationalized in the form of a mathematical model by E. S. Dunn, Jr. (1954). Dunn concurred with von Thunen's basic ideas when considering product type and location. He formalized the von Thunen model in a functional relationship expressed as follows:

$$R = E\,(p - a) - E\,fk$$

where R = rent per unit of land; E = yield per unit of land; p = market price per unit of commodity; a = production cost per unit of commodity; f = transport rate per unit of distance for each commodity; and k = distance from market. Dunn's model, however, summarizes a one-product linear model in which a given land use is capable of producing a net return out to that point where its marginal revenue and marginal costs are equal. Here receipts will be maximized at a site as close to the market as possible and will be minimized at that distance where marginal returns equal zero. However, when considering competitive products, Dunn showed that a variety of situations might occur.

In one area the implications of this theory were extended to gain an understanding of the internal structure of cities and of how the urban land market operates. Ernest Burgess (1925) had used the concentric ring model to theorize about patterns of urban land use and growth in one of America's burgeoning cities (Chicago) and was joined in such a formulation by Homer Hoyt (1933) a few years later. Harris and Ullman (1945) also called attention to the three con-

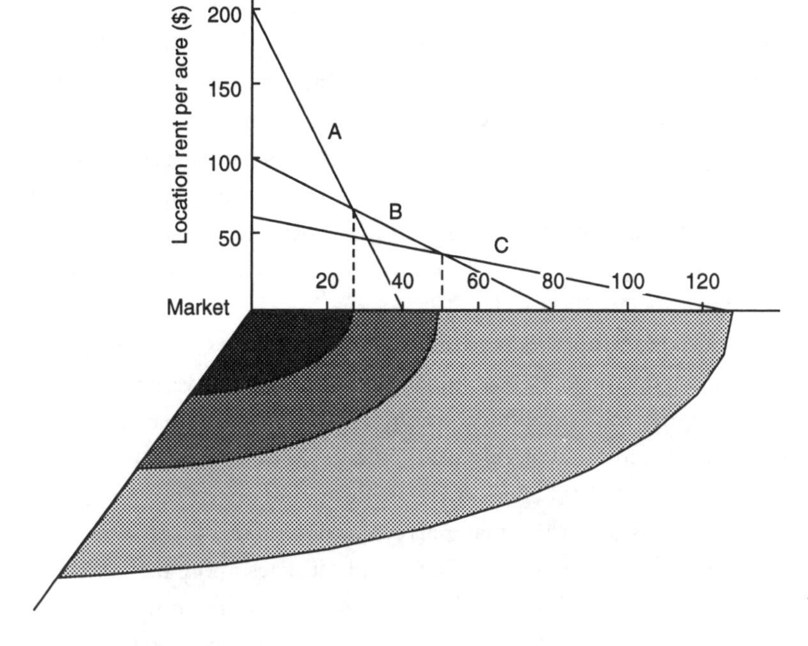

Figure 2.1
Land-use zones from bid-rent curves

ceptual ways of looking at the structure of a city, namely the concentric zonal, the sector and the multiple-nuclei. But it was the regional scientist, William Alonso (1964) who showed that, with the central business district (CBD) operating within a city just as von Thunen's city operated within its isolated state, urban land uses could be illustrated as being in competition around the CBD and generating bid-rent curves on the basis of which land use can pay the highest price per unit of land. The resulting land-value gradients explain the tendency in most cities for multi-storey buildings, occupied by commercial and office establishments, to be found near the city centre and for industrial and residential buildings in different densities of use to succeed systematically as we move to the periphery of the city.

As distinct from this essay into agricultural location theory and its extension to explain the internal structure of cities, a more specialized and systematic location theory evolved from the work of Alfred Weber (1909). Weber too was interested in the tyranny of distance and the role that distance played in location decision processes. As a spatial economist, distance to him had to be translated into a cost unit – the unit he chose was transportation costs. Weber then proceeded to build a series of models that helped to explain what particular combination of regional and local factors influenced the locational decision processes of different industries. He employed the Varignon weight triangle to solve for least-cost solutions, using a mechanical device from physics as an analogue to solve difficult problems in the calculus of variations, a branch of mathematics then underdeveloped. The use of analogies drawn from the physical sciences by these early ordering constructs was hardly surprising in a century strongly influenced by the early writings of Auguste Comte (1830–42) and his 'social physics', but nevertheless it was a surprisingly robust source of analogy which still permeates many theoretical constructs today.

In considering the regional and local factors that influence locational decisions, two dominant lines of thought were noted: the first emphasized cost minimization; the second profit maximization. Weber concentrated largely on the first. He observed that, by definition, manufacturing performs transformation operations on a set of raw materials. These raw materials can be natural resources or the final product of another industry. For most industries, more than one raw material is used. The required raw materials are, however, not uniformly available at all locations. Likewise, the potential markets for the manufactured products were distributed differentially in space. The questions raised by Weber, therefore, were as follows: to what extent is the attraction of raw materials influential in the location decision-making process? To what extent is the physical composition of the materials or the nature of the process using those materials important in the location decision-making process? Do the raw materials used in the production process lose or gain weight, bulk, or volume during manufacture? Is the raw material perishable? How does the value per unit weight of raw material influence the location decision-making process? Are substitute raw materials readily available? In what proportion are different materials used in the manufacturing process? How does the influence of raw materials change as technologies of use and distribution change? And finally, how will the answers to these various questions help decide a geographic location for a given industry?

Weber's reasons for asking some of these questions are straightforward. If a bulky or weighty material can be reduced in bulk or weight by a preliminary manufacturing process, then some savings of transportation cost will be obtained by locating at the material site. This saves payment for the transfer of waste products. If the raw material is perishable but is made more durable during manufacture, economic losses may be avoided by doing this transformation at or near the raw material site (e.g., the production of cheese or butter or vegetable canning). If the value of a raw material is high relative to its weight or volume, it can be transported long distances and the need for a raw material orientation is reduced (e.g., wool). If the value of a raw material is low in its initial state (e.g., copper ore), it cannot be transported too far before transport costs exceed the value of the materials.

Products gain value in the course of manufacturing. The higher this added value, the more likely that the manufactured product can absorb the transportation costs. However, when the sum total of raw materials used to make a product greatly increases the weight, volume or bulk of the product, there is a tendency to achieve cost minimization by locating at the market and bringing the raw materials to that site (e.g., beverage production). These principles are summarized by Weber into two major hypotheses: the weight loss hypothesis and the weight gain hypothesis. In simple cases where one could assume a linear system with a single raw material and a single market, then location was obviously either at the raw material, the market, or some other location usually between the two. If the raw material was weight losing, then the raw material site tended to attract the industry; if it was weight gaining, then the market site proved to be more attractive. If one added alternate sites of labour between the two initial locations, then industries having a very high labour cost component in their total cost structure might move to one of the intermediate labour points if labour costs were low enough.

As one moves to more complex cases with two raw materials and one market, or one raw material and two markets, or as other factors of production become more significant in the total process of manufacturing, they exert an attractive influence on the locational decision-making process. Some manufacturing can be considered to be oriented to sites of cheap power (e.g., aluminium), some to cheap labour (e.g., textiles and fabrics), some to indentured labour (e.g., garment making) and so on.

The alternative approach emphasizing profit maximization paid greater attention to considerations of market areas. It tried to address the situation where demand varies over space and the least-cost location of Weber may not necessarily be the point at which maximum profit accrues to the firm. Thus, assuming a spatially distributed market, instead of the Weberian single market, and identical production costs for all firms, a firm will choose a location to maximize its market area; although, it will take into consideration the locational behaviour of other firms.

In an attempt to integrate these two dominant lines of thought, Smith (1955) came up with a theory which posits that the ideal location for a firm may not necessarily be a universal point but an area. The spatial margin or spatial limits theory, as Smith's formulation is usually called, relates the total cost and the total revenue of a firm in an attempt to determine that zone within which the firm could locate. In many respects, this theory is much more geographic than the two approaches discussed above. Nonetheless, the three theoretical formulations have the same vintage in being derived from neoclassical economics with its assumption of an economic human being. They have been criticized as being unrealistic in their assumptions. These criticisms have led to attempts to make more explicit the behavioural component in location decision making and, subsequently, have led to converging interests with the developing theories of decision making and choice.

Nonetheless, Weberian theory stimulated considerable geographic and regional science research into the industrial locational decision-making process. It fostered investigations of decision-making processes with respect to firms and industries and led to the development of typical profiles that helped explain why some locations are chosen rather than others. More than this, its concern with cost minimization promoted the expansion of locational theoretical ideas concerning issues of allocation of facilities (Teitz and Bart, 1968). Working initially with network models on how to maximize access of network regions to selected network nodes, this theory rapidly developed as an extremely powerful tool capable of handling a wide variety of locational problems. Its general aim is how to locate supply sources for various services so that total costs of transportation can be minimized. Typical examples of problems solved using location/allocation procedures included the location of emergency facilities (e.g., police and fire), location of schools, location of health facilities (e.g., hospitals, medical centres, etc.), location of urban parks and recreational areas, and so on. The basic format for these location/allocation models was expressed by Church and ReVelle (1974), and sets of location–allocation computer algorithms were developed and distributed by Rushton, Goodchild and Ostresh (1973).

The essential form of the location–allocation model is as follows:

$$\textit{Minimize:} \quad \sum_{i=1}^{n}\sum_{j=1}^{m} Iij\, Cij$$

$$\textit{Subject to:} \quad \sum_{j=1}^{m} Iij = Si$$

$$\sum_{i=1}^{n} Iij = Dj$$

where: Iij = unknown allocation from source i to demand j
Cij = known cost per unit allocated (i.e., distance surrogate)
Si = known supply available at source i (capacity)
Dj = known demand at point j (population)

Given Xi, Yi as unknown source locations, Uj, Vj as known demand locations, then cost is equated with linear transportation.

$$Cij = \sqrt{(Xi-Uj)^2 + (Yi-Vj)^2}$$

(i.e., given a destination of demand, the sources with known capacities, the aim of the model is to locate sources so that total cost of transportation required to satisfy the known demand is minimized.) In general, the location–allocation theory applies whenever centres have to be located so that each centre serves a prescribed number of people (e.g., school locations).

Further development of location theories was to explain settlement patterns, especially of towns which were essentially market centres or central places. Explaining the spatial marketing process is central to the development of this variant of location theories. This process consists of three sets of interrelated activities which take place through time and space. The three activities identified in classical economic theory are production, consumption and exchange. However, the marketing process can be conceptualized without any consideration of space assuming that everyone is at the market. This enables one to deduce the occurrence of demand and supply relationships, sets of prices, and market structures. But no spatial distribution of any kind is generated. If, however, we introduce the geographic idea that in any economic system individuals do not all exist at the market, but are distributed geographically over space, then the marketing process becomes more geographic in nature. It can be seen immediately that there is potential for developing different spatial arrangements and distributions of market centres. If we explore the effects of space within this simple type of economic system, then geographic theory concerning the distribution and arrangement, the size and frequency, the functional complexity and distance apart, of the various market centres that make up a system begins to emerge. During this inferential process, one can draw on significant steps that have been undertaken somewhat in isolation in economic theory but which prove necessary for the development of strong geographic theory concerning the marketing process.

For example, Losch (1954) spatialized the idea of a demand curve by inferring that the delivered price of a good consisted of a base price and an increment proportional to the transportation cost that an individual at a given location would have to absorb in order to get to a centre or market to purchase the good. Consequently, by rotating the axis of a standard demand and supply curve and using that part of the price axis above the base price as an indication of the impact that increased distance would have on the purchasing power of (or quantity demanded by) potential consumers, Losch established a demand cone which identified the spatial extent of a potential market area at a given base price and a given rate of transportation cost (figure 2.2). This area is circular in form. As prices changed, the size of this market is increased or decreased. Given that more than one entrepreneur existed in an area, a series of demand cones would emerge identifying the trade areas of all viable entrepreneurs. Through the life and death process of economic survival, which depends on covering costs and obtaining a normal profit, the circular market areas would fluctuate until a type of equilibrium or stable pattern emerges consisting of tightly packed tangent circles of equal size.

What then happens to Losch's solidly packed circular trade areas? It must be obvious that tangent circles would produce areas of non-service, where demand is not met. The existence of such spaces would potentially encourage instability in the system as bold new entrepreneurs become established in these interstitial areas and attempt market penetration of the surrounding areas. If it is assumed that all potential consumers in an area need to be served, then the circular-packing principle produces an inefficient solution. A German geographer, Walter Christaller (1933) had empirically developed a theory of the distribution of urban settlements which included a possible solution to these problems of instability and non-service.

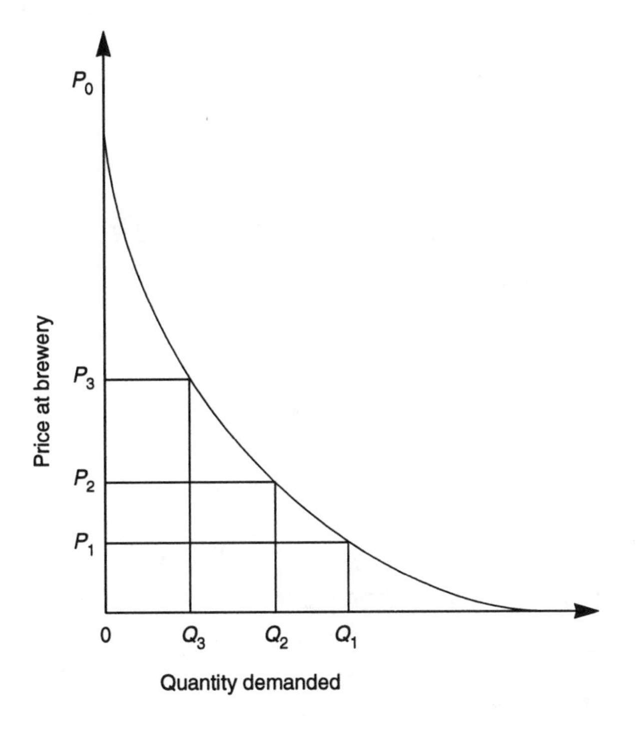

Figure 2.2
Quantity of beer demanded as a function of brewery price

Working with a simplified environment with uniform topography and uniformly distributed populations of identical tastes and preferences, Christaller showed that Fetter-type market area boundaries could be defined between each competing pair of centres. As the distribution of entrepreneurs settled into an equilibrium, the irregular Thiessen polygons formed by the Fetter linear boundaries settled into a uniformly shaped hexagonal market area (the famous Christaller K = 3 system (figure 2.3).

Thus, by combining form-related theories of the potential spatial distribution of market centres with process-oriented theories derived from examining how economic systems work a formal geographic theory emerged. The Central Place Theory, as Christaller called it, has developed as perhaps the single most powerful and influential theory in geography. But, like von Thunen's and Weber's initial theories, it is normative. It tells us what *should* be the case under certain types of constraining assumptions. And that is how many theories are presented. There is no reason to expect them to reflect exactly the world as it is at any particular time; for the world as it is rarely if ever matches the assumptions necessary to build the theory. However, as a normative principle, the theory and its spatial mani-

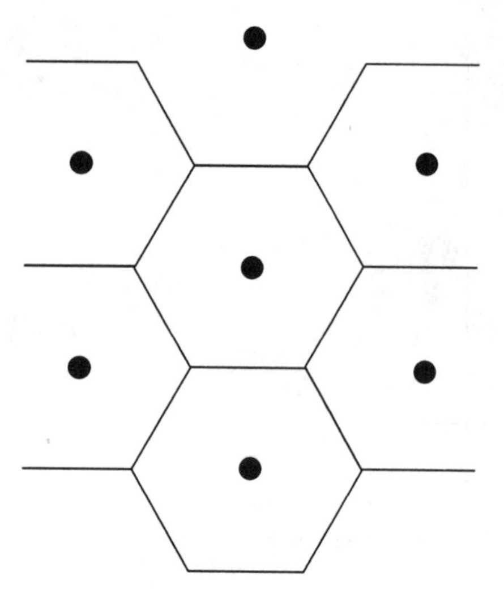

Figure 2.3 Non-overlapping hexagonal market areas

festation, can be used to show departures from this norm. Identifying the reasons for such departures has continued to add significantly to our total understanding of the marketing process and its expression through a system of settlements.

The results of pooling this literature, on market-area analysis and location under various types of competition, together allowed first Christaller and then Losch to produce spatial theories of settlement systems. In Sweden, the geographer-in-exile Edgar Kant (1946) built on the central place theory of Christaller to explain the system of towns in Estonia. Another Swedish geographer, Gerd Eneqvist, from her traditional concerns with problems of cities, explored the central place concept in a manner which came to have strong planning implications among her students (Eneqvist, 1988; Hägerstrand and Buttimer, 1988). As a result, human geography made extraordinary advances in the first decade after the Second World War and its practitioners became catalysts in turn for a new generation of geographers (Ajo, 1955; Godlund, 1951, 1956a, 1956b; Hägerstrand, 1952, 1953) not least through the remarkable *Lund Studies in Geography*, published almost entirely in English since their inception in 1949. Today, the properties of central place systems are well described in many urban geography textbooks (e.g., Berry, 1967; Cadwaller, 1985; King and Golledge, 1978; Yeates and Garner, 1980). Settlement theory is hierarchically as well as spatially organized. Hierarchically, one is able to deduce that as the functional nature of the system of settlement changed from (primarily) marketing to (primarily) transportation to (primarily) administration, there are specific numerical relations between the number of centres at each level of the hierarchy and specific spatial relations concerning the distances separating these centres. For example, in the simple marketing system ($K = 3$), for every centres of one order there is a fixed number of centres at lower and higher orders. In the $K = 3$ system, for every place at one level there would be the equivalent of two *more* centres at the next lowest order. Remember that each centre incorporates the functions of all levels below it and, therefore, acts as a centre at those levels. Consequently, if there was one largest centre, there would be the equivalent of two new lower order market areas plus the larger centre containing a second-order market area, giving the equivalent of three market areas of the next order. Following the same reasoning there would be nine market areas at the next order, twenty-seven at the next order, and so on. Remember, these are equivalent numbers. If at each level we subtract the number of higher-order centres already existing, then the sequence would be as follows: 1, 2, 6, 18, 54, etc. Spatially, the distances apart of centres at each successively lower order would increase by a factor equal to $\lambda\sqrt{3}$ where λ is the distance apart of lower-order centres. Thus, if the lowest order centres were located four miles apart, the next order centres would be ($4 \times\sqrt{3}$); the next highest order centres would be ($4 \times\sqrt{3} \times\sqrt{3}$) and so on.

A probabilistic approach to this problem was also provided by Leslie Curry. Coming into contact with the then new field of operations research at Johns Hopkins University in the late 1950s, Curry's probabilistic thinking was initially informed by queuing theory, a mathematical area of great richness grounded in questions concerning the random arrival of events which had to be dealt with or serviced in an appropriate way. The origins lay in the early papers of a scientist working for the Copenhagen Telephone Company who had tried to estimate the ratio of telephone operators needed immediately to service all incoming telephone calls or, in general terms, the ratio of arrival to service times. As in so many areas of probabilistic analysis, results were often non-intuitive; for example, as the ratio of arrival to service times approached unity, the queue or waiting line went to infinity. The idea of 'randomness' implicit in probabilistic theory informed a wholly original and fresh approach to the rather mechanistic and geometric developments of central place theory (Curry, 1964a, 1964b).

Theories of Spatial Interaction

Theories of spatial interaction are concerned with patterns of movement between places of people (migrations), goods (transportation), money, information, ideas and so on. The relationship between distance and these various types of interaction was noted by many scholars as far back as the nineteenth century. Carey (1858), Ravenstein (1885) and Spencer (1892), among others, observed the attenuating effect of distance on the intensity of human interaction, and Bossard (1932) observed the influence of distance on human behaviour. Stewart, an astronomer, working closely with William Warntz who had been employed by the American Geographical Society as a research associate to investigate the role of distance as one of the basic dimensions of society, had noted certain regularities in various aspects of population distributions which were akin to the laws of physics (1941, 1942). One of these was a tendency for the number of students attending a particular university to decline with increasing distance of the university from the student's home. Stewart's ideas were introduced to geography through his 1947 publication in the *Geographical Review* (Stewart, 1947). This publication sought to establish that human beings, on average and at least in certain circumstances, obey mathematical rules resembling in a general way some of the primitive laws of physics. Indeed, Stewart identified various aspects of population distribution in the United States where such rules could be said to hold. One such rule, for instance, is the rank-size rule for cities. This emphasized that, at least in the United States, the population of a city when multiplied by its rank (measured as 1 for the largest city down to n for the smallest) and standardized by a constant, equalled the population of the largest city. Another rule noted that the distribution of a population could be described by the population potential at a series of points, in the same manner as the potential in a magnetic field is described in Newtonian physics. Stewart and Warntz (1958, 1959) went on to discover other empirical regularities related to the economic and social geography of the United States which were challenging to theoretical formulations. Similar spatial principles which, if not laws, are at least a mark of concern for theoretical thinking, appear in economics (Reilly, 1931; Hoover, 1937) and shortly after the Second World War in psychology (Zipf, 1946; Miller, 1947) and mathematical biophysics

(Rashevsky, 1953), while being extended further in sociology (Dodd, 1950, 1955; Iklé, 1954) and economics (Blumen, Kogan and McCarthy, 1955).

Nonetheless, neither Stewart nor Warntz attempted to give any reasons nor formulate any hypothesis to test the basis of any of these observed regularities. It was left to Zipf (1949) to formulate such an hypothesis in the form of a principle of least effort. According to this principle, individuals organize their lives so as to minimize the amount of work they have to undertake. Movement involves work, and so the minimization of movement is part of the general principle of least effort. To explain this, Zipf elaborated on the empirical regularities already identified by Stewart, especially those relating to the relationship which showed that, with increasing distance, the number of students attending a particular university tended to decrease. He emphasized that, from the point of view of the principle of least effort, going to a university, any university, entailed two aspects of work. The first involved acquiring the necessary information about the university; the other involved the actual travelling there. The greater the distance between any potential student's home and the university, the less likely it is that the student would have enough information about the university and the less likely that he would travel there. The validity of this distance-decay function was tested against many other sets of data. In virtually all cases, it was found that the greater the distance separating any two places, the smaller the volume of contacts or interactions between them.

Zipf referred to this regularity as the P_1P_2/D relationship, which is another way of writing the Newtonian gravity formula earlier identified by Stewart. In Newtonian physics, the formula is of the form:

$$F_{ij} = K\frac{M_iM_j}{d_{ij}^2}$$

where M_i and M_j represent the mass at locations i and j respectively; d_{ij} is the distance separating i and j; k is a coefficient of proportionality (a scaling coefficient); and F_{ij} is the gravitational force between i and j. To use the formula to establish spatial interaction of the type proposed by Stewart and Zipf, all that is necessary is to substitute population (P) for mass (M) and (I) for (F) to indicate interaction.

Much work has been done in geography involving the fitting of this equation to flow or movement between and among places of people, goods and information. Olsson (1965) provides a very detailed review of contemporary work in this field. He observed that, in most cases, to achieve reasonable statistical fits, the various elements of the equation have had to be weighted in the form:

$$I_{ij} = f(P_i^a P_j^b d_{ij}^c)$$

where a, b and c are weights estimated from the data set being analysed. Because of the fact that in almost all studies, different values of a, b and c were produced, considerable controversy raged as to whether the gravity model, in spite of validating the empirical regularity of interaction, actually furnished any theoretical explanation for the pattern observed. In effect, all that the analytical effort was doing was to indicate that the influence of distance on human interaction, particularly on migration from one place to the other, cannot be deduced unambiguously since apparently it varies from place to place, from population to population, and from context to context. The influence was seen to be virtually universal and the theoretical formulation up to then had not been able to account for the variability in the strength of the distance factor.

This, of course, did not deter scholars from seeking to get round the problem in other ways. Borrowing from the work of Stouffer (1940) in sociology, who argued that the difficulties might be in the meaning and measurements attached to each term in the equation and who proposed that in the case of migration the distance function could be better conceptualized as intervening opportunities, Ullman (1956) applied the theory to explain the pattern of commodity flows in the United States. According to him, spatial interaction among American cities on the basis of the amount of commodity movements between them can be explained in terms of three factors; complementarity, intervening opportunity and transferability. Complementarity refers to the degree to which there is a supply of a commodity at one place and a demand for it at the other; intervening opportunity relates to the degree to which either the potential destination can obtain similar commodities from a nearer, and presumably cheaper, source or the potential source can sell its commodities to a nearer market; transferability defines the condition required to be fulfilled for complementarity to exist. It relates to the necessity for movement between any two places to be feasible in terms of the mode of movement, the time required and cost. Nonetheless, not even this conceptual refinement was easily fitted statistically to the gravity model but its relationship to the latter is clear. Undeniably, fitting such models would require accepting that the influence of distance, however expressed, whether as time, cost or any other variable, would vary from place to place and from one period to another.

All of this work on the analysis of spatial interaction and the various movement patterns gained in salience and came to be applied in the planning field, forecasting the growth in traffic flows and the consequent land-use implications. Indeed, land-use and transportation planning (especially intra- and inter-urban) became increasingly sophisticated in a technical sense as they came to rely more and more on spatial interaction theories and their associated quantitative methodology. This was particularly true of the United States and the United Kingdom over the period 1950 to 1970. Initially, data were collected to demonstrate the traffic-generating capacity of various land-use types. Many variants of the gravity model were then applied to these data to evaluate the pattern and intensity of interaction between different parts of the city. The results made it possible to design future land-use configurations, compute their traffic-generating potential, and predict the pattern of flows and the needed road system. One of the most influential variants of gravity models developed in this context was that by Ira Lowry, which was able not only to assess different land-use configurations in terms of traffic flows but also suggest the best 'directions for future urban growth' (Batty, 1978).

The continued demands for sophisticated planning devices stimulated much research based not only on the gravity and Lowry models but also on systems analysis. The use of systems analysis in spatial interaction theories allows dynamic processes to be formally incorporated into the formulation instead of the purely static patterns which are the outcome of such processes. This requires a greater degree of mathematical sophistication and a concern for enhancing the capacity not only to understand change but also to forecast it. Dynamic systems theory and analysis has been the focus of a major research programme conducted at the University of Leeds by Alan Wilson and his associates since the mid-1970s. Wilson, who came to geography after training in physics, noted that many geographical analyses of dynamic systems focused on static spatial patterns or structures, representing these as equilibrium or steady-state situations within ongoing dynamic processes. Change is then handled by forecasting in some other theory or model the independent variables associated with a system and then calculating the new equilibrium or steady state (Clarke and Wilson, 1985:429). Wilson emphasized that such a procedure is untenable when faced with a dynamic and complex system such as a city where there are too many possibilities of transition to different kinds of equilibrium or non-equilibrium states.

Consequently, Wilson questioned the validity of much of the work that had been done with geographical forecasting in the planning field. As a first step, he worked on the gravity models and sought to derive forecasting mathematically so as to give it a stronger theoretical base (Wilson, 1967). He also extended the Lowry model to incorporate a more general set of models concerned with location, allocation and movement in space. He then proceeded, on the basis of the dynamic systems analysis, to introduce the concept of *entropy* into theories of spatial interaction. His use of this concept has been derived from statistical mechanics rather than information theory. Wilson's entropy maximization approach and his derivation of interaction models was thus grounded in 'the most likely state' of a system, focusing explicitly on the constraining effects imposed by relative locations, the connecting costs, and the total amount of money available, as these were measured empirically by Lagrangians. These, in turn, could be interpreted as indices of relative accessibility within the interacting system.

This approach allowed Wilson to conceive of a city as an interactive system in the form of a flow matrix. The flows relate to trips or movements made by groups of individuals on a daily basis from their residential areas to their workplaces. The number of trips originating from a series of residential areas is known, as is the number ending in each of a series of workplace areas. However, the entries in the cells of the matrix with regard to which people move from which residential area to which workplace area are unknown. The question is: what is the most likely flow pattern? Even with only a few areas and relatively small numbers of trip makers, the number of alternatives is very large. Wilson defines three states of such a system. The first is what he called the macro-state. This comprises the totality of trip makers at each origin and the numbers of jobs at each destination. The second he calls the meso-state. This comprises a particular flow pattern, say nine people from zone A going to zone Y and five others from the same origin going to zone Z. What is not known is which nine are in the first category and which five are in the second. This then leads to a consideration of the third or micro-state which represents a particular example of the flow pattern in the meso-state but is only one of many possible configurations of fourteen people moving from zone A, nine to Yand five to Z. Entropy-maximizing procedures find that the meso-state with the largest number of micro-states associated with it has the most probable distribution but is also the one that the analyst is least certain which configuration produces the pattern. The emphasis of this theoretical effort is, thus, about probability and the aim of the theory and of the modelling is to describe the most probable system structure from a given amount of information which is most likely incomplete. The hypothesis that the entries in the flow matrix conform to the most likely distribution can be tested against real data. If it fails to conform, either entirely or in part, it can be refined by building in more contraints. Wilson does this with his intra-urban transport models by introducing, for example, travel-cost constraints, different types of trip makers (identified by class, age, sex, etc.), different types of jobs, and so on.

Wilson has developed both the theory of his modelling and the substantive applications to the theories of spatial interactions (Wilson, 1981). He has produced a whole family of models that can be applied to (and used to forecast) various components of a complex spatial system such as an urban region (Wilson et al., 1981). As a theoretical framework, entropy maximization has proved catalytic to thinking about many problems of spatial interaction even though its balancing Lagrangians point to the hermetically closed nature of the model and, in empirical terms, results in its performing no better than the models based on simple Newtonian analogues (Openshaw, 1989).

In spite of these limitations, much of Wilson's work has been expanded and extended to other areas of spatial interaction. Johnston (1985), for instance, extended its use to the estimation of spatial variations in voting behaviour in England which arose from the flows of information and ideas in different locations. Wilson's impact on development within this field has been aptly described by Gould. In a review of a 1970 presentation of the work of Wilson and others, Gould (1972:689) commented that this was the most difficult contribution he had ever read in geography, but that it planted a number of those rare and deep concepts whose understanding provides a fresh and sharply different view of the world.

Theories of Growth and Development

Notwithstanding the conceptual and analytical sophistication developed in articulating theories of spatial interaction, there grew up within the discipline a strong reaction against their limitations in dealing with problems of growth and development in different parts of the world or even within a single country. Initially, such problems were conceived of in terms of change and spatial dynamics and were addressed in the context of spatial diffusion. Change, it was argued, does not take place uniformly over space. Usually, it originates at one or a few locations from whence it spreads to others through space and time. Diffusion research, thus came to receive considerable attention in the late 1960s. Its basic behavioural postulate, derived largely from sociological research findings, emphasizes the importance of face-to-face communications in the diffusion of innovations. Following the pioneering work of the Swedish geographer, Torsten Hägerstrand (1968), the implications of spatial factors began to be evaluated in such theoretical systems. The attenuating role of distance, the neighbourhood effect on the rate of innovation, and the resultant uneven spatial pattern of spread, were all clearly identified as critical in a theoretical formulation of geographical change. A Monte Carlo simulation technique was initially developed to explicate the process. Such simulation was applied by Morrill (1962) and Hudson (1969) to show how the settlement pattern in a given area develops over time. The latter, in fact, used the results to identify three processes of spatial development which he characterized as colonization, spread and competition. Brown (1981) provides an extensive review of research development in the field and calls attention to the increasing sophistication in analytical techniques for investigating the process.

One such later development, involving a broad range of methods coming under the general umbrella 'expansion models', is associated with the name of Emilio Casetti (Jones and Casetti, 1992). In essence, these models describe changes in some variable of interest as a function of time. They are parameterized from global values and then allowed to adapt their parameters according to relationships with other variables that change from place to place over a region. For example, at the regional level the course of an epidemic may follow the familiar sigmoid function characteristic of a birth–death process. Nonetheless, since it may also be density-dependent, the classical diffusion curve will change if its defining parameters are controlled by geographic variations of population density across the region. The same basic principles underly spatial adaptive filtering (Gould et al., 1991), in which variations from place to place allow the parameters of a function to adapt to local conditions so as to give better local fits. Here the information in both the time and the spatial series is used simultaneously to predict the next maps. Nonetheless, like the positivist stance of gravity models of spatial interaction, these formulations say little about the overriding problems of societal development in the real world.

Indeed, Smith (1989) provides four specific criticisms of the positivist stance of all of these theoretical formulations with regard to their ability to provide an understanding of the problems of growth and development in geography. These criticisms include: first, the pretentious claims of their postulates (and the empirical regularities that they uncover) to scientific status, in the sense of being regarded as laws or virtual laws with universal validity when in fact they are no more than the product of a particular socio-economic situation; second, their inherent lack of historical specificity and therefore their neglect of social inequalities, discrimination and domination which underlie many real-world events and processes; third, their dependence most of the time on neoclassical assumptions of a space-economy that is inherently tending towards equilibrium when in fact crisis, disequilibrium and periodic discontinuities are more the norm in real-life economic situations; and, fourth, their crude conceptualization of geographic space as an inert field of activity, which obscures social relationships between people by treating them as spatial relationships between places.

As a result of this dissatisfaction with the state of theory in the discipline represented by these criticisms, geography went through a period of reappraisal with a search for an alternative theoretical approach. The mid-1970s saw the flowering of a radical school of thought in the discipline that was concerned with critically examining and exposing how much of conventional spatial relationships in geography were undergirded by the inequalities emanating from the dominant and global capitalist mode of production. Problems of growth and development came to be increasingly formulated and better understood in terms of the social relations deriving from this mode of production. Theories of growth and development in geography thus came to lean heavily on the principles of dialectical materialism espoused by Karl Marx and refined by later scholars. Dialectics, in this context, is defined as a way or process of theoretically capturing interaction and change in a historical context. As a process, it emphasizes the struggle between opposites and the need to conceptualize long-term dynamics in human social relations in terms of non-teleological historical laws. Materialism, on the other hand, stresses that, in such relationships, matter (reality) precedes the mind (knowledge), that consciousness results from experience, and that experience results primarily in the material reproduction of life. Dialectical materialism thus provides a theoretical framework for analysing societal development in space in terms of the modes of production; the forces and relations of production; the struggle within each mode by these forces and relations; the consequent growth, transformation and succession of modes of production through time; and the implications of this succession for the eventual achievement of a society characterized by high levels of development, socialized ownership of the means of production, economic democracy, freedom of consciousness and a heightened sense and system of social responsibility.

The particular version of Marxism that influenced earlier works in geography was that of Louis Althusser (1969) and his followers who identified the strong emphasis on structural relations in the works of Karl Marx. This version was particularly relevant because of the many problems it addressed which were of interest to geography. These include the structures of pre-capitalist societies (Meillasoux, 1981; Terray, 1972; Hindess and Hirst, 1975), the articulation and historical transition of modes of production as presented by Rey (Wolpe, 1980); the role of the state in development (Poulantzas, 1975, 1978); and the critical analysis of culture, ideology and consciousness. But it was undoubtedly the publication of Manuel Castells' *The Urban Question* (1977) that indicated the immensely fruitful implications of Marx's ideas for the elaboration of theories of growth and development in geography. Castells saw the city as the projection of society on space. People in their relations with one another give space a form, a function, a social signification (Castells, 1977:115). In consequence, according to Castells, any theory concerning geographical space is an integral part of a general social theory.

Turning to Althusser's conception of modes of production and their constituent elements, he (Castells, 1977:126) argues that:

> *to analyse space as an expression of the social structure amounts, therefore, to studying its shaping by elements of the economic system, the political system and the ideological system, and by their combinations and the social practices that derive from them.*

Under capitalism, the economic system is dominant and is the basic organizer of space. It does this through the three types of economic activities which it promotes. The first comprises of productive activities concerned with producing economic goods and services at particular locations. The second relates to reproductive activities of society as a whole and involves, among other things, the provision of housing and other social services such as education, health care, recreation and so on. The third entails exchange activities such as commerce, transportation and communications. Each of these groups of activities produces their own type of spatial structures. The political system, in its turn, organizes space through its two essential functions of domination/regulation and integration/repression whilst the ideological system marks space with a network of signs loaded with ideological content. Over and above the role of these three social systems in organizing space is the impact of three other factors namely the structural combinations of the three systems, the persistence of spatial forms created in the past, and the specific actions of individual members of social and spatial groups. According to this formulation, then, theories of growth and development concerning any area or region of the world must begin with an abstract theorization of the modes of production and then go on to form a concrete analysis of the specific ways structural laws of contradiction, struggle and transformation are realized in spatial practice.

Within the English-speaking geographical community it is, however, the work of David Harvey that has been most influential in promoting work in geography on theories of growth and development within the Marxist framework. His book, *Social Justice and the City* (1973), written about the same time as the French version of Castells' book, derived its inspirations, apart from Marx's writings directly, from the works of other scholars of structural Marxism notably Piaget (1970) and Ollman (1971). According to Harvey, the contradictory relations and struggle between the constituent social classes in society – notably the classes of capital and labour – govern structural changes and promote or impede growth and development. Structure, in this context, refers to a system of internal relations which is being structured through the operation of its own transformation rules. Contradictions occur between and within these structures. The political structure, for instance, apart from having its own internal contradictions and revolutions, might be in a contradictory relation to the ideological structure, and both of them in turn could be in contradiction to the economic structure of society.

In applying these concepts to the role of cities in development, for example, Harvey notes that the form and functioning of urbanism, generally, is related to the dominant mode of production. In general, cities are socio-economic and spatial forms whose functions are to extract significant quantities of the social surplus created by people in a given politically organized space. They are to be found functioning under different modes of production and economic integration (whether involving reciprocity, redistribution or market exchange). The transformation from reciprocity to redistribution precipitated the development of a separate urban structure with, however, limited power of inner transformation. This transformation itself was based on contradictions between the forces and relations of production, particularly as they relate to property rights. As property rights were not fully developed and pervasive in earlier modes of production and economic integration, cities functioned more as political, ideological and military centres rather than as economic ones. Their powers of transformation both of themselves and society were limited. By contrast, the transition from earlier political cities to the modern commercial and industrial city based on market exchange was catalytic and entailed a fundamental inner transformation of urbanism itself. This involved major reorganization of productive forces and their polarization (between capital and labour). Nonetheless, Harvey argues that modern urbanism must not be conceptualized as simply created by forces and relations of production but must be appreciated as both expressing and fashioning social relations and the organization of space and production.

Other geographers extended Marxist theoretical formulations to other areas of concern in the discipline relating to issues of growth and development. Smith (1984), for instance, explored the phenomenon of uneven development in the world through understanding the role of capitalism in the production of space and the transformation of the environment on a global scale. According to him (1984:xiv):

In its constant drive to accumulate larger and larger quantities of social wealth under its control, capital transforms the shape of the entire world. No god-given stone is left unturned, no living thing is unaffected. To this extent the problems of nature, of space and of uneven development are tied together by capital itself. Uneven development is the concrete process and pattern of production of nature under capitalism.

But as Peet and Thrift (1989:13) observed, with this recognition of uneven development as a structural imperative of capitalism and as a necessary outcome of the unfolding of capital's inherent laws, the structural approach to space reached its ultimate conclusion.

Not unexpectedly, other attempts have emerged in the discipline to reconceptualize and advance the analytical potency of the Marxist contribution to theories of growth and development. One such reconceptualization emphasizes that Harvey's structuralism tended to conceive of geographical space as a passive surface expressing the mode of production rather than an active social force for change and development in its own right. According to Massey (1984:52):

The fact that processes take place over space, the facts of distance, of closeness, of geographical variations between areas, of the individual character and meaning of specific places and regions – all these are essential to the operation of social processes themselves. Just as there are no purely spatial processes, neither are there any non-spatial processes. . . . Geography in both its senses, of distance/nearness/betweenness and of the physical variations of the earth's surface (the two being closely related) is not a constraint on a pre-existing non-geographical social and economic world. It is constitutive of that world.

Massey, consequently, emphasized that as social classes are constituted in places, so class character varies geographically. This implies that the social structure of any economy will develop necessarily in a variety of local forms which she termed the spatial structures of production. Using the example of a multi-locational corporate firm, she observed that such a firm would organize its activities and its space in a hierarchical form, assigning different stages in its production process (organization, research, assembly, parts making, distribution, etc.) in different combinations to various regions, thereby creating different spatial structures in different parts of a country or the world.

Such spatial structures, Massey maintains, not only emerge from the dictates of corporate initiative but are also established and changed through political and economic battles on the part of social groups (that is, through class struggle) in the different places. In turn, spatial structures, through differential employment possibilities, create, maintain or alter class and gender inequalities over space. In short, Massey's main point is to stress that spatiality is an integral and active condition of the production process (1984:44).

One of the important consequences of the Marxist framework for articulating theories of growth and development in geography has been the fact that it has provided opportunities for introducing further refinements to its categories and analytical modalities, and a suitable theoretical orthodoxy, in its structuralist posture at least, against which controversies and contestations of new ideas can be directed. Such refinements and contestations include the ongoing structure–agency debate; the issue of Fordist and flexible production and capital accumulation processes; the new international division of labour; the problem of patriarchy and the subordination of women; the ecological movement; and postmodernist concerns within the discipline. Another important consequence is the enhanced predisposition of geography to be more conscious of the philosophical underpinning of some of its analytical and conceptual constructs. As far as theories of growth and development are concerned, the vigour of the contestations within and against the Marxist framework has meant constant reappraisal and revision of extant theories. It has also entailed a broadening of vision and concerns to account for the role of major 'non-economic' factors, especially those of the state and civil society in emergent spatial structures and organization.

Theories of Decision Making and Choice

One argument against the structuralism of the Marxist framework used to anchor theories of spatial growth and development is that it says very little about the individual decision makers whose aggregate choices determine much in the spatial structure of any country or region. The same criticism has been levelled against the positivist posture of both the theories of spatial location and spatial interaction. Theories of decision making and choice thus begin by challenging the basic axioms of theories of location and spatial interaction which assume that decisions are made on the basis of economic rationality and the implied existence of perfect information. They argue that most decisions are made rationally on the basis of a probably non-random selection of information, are intended to satisfy some goal which is not necessarily to make a perfect decision, and are based on criteria which vary somewhat from individual to individual. Moreover, having learned a satisfactory solution to a given class of problems, decision makers tend to apply it routinely every time such a problem occurs, unless changed circumstances require reappraisal.

One area of geographical research where these theories were first developed was human responses to environmental hazards, especially floods. Based on Herbert Simon's (1957) theories of decision making, Gilbert White and his associates set out in the 1950s and 1960s to investigate the way men view the risks and opportunities of their uncertain environments and the role this plays in their decisions as to resource management (Kates, 1962:1). Kates provided a theoretical construct which he claimed was relevant to decision making and choice over a wide field. This involved the operation of four principles. The first was that people are rational when they make decisions but only in the context of how they observe and interpret the world. This may be quite different from either objective reality or the world as seen by the researcher. Rational decision making, therefore, is constrained and is not necessarily the same as the maximum rationality assumed in neoclassical normative models of economics. The concept of bounded rationality proposed by Simon was appropriate in this context. The second principle was that people make choices. Many decisions are either trivial or are habitual so that they are accorded little or no thought immediately before they are made. Major decisions, such as those concerning the use of environmental resources, may also be habitual but such behaviour usually develops after a series of conscious choices have been made that lead to a stereotyped response to similar situations in the future. Third, choices are made on the basis of information and knowledge. However, only very rarely can all the information required to make a choice be assembled within the time available for taking the decision. Indeed, even if this were possible, people are often not able to assimilate, analyse and use all the information they have. Finally, information for making a choice is usually evaluated according to some predetermined criteria. In habitual choice, the criterion is what was done before, but in conscious choice the information is weighted according to certain rules.

Kates applied his theoretical formulation to explicate why people choose to live in areas they know are prone to flooding. He discovered that their information was based on their knowledge and experience. This could be categorized on the basis of the certainty of their perceptions regarding future floods. In analysing their decisions, it was clear that most of them were boundedly rational and that the people involved had made conscious choices to satisfy specific objectives.

Further development of theories of decision making and choice owes much to the work of Julian Wolpert (1964) who introduced the concept of satisficing to replace that of profit or utility optimization. Comparing the labour productivity of farms in Sweden on the basis of what could be achieved if decision making was optimized, he came to the conclusion that most of the farmers were probably satisficers, acting undoubtedly on the basis of available information and the need to make rational decisions in the context of an uncertain environment. These theoretical formulations were also extended to the field of migration. The objective was to try to understand the decision-making process that lies behind the patterns of migration and so explain the phenomena beyond what the models and analysis of spatial interaction could provide. Wolpert (1965) again found that individuals behaved in a boundedly rational manner, making sequential decisions, first, whether to move at all, and second, where to move to. The final decision to move was then based on their evaluation of the place utility of each potential locational choice, including the one currently occupied. However, the information on which these decisions are based are for many people incomplete, and, in some cases, virtually non-existent. This led Wolpert to form a new concept: action space. Every individual, making a decision about where to move to, is perceived to have an action space. This is the set of place utilities which the individual perceives and to which he responds (Wolpert, 1965:163). Because of its perceived nature, the content of this action space may deviate considerably from that portion of the real world which it purports to represent. Nonetheless, once the decision to migrate has been taken, then the action space undergoes some changes as the potential mover searches through it for potentially satisfactory destinations, in many cases extending the individual's action space in some form of step-wise migration.

The various postulates of human behaviour encapsulated in the theories of decision making and choice encouraged the rise of behavioural geography, with emphasis on investigating the role of social and psychological mechanisms in the development of spatial structures. Behavioural geography helped to extend theories of decision making and choice to other areas of concern in the discipline. One example of this extension is in the area of political decision making. Although the theories till then had emphasized spatial decisions taken largely in relation to environmental issues and primarily within the constraints of uncertainty, utility and problem-solving ability of individuals, political decision making introduced issues of conflict and interpersonal bargaining, and negotiation for conflict resolution. Coping strategies between two interested parties (rather than between human beings and the environment) could entail a mutual exchange of threats and lead to decision making which does not involve the careful and methodical investigation of options until a satisfactory solution is found. In such a situation, decisions are the consequence of conflict between parties with different attitudes and motivations. They are not the product of the joint application to information of criteria that have been mutually agreed upon. Decision making of this variant are usually encountered in situations such as the choice of routes for intra-urban freeways or the siting of community facilities where the positions of different parties have to be ascertained, established, negotiated and resolved.

One aspect of decision making which came to receive some considerable attention is represented by the concept of mental maps, the perceived world or environment that guides the deliberations of decision makers. Drawing initial inspiration from the work of Kevin Lynch (1960), mental maps were construed as an evaluative approach to determining what factors people consider important about their environment and which, having estimated their relative importance, they use in their decision-making process. Although much of the work on mental maps (Gould and White [1974], 1986) has been represented as no more than a study in spatial preferences, it did attract to theories of decision making and choice some consideration of the role of cognitive mapping. Cognitive mapping is a construct which encompasses those cognitive processes which enable people to acquire, code, store, recall and manipulate information about the nature of their spatial environment (Downs and Stea, 1973:xiv). The interest in cognitive processes as it relates to decision making and choice thus came to encourage cooperation with psychologists and sociologists. Such collaboration has produced a geographical alternative to the variety of developmentally based psychological theories of spatial knowledge acquisition and environmental cognition. It incorporates principles such as dominance, hierarchy of decisions and a number of other fundamental principles of geography (Golledge, 1990).

An interesting twist in the development of theories of decision making and choice is provided by the work of a group coming to terms with the problems from the positivist orientation of location theories. This development, initially presaged in the work of Cadwallader (1975), was further elaborated by Wrigley and Longley (1984) and has spawned considerable research interest. It is based essentially on theories of utility maximization. According to these theories, selections are made by decision makers from within choice sets presented to them and which they perceive and evaluate on the basis of the utility which they allocate to each of the alternatives. Consequently, in order to analyse a decision process, it is necessary to know the available choice set, the elements of the choice set considered by the individual, the criteria of each member of the choice set as evaluated by individual decision makers, and the relative importance attached to each criterion. Desbarats (1983) illustrates this with the example of a potential house buyer in a particular town who discards many of the possible houses available through ignorance and then whittles down those perceived as viable buys to a final list from which the selection is made. Given the positivist orientation of this school of thought, sophisticated mathematical modelling and statistical analysis have been introduced to test many aspects of these formulations, although it is recognized that much explication of the models still has to come.

Conclusion

This review of geographical theories has attempted to provide a framework within which to view some of the exciting theoretical thinking going on within geography. I have attempted to capture what could be described as the dominant trends in the theoretical underpinnings of much of the research going on within the discipline. In the process, it has left out some interesting works whose theoretical impact are still far from clear and others which seem marginal to the mainstream of concerns. Clearly, the choice of what to emphasize is personal but the expectation is that what is presented here succeeds in capturing the range of theoretical preoccupations which gives the subject much of its vitality and vibrancy.

The common and dominant element in all of this has been the need for geographers to articulate and make explicit the important role of geographical space in social existence. Geographical space is about the environment within which human beings live. A fundamental assumption of geography, however, is that there is not one single environment. Not only are there various physical and biotic environments but there are a host of social, cultural, political, economic, legal and other environments which daily impinge on human activities, beliefs and values. In addition to these many hidden environments, there are those that are internal to individual human beings. These are the cognized environments, or those transformed by personal experience, by information obtained through personal interaction or via the mass media; or indeed they may be products of our imagination.

All of these environments define the multi-dimensional nature of geographical space. Bringing this space more to the fore in the intellectual understanding of social life everywhere has been a driving force in attempts to develop geographical theories. As has been emphasized within the chapter and will be shown in subsequent chapters, an abundance of regularities can be observed and explained even in the most complex of these different environments. Geographers search for these regularities and try to explain why they occur. But they are also conscious of the important notion that nature-caused or human-caused events occur differentially in these environments and that there are sets of underlying principles and theories which can be formulated to try to understand why these spatial variations occur. This search for an explanation and understanding of places and events in a spatial context remains even in the present, postmodern era, a dominant preoccupation of geographical concern and analysis.

References

Ajo, R. (1955) *Contributions to 'Social Physics': A Programme Sketch with Special Regard to National Planning* (Lund: Gleerup).

Alonso, W. (1964) The historic and the structural theories of urban form: their implications for urban renewal, *Land Economics*, 49:227–31.

Althusser, L. (1969) *For Marx* (translated by Ben Brewster) (London: Penguin).

Amedeo, D. and Golledge, R. (1975) *An Introduction to Scientific Reasoning in Geography* (New York: John Wiley).

Batty, M. (1978) Urban models in the planning process, pp. 63–134 in D.T. Herbert and R.J. Johnston (eds) *Geography and the Urban Environment*, vol. 1 (London: John Wiley).

Berry, B.J.L. and Garrison, W. (1958a) A note on central place theory and the range of a good, *Economic Geography*, 34:304–11.

Berry, B.J.L. and Garrison, W. (1958b) Recent developments of central place theory, *Papers and Proceedings of the Regional Science Association*, 4:107–20.

Berry, B.J.L. (1967) *Geography of Market Centers and Retail Distribution* (Englewood Cliffs, NJ: Prentice Hall).

Blumen, I, Kogan, M. and McCarty, P. (1955) *The Industrial Mobility of Labor as a Probability Process* (Ithaca: Cornell University Press).

Bossard, J. (1932) Residential propinquity as a factor in marriage selection, *American Journal of Sociology*, 38:219–44.

Brown, L.A. (1981) *Innovation Diffusion, A New Perspective* (London: Methuen).

Bunge, W. (1962) *Theoretical Geography* (Lund: Gleerup).

Burgess, E. (1925) The growth of the city, in R. Park, E. Burgess and R. McKenzie (eds) *The City* (Chicago: University of Chicago Press).

Cadwallader, M. (1975) A behavioural model of consumer spatial decision making, *Economic Geography*, 51:339–49.

Cadwaller, M.T. (1985) *Analytical Urban Geography* (Englewood Cliffs, NJ: Prentice-Hall).

Carey, H.C. (1858) *Principles of Social Science* (Philadelphia: J. Lippincott).

Castells, M. (1977) *The Urban Question: A Marxist Approach* (translated by Alan Sheridan) (Cambridge, MA: MIT Press).

Chisholm, M. (1962) *Rural Settlement and Land Use* (London: Hutchinson University Press).

Chorley, R. (1995) Haggett's Cambridge: 1957–1966, in A. Cliff, P. Gould, A. Hoare and N. Thrift (eds) *Diffusing Geography* (Oxford: Blackwell Publishers).

Christaller, W. (1933) *Die zentralen orte in Süddeutschland* (translated by C.W. Baskin, 1963, as *Central Places in Southern Germany*) (Englewood Cliffs, NJ: Prentice-Hall).

Church, R.L. and ReVelle, C.S. (1974) The maximal covering location problem, *Papers of the Regional Science Association*, 32:101–18.

Clarke, M. and Wilson, A.G. (1985) The dynamics of urban spatial structure:

the progress of a research programme, *Transactions of the Institute of British Geographers,* NS 10:427–51.

Comte, Auguste (1830–42) *Cours de philosophie positive*, 6 vols (Paris).

Curry, L. (1964a) The random spatial economy: an exploration in settlement theory, *Annals of the Association of American Geographers*, 54:138–46.

Curry L. (1964b) Landscape as system, *Geographical Review*, 54:121–4.

Dacey, M. (1960) A note on the derivation of nearest neighbor distances, *Journal of Regional Science*, 2:81–7.

Desbarats, J. (1983) Spatial choice and constraints on behaviour, *Annals of the Association of American Geographers*, 73:340–57.

Dodd, S. (1950) The interactance hypothesis: a gravity model fitting physical masses and human groups, *American Sociological Review*, 15:245–56.

Dodd, S. (1955) Diffusion is predictable: testing probability models for laws of interaction, *American Sociological Review*, 20:392–401.

Downs, R.M. and Stea, D. (eds) (1973) *Image and Environment* (London: Edward Arnold).

Dunn, E.S., Jr. (1954) *The Location of Agricultural Products* (Gainesville: University of Florida Press).

Eneqvist, G. (1988) Landscape and life: a personal story, in T. Hägerstrand and A. Buttimer (eds) *Geographers of Norden: Reflections on Career Experiences* (Lund: Lund University Press).

Gale, S. and Olsson, G. (eds) (1979) *Philosophy in Geography* (Dordrecht, Holland: Reidel).

Garrison, W. (1956) Estimates of the parameters of spatial interaction, *Papers and Proceedings of the Regional Science Association*, 2:280–8.

Garrison, W. (1959) Spatial structure of the economy II, *Annals of the Association of American Geographers*, 49:471–82.

Garrison, W. (1960) Spatial structure of the economy III, *Annals of the Association of American Geographers*, 50:357–73.

Garrison, W. and Marble, D. (1957) The spatial structure of agricultural activities, *Annals of the Association of American Geographers*, 47:137–44.

Garrison, W. and Marble, D. (1961) *The Structure of Transportation Networks* (Washington D.C.: US Department of Commerce).

Getis, A. (1963) The determination of the location of retail activities with the use of a map transformation, *Economic Geography*, 39:14–22.

Godlund, S. (1951) Trafik, omland och Tätorter, in G. Eneqvist (ed.) *Tätorter och Omland* (Uppsala: Uppsala University Geographical Institute).

Godlund, S. (1956a) *Bus Service in Sweden* (Lund: Gleerup).

Godlund, S. (1956b) *The Function and Growth of Bus Traffic within the Sphere of Urban Influence* (Lund: Gleerup).

Golledge, R.D. (1990) Applications for behavioural research on spatial problems: I – Cognition, *Progress in Human Geography*, 14.

Gould, P. (1972) Pedagogic review, *Annals of the Association of American Geographers*, 62:689–700.

Gould, P. (1985) *The Geographer at Work* (Boston: Routledge & Kegan Paul).

Gould, P., Kabel, J., Gorr, W. and Gorlub, A. (1991) AIDS: predicting the next map, *Interfaces*, 21:80–92.

Gould, P. and White, R., ([1974] 1986), *Mental Maps*, 2nd edition (London: George Allen & Unwin).

Hägerstrand, T. (1952) *The Propagation of Innovation Waves* (Lund: Gleerup).

Hägerstrand, T. (1953) *Innovationsförloppet ur Korologisk Synpunkt* (Lund: Gleerup).

Hägerstrand, T. (1968) *Innovation Diffusion as a Spatial Process* (Chicago: University of Chicago Press).

Hägerstrand, T. and Buttimer, A. (eds) (1988) *Geographers of Norden: Reflections on Career Experiences* (Lund: Lund University Press).

Haggett, P. (1965) *Locational Analysis in Human Geography* (London Edward Arnold).

Harris, C.D. and Ullman, E. (1945) The nature of cities, *Annals of the American Academy of Political Science and Social Science*, 242:7–17.

Harvey, D. (1973) *Social Justice and the City* (Baltimore: Johns Hopkins University Press).

Heidegger, M. (1977) Science and reflection, in *The Question Concerning Technology* (New York: Harper Colophon Books).

Herbert, D.T. and R.F. Johnston (eds) *Geography and the Urban Environment: Progress in Research and Application*, vol. 6 (Chichester: John Wiley).

Hindess, B. and Hirst, P.Q. (1975) *Pre-capitalist Modes of Production* (London: Routledge & Kegan Paul).

Hoover, E. (1937) *Location Theory and the Shoe and Leather Industry* (Cambridge, MA: Harvard University Press).

Hoyt, H. (1933) *One Hundred Years of Land Values in Chicago* (Cambridge, MA: Harvard University Press).

Hudson, J.C. (1969) Diffusion in a central place system, *Geographical Analysis*, 1:456–85.

Iklé, F. (1954) Sociological relationship of traffic to population and distance, *Traffic Quarterly*, 8:123–36.

Isard, W. (1956) *Location and Space-economy: A General Theory Relating to Industrial Location, Market Areas, Land Use, Trade and Urban Structure* (Boston: MIT Technology Press; New York: Wiley).

Isard, W. (1960) *Methods of Regional Analysis: An Introduction to Regional Science* (New York: John Wiley).

Johnston, R.J. (1979) *Geography and Geographers* (New York: John Wiley).

Johnston, R.J. (1985) *The Geography of English Politics: The 1983 General Election* (London: Croom Helm).

Jones, J. and Casetti, E. (1992) *Applications of the Expansion Method* (London: Routledge).

Kant, E. (1946) Den Inre Omflyttningenk Estland I Samband med de Estriska Städernas Omland, *Svensk Geografisk Årsbok*, 22.

Kates, R.W. (1962) *Hazard and Choice Perception in Flood Plain Management* (Chicago: University of Chicago, Department of Geography, Research Paper No. 78).

King, L.J. and Golledge, R.G. (1978) *Cities, Space and Behaviour: The Element of Urban Geography* (Englewood Cliffs, NJ: Prentice-Hall).

Losch, A. (1954) *The Economics of Location* (New Haven, CT: Yale University Press).

Lynch, K. (1960) *The Image of the City* (Cambridge, MA: MIT Press).

McCarty, H. (1954) An approach to a theory of economic geography, *Economic Geography*, 30:30–101.

McCarty, H. (1959) Toward a more general economic geography, *Economic Geography*, 35:283–9.

McCarty, H., Hook, J. and Knos, D. (1956) *The Measurement of Association in Industrial Geography* (Iowa City: Iowa State University, Department of Geography).

McCarty, H. and Salisbury, N. (1961) *Visual Comparison of Isopleth Maps as a Means of Determining Correlations between Spatially Distributed Phenomena* (Iowa City: State University of Iowa, Department of Geography).

Massey, D. (1984) *Spatial Divisions of Labour: Social Structure and the Geography of Production* (Basingstoke: Macmillan).

Meillasoux, C. (1981) *Maidens, Meal and Money: Capitalism and the Domestic Community* (Cambridge: Cambridge University Press).

Miller, G. (1947) Population, distance and the circulation of information, *American Journal of Psychology*, 60:276–84.

Morrill, R.L. (1962) Simulation of central place patterns over time, *Lund Studies in Geography, Series B., Human Geography*, 24:109–20.

Morrill, R.L. (1963) The development of spatial distributions of towns in Sweden: an historical-predictive approach, *Annals of the Association of American Geographers*, 53:1–14.

Nystuen, J.D. (1963) Identification of some fundamental spatial concepts, *Papers of the Michigan Academy of Science, Arts and Letters*, 48:373–84.

Nystuen, J.D. and Dacey, M. (1961) A graph theory interpretation of nodal regions, *Papers and Proceedings of the Regional Science Association*, 7:29–42.

Ollman, B. (1971) *Alienation: Marx's Conception of Man in Capitalist Society* (New York: Cambridge University Press).

Olsson, G. (1965) *Distance and Human Interaction, a Review and Bibliography* (Philadelphia: Regional Science Research Institute, Bibliography Series No.2).

Openshaw, S. (1989) Computer modelling in human geography, in B. Macmillan (ed.) *Remodelling Geography* (Oxford: Basil Blackwell).

Peet, R. and Thrift, N. (1989) *New Models in Geography,* Vol. 1 (London: Unwin Hyman).

Piaget, J. (1970) *Structuralism* (New York: Basic Books).

Poulantzas, N. (1975) *Classes in Contemporary Capitalism* (London: New Left Books).

Poulantzas, N. (1978) *State, Power and Socialism* (London: New Left Books).

Rashevsky, N. (1953) Imitation effects as a function of distance, *Bulletin of Mathematical Biophysics,* 12:197–234.

Ravenstein, E.G. (1885) The laws of migration, *Journal of the Royal Statistical Society,* 48:167–235.

Reilly, W. (1931) *The Law of Retail Gravitation* (New York: Knickerbocker Press).

Ricardo, D. (1817) *The First Six Chapters of the Principles of Political Economy and Taxation by David Ricardo, 1817* (New York: Macmillan).

Rushton, G., Goodchild, M.F. and Ostresh, L.M., Jr. (1973) *Computer Programs for Location-allocation Problems* (Iowa City: University of Iowa, Department of Geography, Monograph no.6).

Simon, H.A. (1957) *Models of Man* (New York: John Wiley).

Smith, N. (1984) *Uneven Development: Nature, Capital and the Production of Space* (Oxford: Blackwell).

Smith, N. (1989) Uneven development and location theory: towards a synthesis, pp. 142–63 in R. Peet and N. Thrift (eds), *New Models in Geography, vol 1* (London: Unwin Hyman).

Smith, W. (1955) The location of industry, *Transactions of the Institute of British Geographers,* 21:1–18.

Spencer, H. (1892) *A System of Synthetic Philosophy, Vol. 1 – First Principles,* 4th edition (New York: Appleton).

Stewart, J. (1941) An inverse distance variation for certain social influences, *Science,* 93:89–90.

Stewart, J. (1942) A measure of the influence of population at a distance, *Sociometry,* 5:63–71.

Stewart, J.Q. (1947) Empirical mathematical rules concerning the distribution and equilibrium of population, *Geographical Review,* 37:461–85.

Stewart J.Q. and Warntz, W. (1958) Macrogeography and social science, *Geographical Review,* 48:167–84.

Stewart, J.Q. and Warntz, W. (1959) Physics of population distribution, *Journal of Regional Science*, 1:99–123.

Stouffer, S.A. (1940) Intervening opportunities: a theory relating mobility and distance, *American Sociological Review*, 5:845–67.

Taaffe, E. (1956) Air transportation and the United States urban distribution, *Geographical Review*, 46:219–38.

Taaffe, E. (1959) Trends in airline passenger traffic: a geographic case study, *Annals of the Association of American Geographers*, 49:393–408.

Taaffe, E. (1962) The urban hierarchy: an air passenger definition, *Economic Geography*, 38:1–14.

Teitz, M.B. and Bart, P. (1968) Heuristic methods for estimating the generalized vertex median of weighted graph, *Operations Research*, 16:955–61.

Terray, E. (1972) *Marxism and Primitive Societies* (New York: Monthly Review Press).

Thomas E. (1960) *Maps of Residuals from Regression: Their Characteristics and Uses in Geographic Research* (Iowa City: Iowa State University, Department of Geography).

Thrift, N. (1995) Peter Haggett's life in geography, in A. Cliff, P. Gould, A. Hoare and N. Thrift (eds) *Diffusing Geography* (Oxford: Blackwell Publishers).

Tobler, W. (1961) *Map Transformations of Geographic Space* (Seattle: University of Washington Ph.D. Dissertation).

Tobler, W. (1963) Geographic area and map projections, *Geographical Review*, 53:59–78.

Ullman, E.L. (1956) The role of transportation and the bases for interaction, pp. 862–80 in W.L. Thomas (ed.) *Man's Role in Changing the Face of the Earth* (Chicago: University of Chicago Press).

Von Thunen, H. (1966) *Von Thunen's Isolated State* (translated by C.M. Wartenberg and edited by Peter Hall) (London: Pergamon).

Warntz, W. (1957) Contributions toward a macro-economic geography: a review, *Geographical Review*, 47:420–4.

Warntz, W. (1959) *Toward a Geography of Price* (Philadelphia: University of Pennsylvania Press).

Warntz, W. (1964) A new map of the surface of population potentials for the United States, *Geographical Review*, 54:170–84.

Weber, A. (1909) *Theories of the Location of Industries* (translated by C.J. Freidrich, 1929) (Chicago: University of Chicago Press).

Wilson, A.G. (1967) A statistical theory of spatial distribution models, *Transportation Research*, 1:253–69.

Wilson, A.G. (1970) *Entropy in Urban and Regional Modelling* (London: Pion).

Wilson, A.G. (1981) *Geography and the Environment: Systems Analytical Methods* (Chichester: John Wiley).

Wilson, A.G., Coelho, J., Macgill, S. and Williams, H. (1981) *Optimization in Locational and Transport Analysis* (London: John Wiley).

Wolpe, H. (ed.) (1980) *The Articulation of Modes of Production* (London: Routledge & Kegan Paul).

Wolpert, J. (1964) The decision process in spatial context, *Annals of the Association of American Geographers*, 54:337–58.

Wolpert, J. (1965) Behavioural aspects of the decision to migrate, *Papers and Proceedings of the Regional Science Association*, 15:159–72.

Wrigley, N. and Longley, P.A. (1984) Discrete choice modelling in urban analysis, pp. 45–94 in D. T. Herbert and R. J. Johnston (eds) *Geography and the Urban Environment: Progress in Research and Applications*, vol. 6 (Chichester: John Wiley).

Yeates, M. and Garner, B. (1980) *The North American City*, 3rd edition (San Francisco, CA: Harper & Row).

Zipf, G. (1946) The P_1P_2/D hypothesis: on the intercity movement of persons, *American Sociological Review*, 11:677–86.

Zipf, G. (1949) *Human Behaviour and the Principle of Least Effort* (New York: Hafner).

3 Geographical Methodology and the Information Revolution

In spite of its concern with theory and conceptual formation, geography depends essentially on facts and information to improve knowledge of the earth on which we live. Even the early speculations of the ancient Greeks were attempts to derive some plausible information about the nature of the earth. Over the centuries, various individuals and groups have sought to increase or improve upon these facts and information. The rate at which these increases took place accelerated in the great age of explorations and discovery. Men set out deliberately to find out and collect more facts about different parts of the world with a view to presenting these to the enlightened public and thereby improve knowledge and understanding.

The methodology of a subject is, however, not just about facts and information. It is about the set of rules and procedures which defines the types of facts and data that can be collected and provides the framework within which these can be analysed and used to arrive at credible and valid knowledge about particular phenomena or topics. Methodology thus entails not only the methods of data collection but also, and perhaps more important, the different aspects and steps involved in conducting research and argument within the subject. Its value is that it allows the accumulation of knowledge that becomes the distinctive contribution of a particular discipline to human understanding.

Preliminaries

Of preliminary interest, therefore, is the need to appreciate the distinction between words used in relation to the methodology of a subject, in particular such words as data, information and knowledge.

This chapter is based on a contribution by Vladimir S. Tikunov.

Data, for instance, derives from the Latin word *datum* which literally means a fact. The word corresponds to discrete facts about phenomena or a combination of facts formalized either qualitatively or quantitatively and applied in a particular area of scientific inquiry or other spheres of human activities. It can, therefore, be defined as a description of a phenomenon or idea, valuable enough for an attempt to be made to formulate and fix it with some degree of precision (Tsichritzis and Lochovsky, 1985:16–17). Data are thought of as a kind of raw material which can be processed into information. In other words, they are an element in the process of information construction.

Information, on the other hand, is somewhat more difficult to define. Because of its more popular use, its definition can appear vague and ambiguous. For instance, the word is regularly used in newspapers, television and radio programmes, scientific and popular-science fiction, and in each context its meaning may appear very clear. As a result, synonyms for information in common usage appear to be such words as account, insights, signal, data and knowledge. In science, information is a much more rigorously defined concept; it is regarded as a universal feature of all objects and closely akin to energy. That conceptualization of information is central to the present chapter and is deferred for detailed elaboration later.

Knowledge, in its turn, can be said to derive from information. It is the product of a process whereby information is analysed and configured within a particular framework and made to confront the human intellect for its validity and acceptability.

Mapping, Surveying and Cartographic Analytical Methods

As soon as curiosity about the earth reached the point where men and women began to observe specific facts about it, the task of noting or putting down these facts or data began to assume the nature of a challenge. The first challenge was how to represent the earth itself. Indeed, centuries before the Christian era, the Babylonians are known to have produced graphic representations of their parts of the world on clay tablets. Such maps attempt to portray land features as well as identifiable human constructions.

It was, however, the Greeks who first attempted to produce a map of the known world. A surviving map by Herodotus, for instance, was

Satellite dish in a remote area.
Source: Bartholomew/ Liaison Gamma.

said to be an improvement on the delineation of the shape and extent of the then known regions of the world. The greatest figure of the ancient world in respect to the development of geographical methodology was, of course, Claudius Ptolemaeus, better known to history as Ptolemy. He lived in Alexandria, Egypt from AD 90–168. Through his mathematical interest, he was able not only to accumulate sufficient data concerning places in different known regions of the world but also to use these data to provide a method for constructing the globe and projecting it to two-dimensional space.

Much of the work developing from this time helped to underscore the unique characteristics of geographical data. Since any place on the earth's surface exhibits attributes or characteristics such as location (longitude and latitude), distance, area, shape, height above sea-level, orientation or direction, data about them must try as much as possible to capture as much of these attributes as possible. It was in recognition of this fact that quite early in the history of geography, considerable interest was shown in the development of various projections meant to preserve as much of these characteristics as possible. A projection is a systematic method of drawing the earth's meridians of longitude and parallels of latitude on a flat surface. Some projections help to preserve the area characteristics of the surface; others the shape characteristics. Ptolemy's concern with map projection and globe construction thus represented a major milestone in the achievements of geography, leading among other things to the attempt to calculate the areal size of the earth.

Whilst projections were important in the representation of the whole earth surface, generally on a small-scale map, surveying was critical for collecting data and representing these on larger scale maps. Surveying is, thus, a method of undertaking relatively large-scale and accurate measurement of portions of the earth's surface. It entails the determination of the measurement data, the reduction and interpretation of the data to usable form, and the establishment of relative position and size according to given measurement requirements. There is evidence that the history of surveying probably dates back to ancient Egypt where both the accuracy in the division of agricultural fields in the fertile Nile Valley and in the construction and north–south orientation of the Great Pyramid of Khufu at Giza would seem to affirm some command of the science of surveying. Surveying has developed since then both in terms of its instrumentations and its operational methods. Indeed, by the end of the First World War, two other revolutionary changes considerably altered the procedures of both mapping and surveying. The first is the ability to map from photographs of the earth's surface taken from the air, that is, photogrammetry; the second is the electronic distance measurement capability, which include the adoption of laser for this purpose as well as for alignment. Since the 1970s, further technological innovations include the use of satellites as reference points for geodetic surveys and electronic computers to speed up the processing and recording of survey data.

In the early period of the development of geography as a university discipline, considerable attention was paid to ensuring that undergraduates familiarized themselves with the processes of mapping and surveying. The expectation was that, if in the course of their study or research, they had cause to use either methods in collecting data or representing earth's surface information, they would have the necessary proficiency. Today, mapping and surveying have become the central concern of new fields of human inquiry. Although still of some limited interest, they are now seen as associated fields producing information which geography can access when and if needed. But some recent technological revolutions impact on all the fields together.

Until the post-war era of quantitative and theoretical revolution in geography, the basic analytical tool within the subject was cartographic. The important element in geographical analysis was plotting the spatial distribution of data with a view to identifying the regional pattern of concentration and attempting an explanation through assumed relationship between this distribution and those of associated phenomena. Most of the maps used in this connection usually relied on the Transverse Mercator projection, a projection that sought to preserve as much as possible the equal-area properties of maps. On such maps, grid cells of various shapes provide the units of observation. Sometimes, other subdivisions of area, particularly political or administrative subdivisions provided the units of observation. Cartographic analysis thus enabled the analyst through the use of various techniques of representations – dots, choropleths, isopleths, bar charts, pie charts, and so on – to highlight the spatial pattern of distribution of the phenomenon of interest. Where the emphasis of the data was to indicate flow, movement or interaction between places, this was often represented through arrows of varying thickness depending on the volume or the intensity of interactions between the places concerned.

Cartographic analytical methods, however, had little to say about the quality or reliability of the data being analysed. These were usually taken as given. Thus, it required the surveyors to provide the data on the elevation of places for contour isopleths to be drawn. Similarly, climatic data of various types, population, socio-economic and other data are provided, analysed and presented cartographically without much concern about the accuracy and the reliability of the data. Much of the criticism in cartography is about the effectiveness of the particular technique of representation, whether in fact it conveys at a glance the spatial pattern of distribution of the phenomenon of interest.

To proceed to attempt to derive causal, or more correctly, associational knowledge concerning the phenomena, the use of map overlays was the recommended procedure. This requires that for the given area, a series of maps at the same scale is provided. Data for various phenomena which are assumed a priori to have some relationship are then represented cartographically on each of these maps, using mapping paper of a high degree of transparency. These maps are then laid one on top of the other and the congruency among the patterns of distributions of the different phenomena represented is noted. The degree of congruency is assumed to indicate some relationship, or at least some association between and among the phenomena concerned. This type of congruency which, for instance, allows strong association to be noted between tropical areas and a lower level of socio-economic development provided much of the grist for the early emphasis on environmental determinism in the subject.

It has been argued that maps play a vital role in geography not only because they are a device for storing facts areally but also because they are the logical framework upon which geographers have constructed geographic theory. Indeed, it is claimed that maps have been to geography what mathematics has been to some other disciplines. It is a spatial tool of generalization and analysis used by no other science as much as by geography. Yet, there is no way to get round the fact that cartographic methodology does not entail any concern with how the data stored in maps are collected. Nor does it have much to say about the reliability of the process of collection or the accuracy of the data collected. It also does not evince any capacity to squeeze all of the potential information out of the data. All it does is to represent them with a view to establishing the pattern of their spatial distribution. Not unexpectedly, therefore, the inadequacy of this methodology for promoting the theoretical development of the subject was part of the concern of the quantitative and theoretical revolution that took place in the subject in the period after the Second World War.

Mathematics and the Quantitative Revolution

In his publication, *Theoretical Geography*, William Bunge (1962:71) came to the conclusion that in spite of certain advantages of maps over mathematics, mathematics is the broader and more flexible medium for geography. In general, premaps are a subset of maps and maps are a subset of mathematics . . . Premaps, such as aerial photographs, have certain advantages which should be exploited, but premaps are not as selective and as supple a spatial device as maps themselves. And metacartography indicates that maps have not been extended to their fullest potential.

Against this background, the quantitative revolution in geography sought not only to improve the capabilities of maps as a tool of geographical analysis but also to introduce a high level of mathematical and statistical analytical techniques to the subject. Maps, for instance, were recognized as depending basically on the concepts of geometry and topological mathematics. Both elementary and projective geometry have always been important methods in the development of maps. But the latter has only recently started to be used as a general methodological tool in the subject, especially in situations where it is recognized, for instance, that raw distances are not sufficient to convey the right impression and that, instead, cost or time distances are the appropriate measures. Thus, Watson (1955) referred to various informal maps of real distances, such as Harris's cartograms, which bend space into proper economic distances, as revealing how distant places were from one another. Hägerstrand (1957) also showed that both the psychological and economic view of distance of the type that migrants have to entertain when they seek to emigrate from their homes may be generalized within a logarithmic transformation of distance. He used an azimuthal logarithmic projection centred on the place of migration, in this case Asby in central Sweden, to suggest the migrant's impression of distance. The contrast between the conventional map of Sweden and the transformed map on the azimuthal logarithmic projection shows the drastic change in spatial relationships, with places like Göteborg and Stockholm only marginally less distant than the United States.

Topology, on the other hand, is that branch of mathematics that deals with spatial properties that are invariant after being subjected to infinite stretching. It may or may not involve the use of measurement. A good example of the use of topological mathematics in geography is provided by the analysis of transport networks using graph theory. Graph theory deals with connectivity of phenomena, with the connections being represented by lines and the phenomena by points. Its analytical use in geography has enabled a better appreciation of the connectivity advantages or disadvantages of different locations within a given network.

Within geography, the quantitative revolution fostered a methodological trend towards model building. A model in this respect is described as a simplified version of reality, built in order to demonstrate certain properties of reality (Haggett, Cliff and Frey, 1977:17). Usually, models are conceptual and have a high degree of abstraction which enables them to be expressed in mathematical forms. Other models based largely on empirical considerations have a low level of abstraction and are often termed operational. One of the consequences of the shift towards model building in geography is the increasing use of regressional analysis within the subject to establish the dependent or causal relation of a given variable, Y, to one or more independent variables, $X_1, X_2, \ldots\ldots X_k$. That is, we can express the relationship as

$$Y = f(X_1, X_2, \ldots\ldots, X_k).$$

If we now measure the values of the variables, both dependent and independent, on some appropriate scale in each of a set of $i = 1, 2, \ldots\ldots, n$ areas, we can begin to estimate something of the nature of the functional relations (f) that appear to link the variables together. For example, it can be postulated that the variables are linearly related so that

$$Y_i = \beta o + {}^k\beta j Xij + \varepsilon j,\ i = 1, 2, \ldots\ldots, n,$$
$$J=1$$

where the $\{\beta\}$ are unknown parameters and the $\{\varepsilon\}$ are independently normally distributed with mean being zero, the variance constant and the error terms stochastic. This familiar form of the general linear regression model can be estimated by ordinary least squares. In this case, the estimated coefficient values, $\{\beta\}$, allow us to measure the supposed direction and strength of the effect of independent variables of Y. This model has a very wide use in the subject, particularly where the emphasis is to establish some causal or other relationship between one dependent variable and one or more other independent variables, which are assumed to have some direct or indirect effects on the pattern of manifestation of the former.

Mathematical models thus provided opportunities for exploring other areas of spatial relations and spatial interactions. One of the most fruitful of such models, used essentially to analyse the movements of people, goods, information and so on between places is the gravity model. This model draws on the analogy from Sir Isaac Newton's Law of Universal Gravitation propounded in 1687, which states that two bodies attract each other in proportion to the product of their masses and inversely as the square of their distances apart. This gives the familiar formula.

$$F = \frac{G M_1 M_2}{d^2_{12}}$$

where F is the force with which each mass pulls the other, G is a universal constant, the pull of gravity, M_1 and M_2 are the sizes of the two masses concerned and d_1d_2 is the distance between them. In relating Newton's original terms to the geographical field, force is identified with movements or flows between two locations while mass is some measure of the relevant characteristics of these locations regarded as generating the movements or flows. Thus, in the field of population migrations, variants of this model of the following form,

$$Tij = k\ 0_i D_j d_{ij}^{-\beta}$$

have been widely used. In this formula, Tij stands for migration between an origin or source area, *i*, and a destination area, *j*, which is proportional to the product of their sizes, 0_i and D_j respectively, and inversely proportional to the distance between them (d_{ij}), raised to some power and where *k* is a constant and ß is a distance exponent. Models of this form are readily fitted by ordinary least squares method although the value to be given to the distance exponent has remained a source of some controversy with most researchers opting for a figure around 2.

Whichever type of modelling is used, an important feature of the methodology is the assessment of the reliability of the model. In this connection, a distinction is made between the technical accuracy of mapping and the reliability of the correspondence between reality and the cartographic model. In determining the reliability of the model, attention is focused on the whole process of modelling, from the identification and determination of the data set, through map design to the selection of the methods for presenting the results. This is because all stages in the technical process make some impact upon the reliability of the thematic content of the map. Reliability, for instance, can be a function of the multivariate character of the data as well as the particular mathematical algorithms used in the analysis. In the process of modelling, the intermediate results (such as the statistical correlations or the maps produced at different stages of the modelling) should themselves be correlated with reality in order to modify or correct the model, or supplement the existing data set.

The growing emphasis on mathematical and statistical models in geographical methodology has not only considerably enhanced the analytical capabilities but also broadened the scope of spatial problems which the subject seeks to address. These developments have also had the effect of making the discipline more sensitive to the type of data that it uses and the mode and manner of their collection. Censuses are recognized to include the collection of data about a total population whether this is a population of human beings, animals, houses, land units, vehicles, road networks, postal mails, or what have you. It is difficult to ensure, however, that censuses are conducted with the necessary diligence and that the level of accuracy of their results can be assumed always to be reasonably high. Consequently, most geographical investigations and researches tend to depend on samples whose levels of reliability can be determined *ab initio*. Geographical methodology has had to concern itself not only with different types of data and the scale on which they are measured – nominal, ordinal, interval, ratio – but also with the manner of their data collection. It has also had to pay attention to the procedures of the various sampling techniques that it chooses to use in the process of various investigations. One reason for this is that the various statistical techniques to be applied to these data assume that they are not only independent one of the other but that in the aggregate their distributions belong to one of the basic distributions known to statistics, namely normal, binomial or Poisson.

It was in the course of such investigations that the critical implications of contiguity and distance for spatially located data became obvious. These factors have meant that spatially located data cannot be assumed to be independent one of the other, this being one of the basic conditions that data used in inferential statistics must meet. Or as Tobler (1970) would put it, the first law of geography: everything is related to everything else, but near things are more related than distant things would seem to conflict with the fundamental assumption of inferential statistics. Yet, despite this intrinsic quality of (usually positive) dependence between observations in space, which is implied in quite traditional geographical concepts such as the region, geographers have, almost without exception, applied conventional statistical tests which assume independent observations to such data. The null distribution of, for example, the t-test and the F-test, may be very different for correlated observations compared with those for uncorrelated observations. If tests which assume independence to correlated observations were then applied to such data, the risk of reaching quite misleading conclusions is real.

Spatial autocorrelation, as this tendency for spatially located data to display systematic spatial variation is called, is now recognized as a significant problem of geographical methodology. The problem, however, has both technical and substantive issues. The technical issues, relating to the validity of inferential tests with data that are not from independent samples because the value of one observation is related to that of its neighbours, have been successfully tackled (Cliff and Ord, 1980). Methods have been developed for evaluating spatial dependencies in data so as to identify real spatial patterns and relationships and be able to choose appropriate analytical techniques for their investigation (Getis and Ord, 1992). The substantive issues, revolving around the question of the nature of the geographical individual or unit of observation remain and, being more complex, have been largely ignored. In the real world most of these units tend to be polygonal surfaces with very irregular boundaries, shapes or areas physically delimited by lakes, rivers or other earth features, or administratively demarcated along socio-cultural lines or other principles. Mandelbrot (1983) in his publications on fractals has provided some indication of how these problems can be resolved. Algorithms of fractal dimensionality have been developed to provide an index for describing the complexity of curves and surfaces as a means of determining the accuracy of geographical measures, having corrected for the spatial autocorrelation of the data (Lam, 1988).

The Geographical Information System

The growing concern with data characteristics and the increasing sophistication of computer technology has led to a major development in the methodology of geography relating to the capacity to process a wide variety of data from a wide variety of sources. A Geographical Information System (GIS), as this new capability is designated, refers to a process of organizing data with reference to their spatial co-ordinates such that analysis of them and the resulting spatial information can be displayed in combination on a screen or plotter. GIS, however, does not refer to a single programme or methodology of analysis. What is systemic about it is the centrality of the digitization of cartographic information about places and other spatial features, the enormous data-processing capabilities provided by computers both in terms of softwares and hardwares, and the flexibility for storage, retrieval, modelling, analysis, display and animation.

The Geographical Information System thus represents the culmination of various attempts to improve on spatial data analysis. In its crudest form, its history can be said to go back to earlier attempts to provide a series of map overlays for spatial analytical purposes. However, it was the convergence of various events from the late 1950s onwards that made the development of GIS almost inevitable. These include the advances in computer technology especially in the area of graphics, the development of theories of spatial processes, and the increasing awareness of the geographical bases of social and environmental problems. The latter, for instance, was critical whenever major transportation projects or other areally extensive constructions such as dams had to be undertaken. Consequently, GIS came to draw on skills from a number of different disciplines apart from geography and cartography. Such disciplines include, for instance, remote sensing, photogrammetry, surveying, geodesy, statistics, operations research, computer science, mathematics and civil engineering. As an analytical methodology, therefore, GIS is multifunctional and multipurpose. For this reason, the GIS usually comprises several structural units called modules or subsystems, each having more or less definite functions. These functions, in their turn, are designed to resolve four types of problems: data collection; data processing; data modelling and analysis; and data use for decision making (H. Calkins, 1977).

In essence, then, there are *four* features to a Geographical Information System. These are:

1 The collection or acquisition of the raw information or data of interest;

2 The processing, modelling and analysis of these data (the data base);
3 The expert system that determines how the data is manipulated; and
4 The use of processed information for decision making.

1 The collection or acquisition of information

The term information in this context is defined as the product of a communication process involving a number of individuals (Blumenau, 1989). In systems theory, information is also defined as the potential of interaction among objects and is regarded as a function of the degree of complexity, organization and ordering within a given system (Anon, 1979:114–15). As against this qualitative definition of information, there is the quantitative definition which sees information as the energy in a system and therefore as information able to be quantified, especially in the context of entropy theory.

Whichever definition of information is appropriate in a given situation, it is evident that for most human beings information is accessed conventionally by our senses of sight and, to some extent, hearing. Our awareness of information is, thus, greatly limited by the scope of these senses. To extend the information base upon which our conceptualization is founded, we have traditionally sought to gather data which make it possible to access other times (history) and other places (geography). The viewpoint adopted in these temporal and spatial excursions, however, has always been human, projecting our eyes and ears to other times and places. An invaluable contribution of the sensing and recording technologies developed has been to extend the spatial and temporal scope of data acquisition. These have provided human beings with the capacity to explore dimensions of the world that would otherwise have been inaccessible to the unaided human senses. Indeed, so restricted is the range of the electromagnetic spectrum, which is capable of perception by human eyes, that our patterns of living are structured so that we are active in the daytime and are constrained to rest at night.

One of the primary advances in information technology is the capacity to sense and record information about physical entities and events through the non-visible parts of the electromagnetic spectrum. In non-military applications, the potential is best seen in the science of remote sensing. Moving beyond visible light, the satellite or aircraft scanner sees first the near infra-red with its greatly improved ability to achieve enhanced discrimination of geographically relevant phenomena such as vegetation, soil and moisture. Further on, through the dimension of the thermal infra-red, it adds still more information and transgresses for the first time the necessity for light in order to permit sight. Equal potential lies elsewhere in the spectrum which, however, can be accessed only through the radar scanner. Each new sensor reveals a different aspect of the world. Each makes some sacrifice in order to do so, but in combination their products build up an ever more complete and complex picture of reality (Clark, 1989). These developments, involving among other things, operational satellite environmental remote-sensing and monitoring capabilities have had the effect of breaking down the scale/resolution link and provide frequent detailed regional and global coverage. Scanning electron microscopy has also begun to access the fundamental microprocesses of physics, chemistry, physiology and (perhaps) psychology. Clearly, local-scale and micro-level studies stand to gain much from these improvements in the spatial resolution capabilities of remote-sensing systems.

Consequently, for geographical analysis, information can come from a wide variety of sources. They can come from publications, statistical and cartographical materials as well as from aerial and satellite images. The most revolutionary development in information or data acquisition has come through remote sensing. Modern sensors, however, can either be passive or active. Passive sensors catch the reflected or emitted natural radiation while active ones send the necessary signal themselves and record it after it has been reflected from the object. Passive sensors include optic and scanner devices operating in the interval of reflected solar radiation (ultraviolet, visible and near infrared bands); active sensors comprise radars, scanning lasers, microwave radiometers and a number of similar devices. The modern trend in the development of operational space electronic systems of remote sensing is to combine various multichannel and multipurpose sensors with high resolution, including all-weather equipment. Other non-operational space systems still in use include panchromatic photo equipment and multispectral cameras providing high resolution and geometrical precision (Permitin and Tikunov, 1991).

In short, technology has made it possible to collect large amounts of raw information about the world in which we live using a wide range of sensors, both human and non-human. There are thus two basic sources of information: terrestrial-based sources and remotely sensed imagery. The former comprises of analogue maps, statistics and digital data recording usually collected for a spatial unit of differing dimensions. These data may be of census aggregates or averages, climate measures, vegetation or soil classes, or perceptual responses from a statistic investigation. The second type can be captured either in analogue form (photographs) or in digital form as numerical arrays of data transmitted to a ground station or recorded on board. These arrays are in bulk form and have not been classified to isolate specific features or categories. The onus is thus on the analyst to identify, interpret, measure and classify these data into meaningful categories so that they can be input into the system.

2 The database – processing, modelling and analysing data

Once the raw information or data has been collected or acquired, there is the need to input it into the digital database. A database is a collection of spatially referenced, digitized data that acts as a model of reality in the sense that it represents a selected set or approximation of real-life phenomena. These selected phenomena are referred to as entities and their database representations as objects. Objects comprise geometrical elements such as points, lines, areas and volume which can be grouped into layers, also called overlays or themes. A layer may represent a single entity type or a group of conceptually related entity types; for example, a layer may have only stream segments or may have streams, lakes, coastlines and swamps. With the help of the computer software and hardware, the database is designed to be able to reconfigure reality in conceptual terms and provide logical systems for analysing the conceptual form so produced.

The database thus refers to the automated and digitized cartographic data together with other tools and techniques which are being used to capture, reconfigure, analyse and display processed data. There are various means of automating cartographic data and consequently different types of databases exist. Three types are generally recognized. These are the network, the hierarchical and the relational databases. Although all three are used, the relational type has been the most successful within the GIS. Whatever the type, they all tend to include Digital Line Graph (DLG) Data, DIME files,

Coordinate Geometry (COGO) data, Standard Interchange Format (SIF) data, and a series of variously formatted polygon data, image or cellular data, and digital elevation data. Some also include various procedures for improving map accuracy before, during and after data acquisition as well as other means for address-matching of records and for acquiring and entering tabular data, text, and data on digital laser disks.

Data processing is, however, the first major task of a database. This requires the application of various mathematical models, with the spatial aspect (topological mathematics) being of primary importance in geography. One can distinguish three types of mathematical models in the methodology of geography. The first type comprises mathematical models which are elaborated without any account taken of the spatial coordinates of the phenomena concerned. The results of such models are generally not meant to be mapped. The second type consists of those mathematical models the results of which are meant to be mapped but whose spatial aspects were not taken into account in realizing the mathematical algorithm used for their analysis. The third type comprises models, the mathematical calculations of which cannot be performed without account being taken of the locational characteristics of the phenomena (Tikunov, 1986a). The first type deals essentially with spatial statistics (Vasilevskiy and Polyan, 1977; Griffith, 1987). The second type relates to attempts at simulation or optimization in geography, whilst the third type belongs properly to attempts to unify mathematical and cartographic models (Zhukov, Serbenyuk and Tikunov, 1980) and uses these to design or analyse the thematic content of maps. Such mathematical–cartographic modelling makes it possible within GIS to construct both elementary one-link models and complex combinations such as chain, network and tree-like models in which alternating mathematical models and maps provide opportunities to optimize the process, to modify it and to correct likely mistakes (Tikunov, 1985).

Within GIS, the technology of modelling geographical phenomena has been developed to have a capacity for dealing with multivariate variables since these can manifest themselves at all stages of the process starting from the stage of information or data acquisition, through data processing to the display of modelling results (Tikunov, 1990). The stage of data entry can involve the acquisition and use of various information blocks to describe a particular phenomenon such as, for example, the level of social and economic development of countries. It is possible to use different systems of initial parameters to process these data using, however, the same algorithm and representing the results on maps in much the same way. In this case, the reliability of the results will depend only on the source of the information used in the modelling.

Multivariability may, however, also manifest itself when an information block is processed using different algorithms. In this case, one should be conscious of the fact that not all algorithms correctly reflect the nature of the modelled phenomenon and the accuracy of the results obtained through using different algorithms needs to be seriously evaluated. The results should be approximately the same if one is to have confidence in the output; otherwise, the result will have to be given different weighted indices, even though such assessment of results is itself often very complicated. Application of different mathematical methods to obtain a single final result is becoming more and more widespread and is being promoted by the expansion in use of computers and the improvements in programme libraries devoted to modelling.

The third aspect of multivariability in modelling is connected to the different ways of mapping the results of an exercise. The language of maps is very rich and profound. Although it has been used for centuries, it still remains possible to devise newer and newer ways of graphic representation of phenomena on maps. The automated reproduction of cartographic images seems to have given a remarkable impetus to the rich possibilities of elaborating new types of data presentation.

An important feature of any modelling is the assessment of its reliability. In this connection, one must distinguish between the technical accuracy of the cartographic representation and the degree of reliability in the correspondence between the real-world phenomena and their cartographic models (Tikunov, 1982). In determining reliability, it is necessary to evaluate the whole process of modelling from map design to the selection of ways of presenting the results. This is because all stages in the technological ensemble exert a certain impact upon the reliability of the thematic content of maps. Thus, one way of ensuring a high degree of reliability in the final result is to use simultaneously various information blocks, different mathematical algorithms and different systems of displaying the results, bearing in mind that there are diminishing returns as the number of different versions used increases. Notwithstanding, one must observe that presenting the results of reliability tests helps to indicate the level of accuracy of various parts of the model.

3 The role of the expert system

Although the computer in the GIS is expected to process, model or analyse the data fed into it, the issue of the purpose or meaning of all of these manipulations remains embedded in the human mind. Part of the challenge of the GIS is to create programmes capable of analysing the semantics (the meaning) that can be extracted from the data collected and input. Such programmes are meant to be able to create chains of inferences based on earlier conclusions through, as it were, filtering knowledge through these conclusions, verifying their logic, refining it and creating more elaborate constructs. If the result obtained was to contradict common sense, this would, of course, make no difference to a computer employing a specific algorithm. It is to make it easy to detect such obvious contradictions that any geographical information system has to depend on an expert system.

An expert system is characterized by three things: its domain, its knowledge base and its control mechanism (Heikkila et al., 1989). The subject area of an expert system such as transportation or infrastructure planning is called its domain, while its knowledge base consists of the collection of facts, definitions, rules of thumb and computational procedures that apply to the domain. The control mechanism consists of the set of procedures that are used for manipulating the information in the knowledge base. This control mechanism could be based on logical deductions from a set of rules or facts, or could include the procedures for determining which rules to examine first and which facts to obtain by querying the user. Expert systems are distinct from conventional programming in that the control mechanism, that is, the thinking engine as some people call it, can be disembodied from the rules that it processes (Ortolano and Perman, 1989). Given an expert system shell, the expert system developer translates expert knowledge for the problem domain in question into precise rules that conform to the logical and syntactical requirements of the shell. It is this that has encouraged efforts to develop systems of artificial intelligence capable of self-training using experience in analysis, control and decision making accumulated in the process of studying the real phenomena and therefore are able to produce from the data second generation knowledge or meta-knowledge (Waterman, 1986). It is no wonder that commercial development of expert shells is now a flourishing industry.

Besides the knowledge base, the control mechanism and the user interface, an expert system has a working memory that contains information generated during a particular run. A key feature of expert systems is that the knowledge base is coded separately from the control mechanism and, in this sense, both are independent one of the other. This makes it possible for the user to add to the knowledge base without having to rewrite substantial portions of the computer programme. It is this ease of expanding or modifying the knowledge base that makes it possible for expert systems to solve problems through rapid prototyping (Turban and Watkins, 1986).

Expert systems not only allow the user to ask questions during an interactive session to learn how certain conclusions are reached, they also show the chain of reasoning used to arrive at a particular conclusion. While expert systems may be applied in many subfields of geography for purposes of data interpretation, problem diagnosis, prescription, monitoring, control and instruction, they can also assist in the creation of intelligent data bases by combining expert systems with traditional data base management programmes.

There are several kinds of expert system used in geography. Among them, the cartographic expert systems are the more important. This is because of the complicated nature of producing cartographic images in a GIS. Usually, this requires the logical co-ordination between elements, their correlation with real-life phenomena, and the complex formalization of their compiling techniques. It also entails an ability to determine the type, the size and the colour of conventional signs, while co-ordinating the content elements, designing new signs, revealing mistakes in map digitalization, discerning cartographic images, and interpreting photographs. Cartographic expert systems also make it possible to achieve automated cartographic generalizations (Zhao, 1988; Zhang, Li and Zhang, 1988) and to select appropriate map projections (Nyerges and Jankowski, 1989; Smith and Snyder, 1989). They have also been used to model the thematic content of maps used in the evaluation of the environmental and geographical impact of locating an industrial plant in alternative sites in the polar regions of Russia (Bogomolov et al., 1990). Expert systems have also been used in other subfields of geography, for instance, in forecasting the weather and the yield of crops, in soil assessment and in monitoring landscape conditions, in predicting fire-hazardous situations, and in town planning.

Perhaps the more important feature of expert systems is their potential ability to work with what is now referred to as fuzzy information. Fuzziness generally results from imprecise definition or loose logic, or blurred boundary conditions. For example, the notion of a wide river must mean different things to different people, between someone in a desert and someone in the Amazon valley. Many remote-sensing materials, statistical data, territorial units and classificatory schemes tend to have this characteristic of being fuzzy. Attempts are being made based on the theory of fuzzy sets to develop software programmes which would enhance the capacity to process and utilize such information effectively. Already, Zadeh (1965) and a number of other scientists are using the theory of fuzzy sets to resolve the problems of territorial units with transitional or fuzzy boundaries in the process of mathematical modelling.

4 GIS information and decision making

Information processing through the GIS has the potential of becoming a very powerful tool in facilitating decision making and conflict resolution based on multiple criteria. This is because many GIS programmes already have integrated into their system with reasonable success multiple criteria methods. Such methods begin by having programmes which can define the relationship between objectives by quantifying them in commensurate terms, developing procedures for resolving conflicts between them, and finding an optimal solution. It is, of course, realized that in real-world situations, decision making rarely ever collapses into a neat single goal. More often than not, real-world problems are inherently multi-objective in nature. Their analysis and decision making must inevitably seek to identify and maintain this multiple-criteria characteristic of reality. It must also appreciate that what decision making entails in such a situation is the capacity to be able to evaluate the implications of the trade-off relationship between the various criteria. In other words, the strength of the GIS is to be able to provide decision makers with a number of decision options from which they can make a choice.

For this reason, the GIS generally provides opportunities for users to engage in active interaction with it, with a view to modifying solutions or performing sensitivity analysis on the results of any investigations. Since GIS databases combine spatial and non-spatial information and have ideal viewing capabilities, it is possible for the user to secure efficient and effective visual examinations of a wide range of solutions. This also entails that the GIS, almost by definition, contains spatial query and analytical capabilities such as measurement of distances and areas, overlay capability and corridor and network analysis.

Notwithstanding, it is important to recognize that many spatial decision-making problems are complex, semi-structured and ill-defined partly because not all of their aspects can be measured or modelled. As a result, spatial decision support systems (SDSS) of various types are being developed to facilitate the process of decision making with GIS. Most of these systems tend to have the characteristics of being iterative, participatory and integrative. They are iterative in the sense that they generate a set of alternative solutions which the decision maker is able to evaluate, and use the insights gained thereby as input into a further round of analyses. They are participatory in that the decision maker plays an active role in defining the problem, carrying out the analyses and evaluating the outcomes. And they are integrative in that the value judgements that materially affect the final outcome are made by the decision maker who has the expert knowledge that must be integrated with the quantitative data in the models.

Geographic Visualization (GVIS) and Multimedia Technology

According to MacEachren et al. (1992:101) geographic visualization (GVIS) can be defined as the use of concrete visual representations to make spatial contexts and problems visible so as to engage the most powerful human information-processing abilities, those associated with vision. This is the basis for the emphasis in geographical studies for the display of various output of data such as tables, summary diagrams, files on machine carriers, prints on papers, schemes and maps. Both paper and without paper technologies are used to achieve this objective. The latter technology, referring to the reproduction of maps on the display screen, requires the elaboration of special means of computer graphics as well as the adaptation of mapping techniques. Map compilation on the display screen makes it possible to check the appropriateness of numerous versions of map content and design.

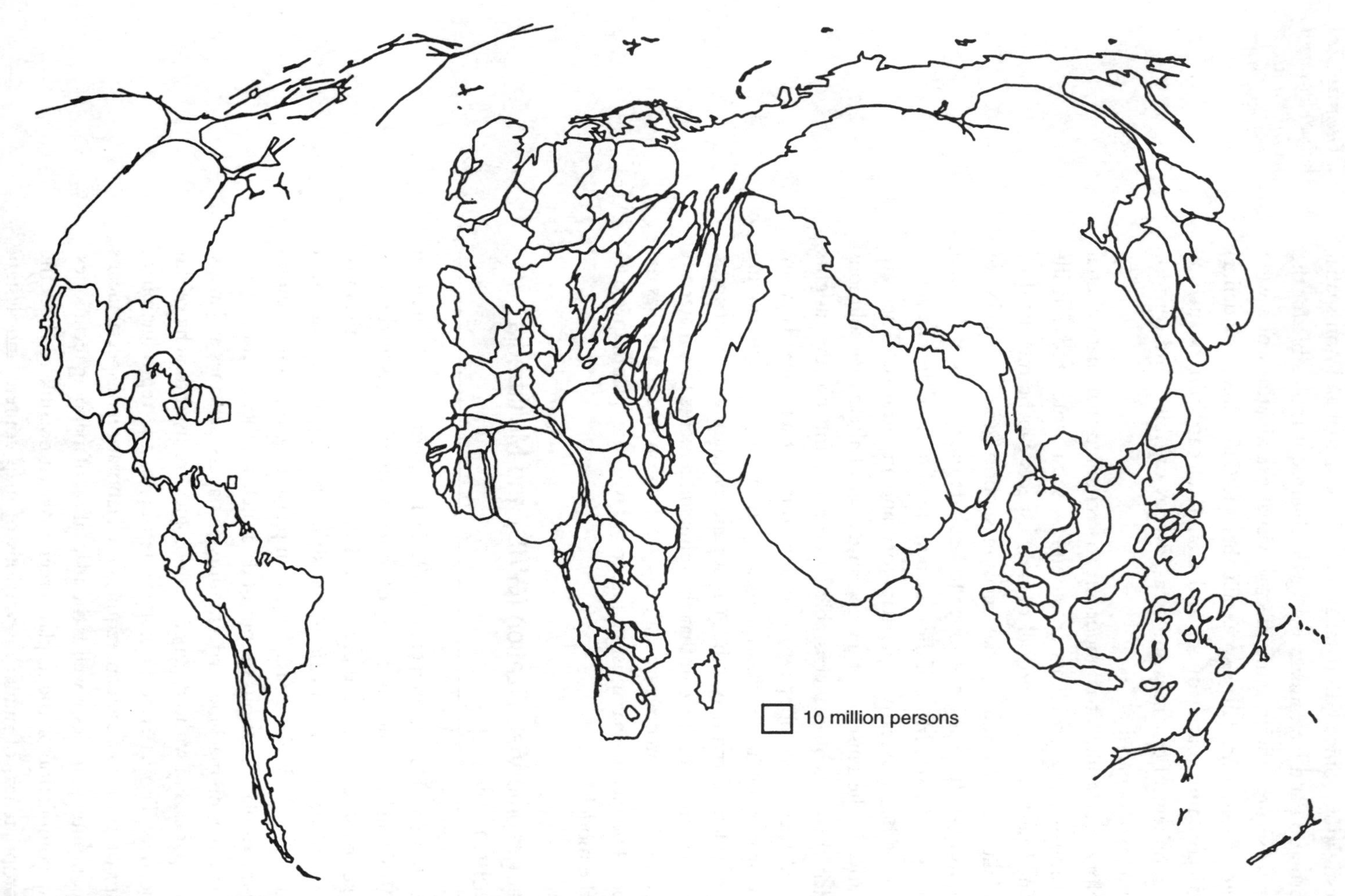

Figure 3.1a Projection of the countries of the world on the basis of population (numbers)

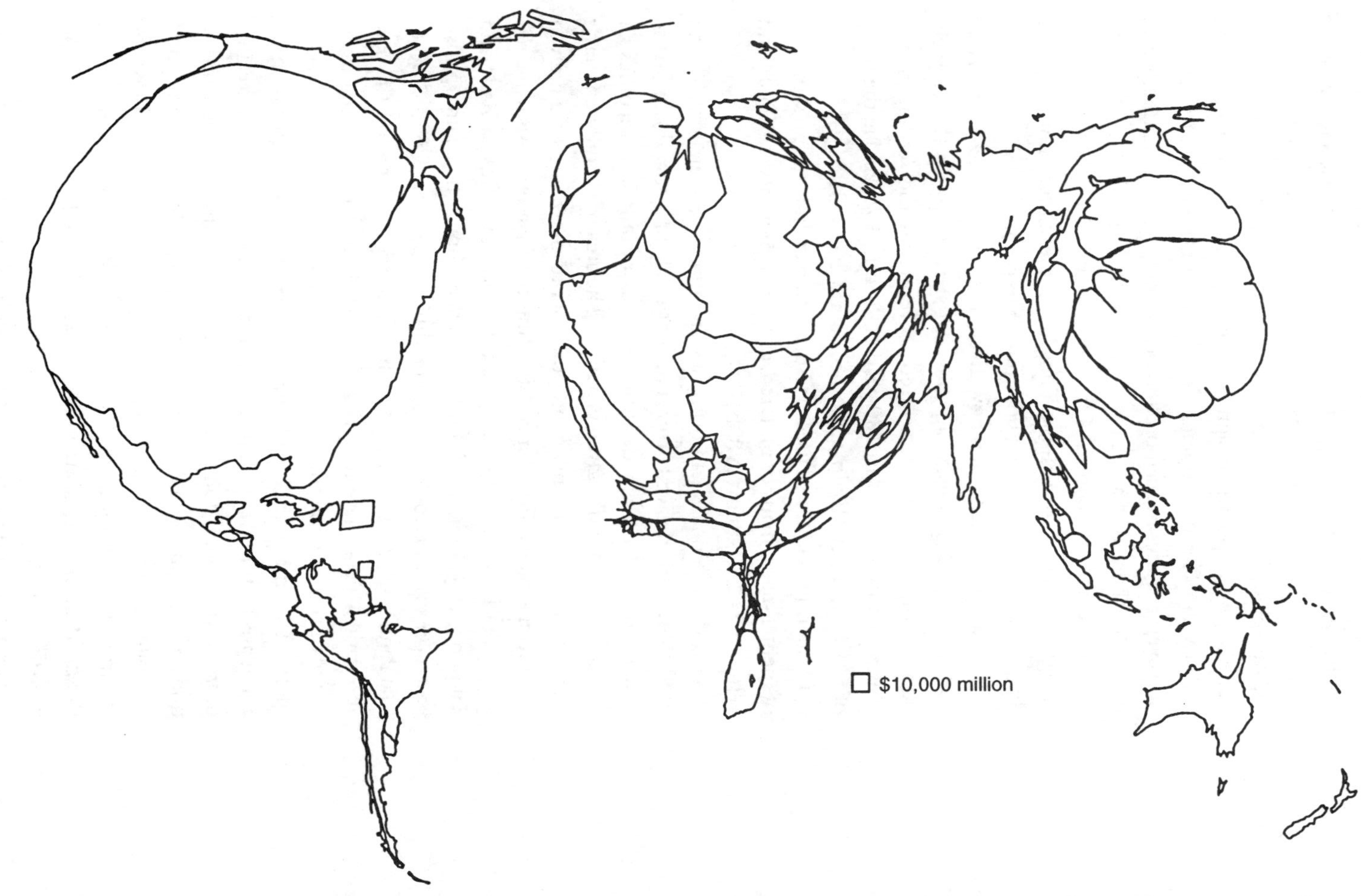

Figure 3.1b Projection of the countries of the world on the basis of gross national product. The square on figure 3.1a represents 10 million persons. The square on figure 3.1b represents $10,000 million.

More importantly, it makes it possible to inject a dynamic dimension to the presentation of phenomena through a quick succession of maps. This projects the course of a process almost in the form of a cartographic film or through animation of individual symbols such as the blinking of points or signs, or through their movement around the screen. The development of dynamic computer cartography, however, requires the elaboration of special symbols and principles of image generalization as well as the psychological and physical aspects of their visual perception. Interesting results have been obtained using animated cartographic films in, for example, imitating the dynamics of environmental pollution (Molotchko, 1987).

Considerable attention is now given to how to visualize 3D images such as block diagrams, anaglyphs and stereomaps. In this connection, holographs have been used to represent the settling of the United States over the last 180 years (Dutton, 1979) and in the study of forest typology (Pavlova, Korneev and Chalov, 1984).

Other non-traditional uses of visualization include the presentation of mental maps, maps of preferences and anamorphoses (Gould and White, 1974; Tikunov 1987b).

Two-dimensional anamorphoses are intensively elaborated and used. They are derived from conventional mapping through the equalization of certain densities such as population, incomes or consumption levels for given products. The area of the different territories represented are made proportional to the values of the given parameter which is used for constructing the anamorphosis while the shape and location are kept up as far as possible (Gusein-Zade and Tikunov, 1990). Figure 3.1a, for instance, shows the anamorphosis based on size of population for the different countries of the world (data and state areas as in 1989) while figure 3.1b represents the data about their gross national product. The difference in the representation of countries on the two maps is quite striking. The comparison of the two maps indicates that anamorphosis can give a more adequate picture of the incidence of a disease since this is related more to the population than to the area of territorial administrative units. Geographic visualization thus combines display with analytical capabilities to enable the search for patterns and relationships, the identification of anomalies, and analysis of directions and flows, the delineation of regions and the integration of local, regional and global information.

The application of multimedia technology enhances visual capabilities through displaying not only maps, schemes and tables but also colour photographs, realistic 3D images and fragments of cartographic and other films, along with text and sound, such as music. Ralf (1994) defines multimedia as computer-based systems for integrated acquisition, storage, representation, transmission and processing of independent data from multiple time-definite and time-indefinite media. These data are synchronized according to spatial, temporal and computer-specific relations. Of course, elements of what is now called multimedia have long been known in cartography, for instance, when maps in atlases are supplemented with explanatory texts, graphical representations, photographs and so on. But the use of sound, animated images and films in this regard has brought us to another qualitative level of combining various capabilities of representing and presenting spatial phenomena. It is thus not surprising that many studies in multimedia are devoted to the compilation of so-called computer atlases (Armenakis and Siedierska, 1991; Ormeling, 1993; Raveneau et al., 1991; Siekierska, 1993). Multimedia tools are also being developed to link maps, graphics, text and data for the purpose of understanding the complexities of geographic processes involved in human–environment interactions.

Conclusion

Geographical methodology has come a long way since the days when all that was important was the ability to map data; but so have the complexity of the problems which geography as a discipline now confronts. All over the world, there is increasing concern over the quality of the earth's environment and the role human beings play in increasing the threat of unsustainability. The issues of global warming, the greenhouse effect and the ozone hole, the acceleration in the rate of land transformation, deforestation and environmental pollution as well as the growing incidence of natural and technological disasters are all setting a new agenda of concern for the human race in the twenty-first century. At the level of individual nation states, these translate to the burning preoccupation with issues of equity and distributive justice in the allocation of and access to national resources by different groups on the basis of race, gender, ethnicity or social class. There is clearly an ever greater need to monitor physical and social processes at a local, national, continental and global scale in order to enhance human knowledge about the increasingly intricate nature of the human–environment relationship.

Geographical Information System (GIS), Geographic Visualization (GVIS) and Multimedia Technology have certainly marked a tremendous advance in the methodology of geography and in its capability to monitor, analyse and gain increased understanding of the human–environment relationship. But even the most enthusiastic proponent of these developments would admit that they are still a long way from giving us the capability for a full understanding of how social processes in different parts of the world do impact on the resources of the earth's environment. It is, of course, true that the situation continues to improve with constantly advancing technological frontiers, especially the increasing speed and power of super-computers and the ever expanding sources of data with fine enough resolution for analysis and modelling. But, our experience in handling very large data bases do not match the rate of their production and raise issues of structure, access and administrative support.

But the future remains challenging and full of promise. The information revolution has served in a way to integrate the various methodological tendencies within geography and to provide the discipline with a sophisticated analytical system whose values transcend the needs of the subject. The Geographical Information System is already proving of immense importance in a variety of management and policy situations both in the public and private sectors. Increasingly, it is becoming a major tool of planning and locational decisions, as well as a tool of monitoring natural resources and environmental conditions. The global span of its information sources, especially through remote sensing also makes it inevitable that its role in the area of international policy and international economic relations is destined to grow with time.

References

Anon (1979) *Concise Dictionary in Philosophy* (Moscow: Politizdat Publ.) in Russian.

Armenakis, C. and Siedierska, E.M. (1991) Issues in the visualization of time-dependent geographical information, *Proceedings of the Canadian Conference on GIS*, 584–92.

Blumenau, D.I. (1989) *Information and Information Service* (Leningrad: Nauka Publ.) in Russian.

Bogomolov, N.A., Borisov, V.M., Krasovskaya, T.M., Tikunov, V.S. (1990)

Expert system for selection of thematic content versions for ecological-geographical maps of industry location, pp. 17–18 in *Problemy kompleksnogo territorialnogo planirovaniya i geografo-ekonomitcheskogo analiza prirodopolzovaniya v avtonomnykh respublikakh i perspecktivy ikh resheniya v novykh uslovivakh khozyaistvovaniya i upravleniha* (Saransk) in Russian.

Bunge, William (1962) *Theoretical Geography*, Lund Studies in Geography: Series C. General and Mathematical Geography No. 1 (Lund, Sweden: C.W.K. Gleerup).

Calkins, H.M. (1977) Information system development in North America, *Proceedings of the Commission on Geographical Data Sensing and Processing* (Ottawa).

Clark, M.J. (1989) Geography and information technology: blueprint for a revolution?, pp. 14–28 in Derek Gregory and Rex Walford (eds) *Horizons in Human Geography* (London: Macmillan).

Cliffe, A.D. and Ord, J.K. (1980) *Spatial Processes* (London: Pion).

Dutton, G.H. (1979) American graph fleeting a computer-holograph map animation of United States population growth, 1790–1970, pp. 53–62 in *Computer Mapping in Education Research and Medicine* (Cambridge, MA: Harvard University Laboratory for Computer Graphics and Spatial Analysis).

Getis, A. and Ord, J.K. (1992) The analysis of spatial association by the use of distance statistics, *Geographical Analysis*, 24:189–206.

Gould, P. and White, R. (1974) *Mental Maps* (New York: Penguin Books).

Griffith, D.A. (1987) Toward a theory of spatial statistics: another step forward, *Geographical Analysis*, 19, N1:69–82.

Gusein-Zade, S.M. and Tikunov, V.S. (1990) Numerical methods of creation of anamorphoses, *Geodeziya I kartografiya*, N1:38–44, in Russian.

Hägerstrand, T. (1957) Migration and area: survey of a sample of Swedish migration fields and hypothetical considerations on their genesis, Lund Studies in Geography, Series B, *Human Geography*, 13:27–158.

Haggett, P., Cliff, A.D. and Frey, A (1977) *Locational Analysis in Human Geography*, vol.1, 2nd edition (London: Edward Arnold).

Heikkila, E.J., Moore, J.E. and Kim, T.J. (1989) Future directions for EGIS: applications to land use and transportation planning, in T.J. Kim, L.L. Wiggins and J.R. Wright (eds) *Expert Systems: Applications to Urban Planning* (New York: Springer Verlag).

Lam, Nina Siu-Ngan (1988) The measurement and description of polygonal surfaces using factals, paper presented at the 26th Congress of the International Geographical Union, Sydney, Australia.

MacEachren, A.M., Buttenfield, B., Campbell, D., DiBiase, D. and Monmonier, M. (1992) Visualization, pp. 99–137 in R. Abler, M. Marcus and J. Olson (eds) *Geography's Inner Worlds: Pervasive Themes in Contemporary American Geography* (New Brunswick, NJ: Rutgers University Press).

Mandelbrot, B (1983) *The Fractal Geometry of Nature* (New York: Freeman).

Molotchko, A.N. (1987) Animated principles in cartographic modelling, pp. 214–15 in *Kartograficheskoye obespecheniye osnovnykh napravieniy ekonomicheskogo i sotsyalnogo razvitiya USSR i eyo regionov*, Tez.dokl.6 Resp.nauchn.konf. (Chernovtsy) in Russian.

Nyerges, T.L. and Jankowski, P. (1989) A knowledge base for map projection selection, *American Cartographer*, 16, N1:29–38.

Ormeling, F. (1993) Ariadne's thread – structure in multimedia atlases, *Proceedings of the 16th International Cartographic Conference*, Cologne (Bielefeld) vol. 2:1093–100.

Ortolano, L. and Perman, C.D. (1989) Applications to urban planning: an overview, in T.J. Kim, L.L. Wiggins and J.R. Wright (eds) *Expert Systems: Applications to Urban Planning* (New York: Springer-Verlag).

Pavlova, Z.G., Korneev, A.A. and Chalov, V.P. (1984) Holographic methods used for the problems of forest typology, pp. 171–2 in *Aerokosmic-heskiye metody issledovaniya lesov.*, Tez. Dokl. Vses. Konf., 7–9 iyulya, (Krasnoyarsk) in Russian.

Permitin, V.E. and Tikunov, V.S. (1991) Environmental monitoring in the USSR: present state and new tasks, *International Journal of Environmental Studies*, N2:1–11.

Ralf, B. (1994) Multimedia GIS – definition, requirements and applications, *The 1994 European GIS Yearbook*, 151–4.

Raveneau, J-L., Miller, M., Brousseau, Y. and Dufour, C. (1991) Micro-atlases and the diffusion of geographic information: an experiment with HyperCard, *Geographic Information Systems: The Micro-computer and Modern Cartography* (Pergamon Press) vol. 1, pp. 201–23.

Siekierska, E.M. (1993) From the electronic atlas system to the electronic atlas products (electronic atlas of Canada from the beginning to the end), *Proceedings of the Seminar on Electronic Atlases*, Visegrad, Hungary, 103–11.

Smith D.G. and Snyder, J.P. (1989) Expert map projection selection system, *United States Geological Survey Yearbook*, Fiscal Year 1988, 15 (Denver, Colorado).

Tikunov, V.S. (1982) Reliability evaluation techniques for mathematical-cartographic modelling, *Vestn. Mosk. un-ta Ser.5. Geogr.*, N4:42–8, in Russian.

Tikunov, V.S. (1985) *Modelling in Socio-economic Cartography* (Moscow: MGU) in Russian.

Tikunov, V.S. (1986a) *Mathematization of Thematic Cartography*,

Vladivostok, in Russian.

Tikunov, V.S. (1986b), Anamorphoses: history and techniques of compilation, *Vestn. Mosk, Un-ta*, Ser. 5 geogr. N6:45–52, in Russian.

Tikunov, V.S. (1990) Multivariability of geographical system modelling, *Izvestiya AN SSSR, ser. geografich*, N5:106–88, in Russian.

Tobler, W.R. (1970) A computer movie simulating urban growth in the Detroit region, *Economic Geography*, 46:234–40.

Tsichritzis, D.C. and Lochovsky, F.H. (1985) *Data Models* (Moscow: Finansi I statistika Publ.) in Russian.

Turban, E. and Watkins, P.R. (1986) Integrating expert systems and decision support systems, *MIS Quarterly*, June.

Vasilevskiy, L.I. and Polyan, P.M. (1977) Mapping of parameters of territorial structures, *Teoriya i metodika ekonomiko-geograficheskikh issledovanity*, pp. 34–47 (Moscow), in Russian.

Waterman, D.A. (1986) *A Guide to Expert Systems* (Reading, MA: Addison-Wesley).

Watson, W. (1955) Geography – a discipline in distance, *Scottish Geographical Magazine*, 71. 1–13.

Zadeh, L.A. (1965) Fuzzy sets, *Information and Control*, 8:338–53.

Zhang, W., Li, H. and Zhang, X. (1988) MAPGEN: an expert for automatic map generalization, *Proceedings of the 13th International Cartographical Conference*, Morelia, vol.4, Aguascalientes: 151–7.

Zhao, X-C (1988) La generalisation cartographique par l'intelligence artificielle, *Cahiers CERMA*, N8:91–126.

Zhukov, V.T., Serbenyuk, S.N. and Tikunov, V.S. (1980) *Mathematical-cartographic Modelling in Geography* (Moscow: Mysl), in Russian.

Part II

Environment and Society

4 The Human Transformation of the Earth's Surface

From very early times, human beings have shown an incomparable capacity to change or transform the character of the earth's surface at local or even regional scale (Sauer, 1956). The use of fire by hunter-gatherers altered flora and fauna; incipient cultivators cut down forests and spread domesticated crops; whilst early civilizations transformed deserts through irrigation. Some of these early activities of human beings may have led to wide and lasting transformations, as is possibly the case in the eradication of megafauna in the western hemisphere by Pleistocene hunters (Martin and Klein, 1984). Others produced major alterations which were subsequently reversed through natural regeneration) as in the reforestation of the central Maya lowlands after its abandonment between AD 800–1000.

The human role in changing the face of the earth has grown continually, reaching to quite unprecedented proportions in recent times. Consequently, there is great concern about global environmental resources and their sustainable utilization. Because of the dramatic acceleration in the rate of transformation, problems such as deforestation, land degradation, and pollution of freshwaters are rightly viewed with anxiety by the scientific community, by politicians and by society at large. In addition, there is the prospect of increased concentrations of greenhouse gases causing global warming during the course of the twenty-first century.

Nonetheless, there is need to stress the considerable scientific uncertainty that surrounds any predictions about the global environmental future. One important reason for this uncertainty is that the earth is not, and never has been, free from change. Natural variability is an intrinsic part of all environmental systems. Only the short time span of direct human observation deceives us into thinking of the

This chapter is based on a contribution by Neil Roberts.

earth as being in a steady state. In some environmental systems, it is now difficult to be sure what their natural, pre-disturbance condition was like. For instance, acid deposition as a result of industrial emissions has meant that there may be no pristine lake ecosystems remaining in the boreal forest zone of the Northern Hemisphere. Even remote mountain lakes in Canada and Europe have had their pH levels reduced since the time of the industrial revolution (Battarbee, 1994). Before estimations can be made about sustainable yields of water, crops, fish or other natural resources, it is necessary to establish the earth's stock of natural resources. For renewable resources, in particular, this resource base has changed through time. In order to understand its natural variability, the earth's environmental and cultural history needs to be investigated.

Time-scales and Data Sources

Human-induced transformation of the natural world is proceeding at different rates and in different ways over the face of the earth, but its overall pace has certainly accelerated during recent centuries (Meyer and Turner, 1994). There are at least three types of techniques available for studying these transformations. In the first place, over time periods of a few decades or less, such transformations can be studied through the methods of observation and monitoring, either through direct measurement or via remote sensing. Satellite imagery is proving to be an especially vital tool in monitoring global and regional changes in forest cover, urbanization and other forms of land use (Haines-Young, 1994). Nearer to the ground, observations of residual, non-westernized societies can be extremely illuminating about traditional relations between nature and society. Ethnobotanical and anthropological studies have revealed a remarkable indigenous knowledge of the natural environment and its uses by modern non-literate hunter-gatherers, peasant farmers and nomads (Richards, 1985). By their very nature, many of these cultural traditions are long-established and have therefore proven themselves to be ecologically sustainable.

In the second place, many environmental and cultural systems change over time scales longer than those which can be observed directly. It is, therefore, necessary to turn to human and natural archives to study them (figure 4.1). The time span and precision of these different data sources strongly influence the ways in which indi-

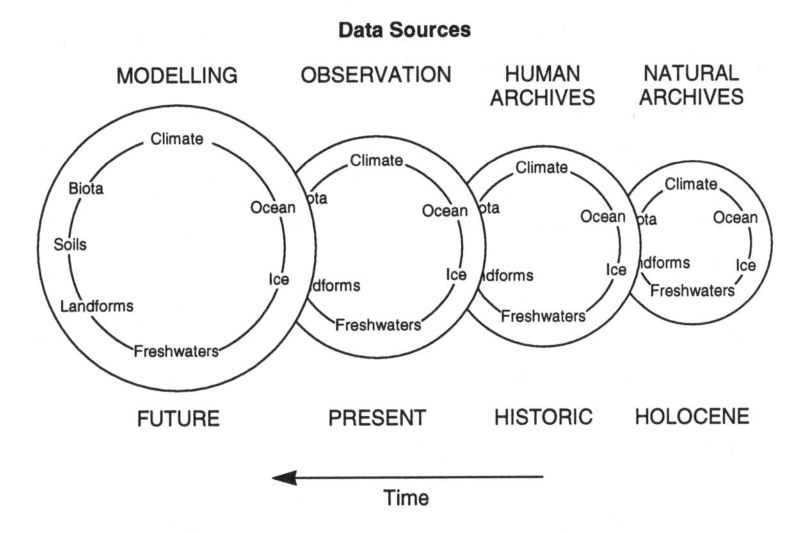

Figure 4.1 Changing sources of environmental data over different time-scales (after Roberts, 1994)

vidual human and environmental histories can be reconstructed. Over time scales ranging from decades to centuries, documentary archives provide the primary means of accessing the past (Brimblecombe and Pfister, 1990; Worster, 1988). Historic maps, diaries, tax records, explorers accounts, and similar archival sources have provided important foci for research on landscape changes over these intermediate time scales (Sheail, 1980; Hooke and Kain, 1982), although these data were not always collected in a form ideal for the analysis to which they were often subjected. Written records are, in any case, restricted to literate cultures. As late as AD 1500, these existed only in Europe, Asia and North Africa. Whereas we have written records from the ancient Near East, including data on early

Australian Aboriginals lighting a fire by rubbing dry sticks together. Evidence from long pollen records suggests that Australian aboriginals used fire as a method of encouraging the growth of new shoots and grazing animals. This may be part of the source of the fire adaptation of most of Australia's pre-European flora. *Source:* Roger Viollet.

problems of soil salinization, dating back to 3000 BC, prehistory in the highlands of Papua New Guinea ended only in the 1930s. There are consequently long periods of human history for which recorded observations are absent, termed by archaeologists prehistory.

Prehistory has come to be associated in the European mind with all that is remote both in time and in cultural affinity to twentieth-century life. On the other hand, to a New Guinea highlander (or a Maori or black Zimbabwean), prehistory represents a direct cultural heritage which ended only a few generations ago. All of this means that the attitudes towards the natural world of many past societies have either gone unrecorded, or have appeared in written form only through the eyes of others.

For still longer time periods – such as the Holocene (or post-glacial) or in regions with short historical records, a third group of techniques becomes especially important. These are proxy methods of reconstructing environmental and cultural change, like tree-ring analysis and radiocarbon dating (Roberts, 1989). A Holocene time scale is valuable if we want to establish how typical present-day rates of environmental change are compared to those of the past. While the main applications of these techniques remain on a millennial time scale, they can also produce valuable insights into shorter term environmental processes, such as cultural eutrophication (Oldfield, 1983). As a result, palaeoecology and archaeology have a potentially significant role to play in elucidating relatively recent changes in culture–environment relations, especially in regions such as sub-Saharan Africa and the Americas, where the depth of the historical record is less than it is in Europe and Asia. In any case, it is important that attempts to reconstruct the major human and environmental features of past times should employ as diverse a range of sources and perspectives as possible (Butlin and Roberts, 1995).

Human Colonization of the Earth's Surface

The probable birthplace of the human race lies somewhere in East Africa and dates back to around five million years ago. From here, by the beginning of the Holocene period, that is, around 12,000 years ago, the genus *Homo* spread to all other continents except Antarctica. For the first few million years of existence, the hominids appear to have been restricted to eastern and southern parts of the African continent. Around a million years ago, they began to spread into the adjacent continents of southern Asia and Europe. A second wave in the expansion came significantly much later and involved the peopling of Australia and the Americas. For both groups of migrants, there must have been formidable barriers to overcome.

To reach Australia–New Guinea meant a sea-crossing by raft or canoe of at least 10,000 km. Remarkably, this was achieved at least 40,000 years ago as indicated by burial sites such as Lake Mungo. The palaeo-Indians arrived somewhat later in America than the Aboriginals did in Australia. It was possible to walk from Asia to America via a land bridge across what is today the Bering strait. Once in Alaska, any potential colonizers would have found their way blocked by ice sheets. It is, therefore, not entirely surprising that the main expansion of human population in the Americas did not take place until the end of the last Ice Age about 14,000 years ago. On the other hand, there is increasing evidence that small numbers of pioneering human groups reached the Americas before this time. The third and final stage in the colonization of the earth took place in the Holocene period and involved the inhospitable terrain around the ice sheets of Greenland and northern Canada, and many of the world's islands.

This initial colonization of the earth by human beings was an extraordinary achievement. It was undertaken with subsistence economies entirely based on hunting of wild animals, fishing and gathering of wild plants. This represents the earliest and, yet in many ways, one of the most subtle human uses of the natural environment. While hunters, fishers and gatherers (h-f-g s) may have been important components within ecosystems, for the most part they changed the environment in degree rather than in kind. Few such traditional societies remain today. Those which have survived – such as the Kalahari Bushmen and the pygmies of Zaire – have been pushed into the remoter habitats of the globe such as drylands and dense tropical forests. However, viewed over a longer time perspective, hunting, fishing and gathering had been a highly successful mode of production. Its success can be gauged from the fact that for 80 to 90 per cent of their existence on earth, it is estimated that human beings have depended on this mode (Lee and De Vore, 1968).

Anatomically, modern human beings (*Homo sapiens* var. *sapiens*) probably emerged between 100,000 and 40,000 years ago and are represented archaeologically by Early Stone Age (Paleolithic) cultures. This was the time of the last Ice Age, when most of Eurasia was covered by steppe tundra rather than by forest and when wild reindeer, bison and mammoth roamed over the open landscape. The Upper Paleolithic peoples of Eurasia were primarily hunters, as the low-biomass glacial vegetation provided few edible plants. Hunting groups were able to follow animal herds as the Saami (Lapps) follow reindeer herds today, or wait for them at mountain passes or at crossing points on rivers. The importance of wild animals for our Stone Age ancestors is reflected in their cave art. Animals dominate the vivid wall paintings at sites such as Lascaux in south-west France. Wall paintings of animals are also a feature of extant hunting groups such as the Bushmen of southern Africa.

With the end of the last Ice Age came dramatic changes in the natural environment. Sea-levels rose; the climate became warmer and wetter; and forests expanded from their glacial refuge to displace the open steppe tundra. This new biome, along with its animal herds, pushed northwards into the polar regions of Eurasia and North America. Not surprisingly, human cultures had to adapt to the new environmental conditions. Woodland supported fewer large mammals and they were better camouflaged from the hunters. Instead, it offered a wealth of new plant resources ranging from hazelnuts and berries to fungi and fruit. Similarly, the many lakes, rivers and coasts offered fish and shellfish. It is probable, therefore, that the Mesolithic (Middle Stone Age) people of the early post-glacial period had a more balanced diet than their Paleolithic predecessors, with meat no longer accounting for the bulk of their food supply.

Hunter-fisher-gatherer peoples need to have an intimate knowledge of their natural environment upon which they depend for food and other resources. The indigenous environmental knowledge (IEK) of these cultures has enabled them to survive in places which a western urban dweller would find hostile and dangerous. To the BaMbuti pygmies, for instance, the tropical jungle is not a threat. It is a benevolent environment, providing all their needs. They call the Ituri forest their mother. This traditional knowledge about the uses of wild resources is rapidly being lost as indigenous groups become acculturated to modern life.

Because all of their material needs are provided by their natural habitat, h-f-g s have every incentive to preserve rather than over-exploit that habitat. Flora and fauna are, therefore, left largely intact under this mode of production. Human beings form only *part* of the natural environment. Of course, such a relationship with the environment can be achieved only at low levels of population density (< 3 km^2 per person). H-f-g bands normally move camp throughout the year following game, water or other resources. Consequently they need a relatively large territory to support them. The resulting low population densities mean that even at the beginning of the Holocene period, world population was unlikely to have been above ten million.

H-f-g s demonstrate that it was possible for humankind to live in more or less stable harmony with the natural world over long time spans without major demographic or resource crises. But even at that time, the human role in nature was far from passive. There may have been little benefit from cutting down whole forests but h-f-g groups have been well capable of manipulating those natural resources of use to them. The most potent of the tools that were used and the earliest form of extra-somatic energy was *fire*. Burning of vegetation to encourage regrowth of new shoots and hence to attract grazing animals is one of human's oldest tools for environmental management (Lewis, 1972). Significantly, in long pollen records from northern Australia, charcoal frequencies increase many times during the later part of the last Ice Age, plausibly due to the arrival of Australia's aboriginal population (Kershaw, 1986). Most of Australia's natural flora is fire adapted. If most fires have been of human rather than natural origin, it raises fundamental questions about the naturalness of this flora. Fire has also been traditionally used to prevent trees from invading grassland ecosystems, for example, in the North American Great Plains where the native Indians burnt the prairie as they hunted the bison.

The other major form of human impact under the h-f-g mode of production has been on large animals and birds which were the prey or competitors of *homo sapiens*. Towards the end of the last Ice Age, there occurred a devastating wave of animal extinctions. The most obvious victims were large mammals such as the mammoth, the woolly rhino and the native American horse. The extinctions varied in their timing and severity across the globe. The most widespread phase affected the Americas, Australia and northern Eurasia. The end of the last Ice Age was a period of rapid climatic change and it has been suggested that difficulties of adapting to the new environmental conditions were the cause of extinction for most of the mammals.

However, researchers such as Martin (1967) have instead been struck by the coincidence between cultural changes and the timing of megafaunal extinctions. This is most clearly the case in North America where the appearance of the palaeo-Indians about 14,000 years ago was followed within less than a thousand years by the demise of many large mammals. Four out of every five genera of large terrestrial mammals became extinct at this time in the Americas (compare the depauperate American fauna with the diverse animal population of Africa, for instance). It has thus been argued that the extinctions could equally be the work of prehistoric big game hunters who butchered in vast numbers animals that were quite unused to human predation (Martin and Klein, 1984). Later Holocene island extinctions, such as that of the giant flightless moa bird, similarly followed within a few centuries of the arrival of human beings, in this case the Maori peoples of New Zealand. The human role in the megafaunal extinctions remains a matter of controversy. It may have involved competition for resources, such as water, as much as direct overkill. Nonetheless, it is likely that human cultural impact on ecosystems was substantial even before the end of the last Ice Age.

From Wilderness to *Oecumene*: The Emergence and Development of Agriculture

For much of human history (or prehistory), *Homo sapiens* was simply one of several large animals adapted to savanna ecosystems (Foley, 1987). The influence of the natural environment on human beings was much greater than the other way round. Starting in the early Holocene period, however, this balance began to change decisively as a result of the emergence of farming as a mode of production. Agriculture was made possible by the domestication of plants and

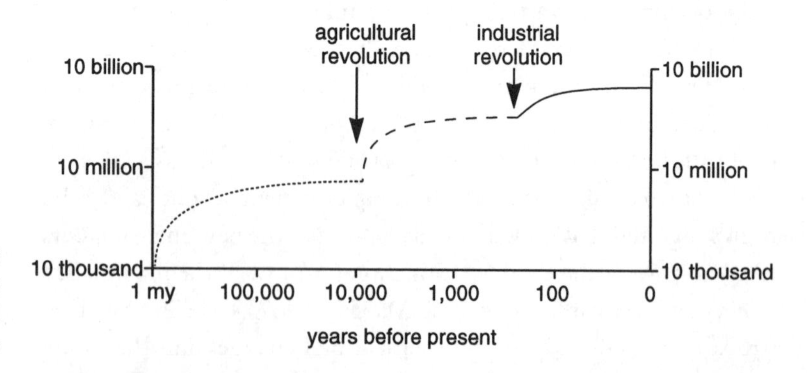

Figure 4.2 Changing human population, plotted on a log–log scale

animals, a process which occurred independently in a number of different parts of the world. Notable among these were three sub-tropical centres – the Near East, Mesoamerica and South-East Asia. In the Near East, the so-called Fertile Crescent of Mesopotamia is probably the earliest and arguably the most important of all the global centres of domestication. It is from the Neolithic (New Stone Age) farmers of this region that we have inherited cereal crops such as wheat, barley, rye, pulses including pea and lentil, and a range of domestic animals including cattle, sheep, goat, pig and dog. The capacity to settle in a place for an extended period as a result of this Neolithic revolution in agriculture some 10,000 years ago led to increased economic productivity and new social organization and was accompanied by a growth in human population and great developments in material culture (figure 4.2).

The major cereal crop of New World origin is maize,whose probable place of domestication is Central or Mesoamerica. Significantly, and in marked contrast to the situation in the Near East, there was no rush in Mesoamerica to abandon hunting and gathering and take up farming. Garden cultivation of maize and other plant domesticates such as squash, chili pepper and avocado were simply added to existing forms of food procurement. A critical threshold may have been reached around 5,000 years ago when, with yields of 200 kg/ha, maize became more productive than any other wild-plant resource. It is suggested that only at this point was maize-based farming capable of supporting settled farming communities, and it is at this date that fully developed agriculture first appeared in the New World. This slow take-off of agriculture may also have been because there were few animal domesticates in the New World.

A third of the world's great cereal crops is rice, of which the most important type is that indigenous to eastern and southern Asia. The earliest known communities with economies based on rice cultivation were those of the early Neolithic culture of China's lower Yangtse river valley, dated to around 7,000 years ago. It is probably that the origins of rice domestication are to be found to the south of this area in a belt which includes Burma, Thailand, Vietnam and southern China (Chang, 1976).

The tropics have also been home to a wide variety of different types of domesticated plants, including tree, fibre, root-tuber and seed crops. However, in tropical areas, domestication appears to have been spatially diffuse rather than concentrated in a few core areas.

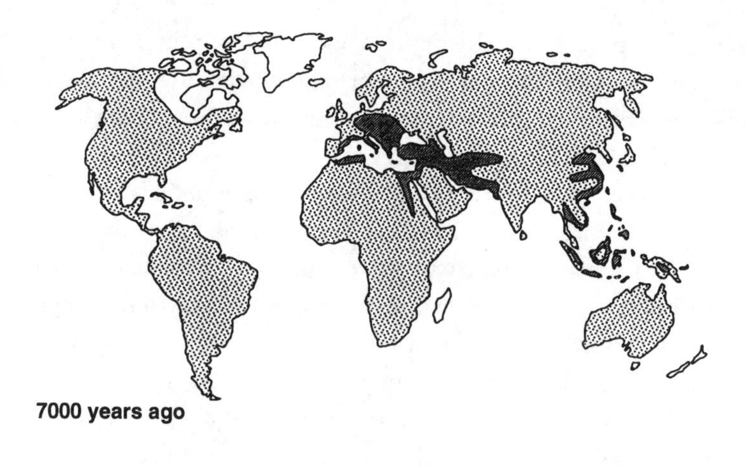

2000 years ago

Uninhabited (mainly ice)
Wilderness, HFGs and/or pastoralists
Small-scale agriculture
'Civilized' lands

Figure 4.3 Changing extent of the wilderness and the oecumene through time (based partly on maps in Sherratt, 1980)

Domestic crops, for instance, appear as early as 9,000 years ago in South America and in Papua New Guinea. The first farmers on the south side of the Sahara grew locally domesticated crops such as sorghum (or guinea-corn), not imported ones. In South America, manioc is a crop of lowland origin while the potato is native to the Andean mountains. Thus, unlike the Near East or Mesoamerica, the tropics did not produce any major centre of diffusion for new plant and animal domesticates.

Domestication started the long transformation of many natural landscapes into cultural landscapes, fashioning what Glacken (1967) has termed a second nature. This taming of nature was the work of agricultural societies which have since developed and spread over

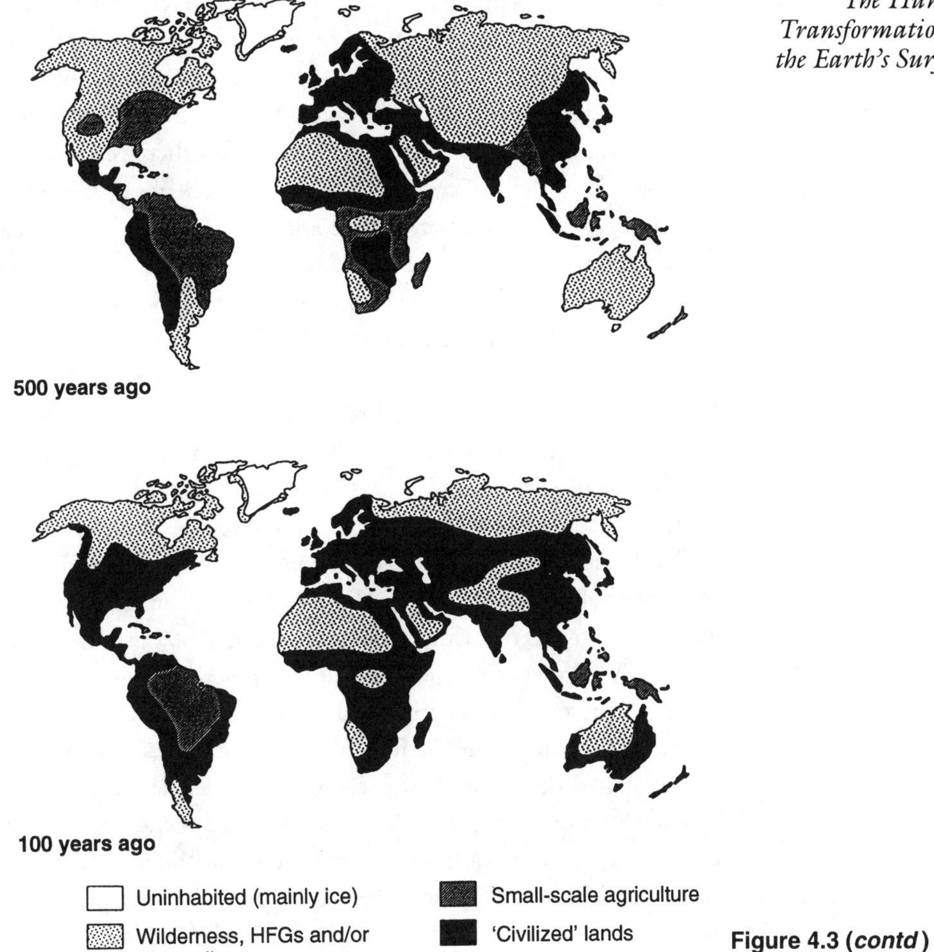

Figure 4.3 (*contd*)

large parts of the earth's habitable land masses. They have created the *oecumene*, that part of the earth's surface settled and transformed by people to create a landscape of farms, fields and cities (figure 4.3). From a common starting point in peasant farming, there has evolved a diverse range of societies with cultures based on agriculture or pastoralism. The most important of these societies were socially stratified and were engaged in what Karl Marx distinguished as the Asiatic, classical or feudal modes of production (Hobsbawm, 1964). All of these shared in common the production of an economic surplus by agricultural labour that was then taken and used by another social class.

This new form of social relations also changed human–

environment relations in several important ways. Agricultural output was maximized by improving the productivity of existing farmland or by increasing the cultivated area. The latter led to clearance of woodland and draining of marshland. It also included cultivation of marginal land susceptible to soil erosion and other forms of degradation. The advent of complex agricultural societies distanced and often weakened the link between people and nature. Nature became less the habitat for the farmer than a set of economic resources to be exploited, managed and manipulated by the controlling group. Although these complex agricultural societies shared certain key characteristics in common, such as urbanism, they each had their own distinctive set of environmental relations. This is well illustrated by the Asiatic mode of production, or as Wittfogel (1956, 1957) termed it the hydraulic civilization.

The ancient civilizations of Egypt, Mesopotamia, India and China all developed in a specific physical context. They emerged within major alluvial river valleys under climates of low or unreliable rainfall. Wittfogel suggests that the agricultural potential of these fertile lands could only be realized by large-scale manipulation of their soil and water resources. This encouraged the development of hydraulic irrigation schemes, constructed under a centralized organization. The agrarian economy in these great irrigated river valleys became the function of the state, which was also the sole owner of the land. The timing and distribution of irrigation water, the maintenance of canals and the collection and storage of surplus food were the responsibility of a professional bureaucratic ruling class which held total, despotic power. The combination of organized human labour and a productive environment created a distinctive and potent mode of production which transformed nature so completely that almost nothing now remains of the original alluvial landscape.

Complex agricultural societies, such as existed in Egypt and China, were not the only form of cultural development to emerge from communal peasant farming. One of the most distinctive cultural developments was nomadic pastoralism, an extensive form of production adapted to arid or rugged terrain. Like hydraulic irrigation, pastoralism involves a special set of ecological relationships between human groups and their environments, in this case mediated not through water use but through domestic animals. The domestication of the horse led to the emergence of pastoralist groups on the steppes of central Asia some 4,000 years ago. Nomadism developed around the same time in the deserts of Arabia and the Sahara where the camel was found to be better suited to the arid environment. In sub-Saharan Africa, pastoralists like the modern Masai were above all reliant on cattle. Some 5,000 years ago, they began to spread progressively across the African continent. Like the plough, pastoral nomadic societies such as these were effectively confined to the Old World prior to the era of European expansion from about AD 1500 onwards. They occupied land that was at best climatically marginal for settled agriculture and where environmental conditions were inherently prone to great fluctuations in grazing, water and food supplies. Livestock numbers were traditionally held in check by these constraints of the carrying capacity of the environment.

Agricultural Impact and Land Degradation

The advent of agriculture brought with it tremendous potential for modifying the natural environment. The most immediate impact was on the domesticates themselves. Cultural pressures did not stop with domestication but continued to select the most productive crop and livestock forms. Plant domesticates often lose their natural means of dissemination and become dependent on cultural practices for their propagation. But if cultigens could not survive without human help, neither could humankind survive without cultivated crops. Agriculture thus strengthened the mutual dependence between people (as farmers) and a limited range of (domestic) plants and animals. In so doing it also brought human beings into conflict with other elements of nature. The artificial environment created by agriculture protected crops and farmyard animals from predators and gave them great advantages for survival in relation to natural competitors. The ecological consequences of such selective nurturing were sometimes greater for the wild species than they were for the domesticates themselves. Sheep and cattle are numerous today, while predators such as the bear and competitors such as the aurochs are now either rare or extinct. Extinction has occurred both directly through hunting kill-off and indirectly through loss of habitat. The latter relates to the most important aspect of all in the new human relationship with nature, namely the use of the land for agriculture.

Cultivation of crops requires at least partial clearance of the existing vegetation cover and its replacement by field plots. Consequently, early agriculture is associated with the first substantial human impact upon the soil. This impact was all the more permanent because of agriculture's association with a settled or sedentary way of life. During the last 10,000 years, the progressive adoption and intensification of agricultural modes of production have led to widespread conversion of natural woodland and grassland vegetation into farming land. Clearance for crop cultivation need not leave the soil completely bare. Some of the original flora may remain, for example, in hedgerows while the crop itself affords considerable protection from raindrop impact. The invasion of weeds during fallow periods or of secondary forest after abandonment by tropical shifting agriculture, has a similar protective effect. On grazed grassland, the ground cover may hardly be reduced at all. Nonetheless, agricultural clearance has certainly increased soil erosion losses well above the long-term geological norm (Douglas, 1967).

The environmental consequences of soil loss for agro-ecosystems are complex and variable. Nutrient depletion and other forms of soil degradation can crucially affect conditions for plant growth, not only changing crop yields but also restricting the range of wild plants that could eventually reoccupy the site. One useful indicator of environmental impact is soil life, or the balance between the rate of soil loss and the creation of new soil by substrate weathering. Soils on alluvial floodplains, for example, may take only 50 years to form and are generally subject to low rates of erosion, making their soil life effectively infinite. In contrast, soils on the ancient land surfaces of Africa have taken millions of years to form and in some cases have lives as short as the rate at which they are being lost, which can be as little as ten years on soils actively eroding today (Elwell and Stocking, 1984).

Human beings are not passive to land degradation. Conservation measures to maintain soil fertility are surely as old as agriculture itself. Traditionally, tropical-shifting cultivators have left tree stumps in their fields and intercropped to reduce the impact of violent convectional rains on the soil surface. Manuring, which provides nutrients and helps maintain the soil's organic-matter levels, is another very ancient practice. The laborious work of terracing China's loess plateau by peasant farmers has greatly reduced erosion in an environment where natural soil losses were amongst the highest in the world. However, there may be a long-term price to pay for cultivating marginal land such as this. Soil conservation systems need maintenance. If they break down, it can have environmental consequences worse than if the land had been left alone in the first place.

Around the Mediterranean basin, there is evidence from denuded hillsides and from valley alluviation that soil erosion has been widespread during recent millennia. Many Mediterranean river valleys contain alluvial deposits of historical age, and the pottery and other datable remains which they contain indicate that a main period of deposition was from late Roman to early medieval times (Vita-Finzi, 1969). It seems as if the replacement of forest by terraced vineyards and olive groves during classical times may have conserved the soil initially. So long as this agricultural system was maintained, land degradation was avoided. But when the Roman Empire was invaded by barbarians, rural security declined and soil terraces fell into disrepair. The equilibrium was upset and soil erosion followed, valleys became choked by sediments.

The intensity of soil erosion provides one of the clearest and simplest indications of landscape stability or instability (Boardman, Deaning and Foster, 1990). While land degradation through soil erosion is not new, it has certainly accelerated in recent decades in many parts of the Third World, especially in the tropical areas where it is now a pressing environmental problem (Douglas, 1994). The spatial linkages within river basins mean that the human and environmental effects of land degradation are not confined to the eroding headwater zones but extend downstream to lowland river floodplains. For example, the deforestation of Himalayan watersheds also affects the Ganges floodplain. This is home to some of the largest rural populations anywhere on earth and is now constantly threatened by river aggradation and flooding (Haigh, 1994).

Imperialism: Ecological and Political

The modern pattern of human–environment relations began some five centuries ago when Vasco da Gama rounded the Cape of Good Hope and Columbus set foot on the New World. In AD 1450, Eurasia, the Americas, sub-Saharan Africa and Australasia were unknown to each other. By AD 1550, European explorers had brought all except the last within a system of global contact. The explorers returned not only with gold but also with a plant to be smoked and fine bird feathers for ladies hats. The discovery and subsequent interchange of plant and animal domesticates brought to an end the discrete crop cultures based around wheat, rice and maize that had existed for thousands of years in western and eastern Eurasia and in the Americas. Indeed, so radical was this transformation that Irish potatoes, Australian sheep and Canadian wheat are now accepted as native. All were introduced and adopted following the age of European discovery.

For plant crops, there was two-way traffic between the Old and the New World but the situation was different for animals. The Americas had almost completely lacked indigenous domestic animals before AD 1500. Even the wild-mammal fauna was species-poor following the wave of megafaunal extinctions at the end of the last glaciation. The ecological potential for intrusive animal species was immense and the colonization of the New World was consequently accomplished by four-legged as much as by two-legged Europeans. The omnivorous domestic pig was a particularly effective primary colonizer. It was followed by cattle, horses and sheep, some of which escaped to become feral species.

It has often been the unintended rather than the intended consequences of European contact that have had the most far-reaching ecological effects in different parts of the world (Crosby, 1986). Species which had evolved separately over millions of years suddenly came into contact to produce a homogenized, melting-pot biota. Weedy plants such as plantain (Englishman's foot) expanded their range to disturbed habitats throughout the temperate world. But while some creatures profited, others lost out. Before European contact, millions of passenger pigeons were to be found in the woodlands of eastern North America. Millions of bison (buffalo) grazed upon its prairie grasses and herbs (McDonald, 1981). By the end of the nineteenth century, the buffalo population of the entire continent had been reduced to a few hundred heads and the passenger pigeon was extinct. Extinction rates were highest on island ecosystems, especially on those few – such as St Helena – which remained uninhabited prior to European contact (Grove, 1995). But the greatest unplanned impact of European exploration and discovery was in the transmission of human disease pathogens which devastated the indigenous population of the Americas.

The Americas were certainly no pristine Garden of Eden at the time of first contact. Human-induced landscape changes were already well underway before the European encounter. In regions like central Mexico, pre-Hispanic land degradation is now well attested (O'Hara,

Table 4.1 Net conversion of land to crops by region, 1700–1980

World region	*Area (million ha)* *1700*	*1980*	*% increase*
Tropical Africa	44	222	405
North Africa/Middle East	20	107	435
North America	3	203	6,667
Central and South America	7	142	1,929
South and East Asia	86	399	464
Former Soviet Union	33	233	606
Europe (exc. FSU)	67	137	105
Australia/New Zealand	5	58	1,060
Total	265	1501	466

Source: Data from Richards (1990).

Street-Perrott Arandol Burt, 1993). However, the demise of much of the native Amerindian population in the years following AD1492 led to the demographic and cultural collapse of several New World civilizations (Butzer, 1992). This in turn caused an end to indigenous woodland or grassland management systems. In many regions, the resultant ecosystem shock in the decades following 1492 led to a re-expansion of forests, leaving European settlers with the false impression of entering an empty land.

European discovery usually led on to economic exploitation and to political control. During subsequent centuries, European colonialism brought important changes in land use to many parts of the world (table 4.1). And where colonialism did not lead to the eclipse or demise of indigenous peoples, it disrupted their established mode of production and threatened their traditional relations with nature. Many contemporary problems, such as the desertification of Africa's drylands, can be understood only by reference to changes introduced during the colonial era.

Drylands are characterized not simply by low rainfall but also by a large rainfall variability and low reliability. Traditionally, societies coped with the threat of drought and famine by a series of ingenious socio-economic and technical strategies. They included inter-cropping and cultivation of drought-tolerant staples (e.g., sorghum), small-scale irrigation schemes such as rainwater harvesting, the exploitation of famine foods (in many cases returning to the gathering of wild plants), and redistributive and food storage mechanisms (e.g., Adams and Anderson, 1988; Mortimore, 1989; Watts, 1983).

There were, of course, food shortages but the socio-economic system acted to minimize their adverse effects and ensure that the rural population as a whole survived through times of hardship. If societies in drylands, such as the Hausa of northern Nigeria, were traditionally well attuned to their variable natural environment, is it reasonable to attribute desertification – as some have – to squandering of natural resources by ignorant peasants and pastoralists?

In fact, what colonial rule did was to introduce an alternative form of knowledge, about how to cope with Africa's environment. The methods of western science were applied to farming problems, for example, at agricultural research stations, at which there were experimental plots with controlled conditions of crop growth. These built up a corpus of recommended strategies for managing the natural environment, which were often at odds with those strategies established by trial-and-error experience by African farmers. It was impossible on agricultural research stations to simulate problems of seasonal labour shortage and how these could best be avoided (e.g., by spacing out the harvest season). The new scientific knowledge was, in any case, applied to cash crops, such as tobacco and coffee rather than to subsistence ones like sorghum and cassava. It was consequently more appropriate to European settlers than to native African farmers.

Conflict between the two forms of knowledge can be illustrated by rice cultivation in Sierra Leone (Richards, 1985). Here, colonial policy had opposed the traditional mix of upland and swamp rice and discouraged shifting cultivation which, while 'wasteful' of land, avoided soil nutrient depletion. Colonial-farming specialists instead tried unsuccessfully to introduce swamp rice monoculture to low-lying boli-lands. Here, as elsewhere, colonial (and, after political independence, post-colonial) policy led to the modification or even abandonment of traditional-farming methods, the introduction of cash crops, and the associated enforcement of colonial conservation methods which were often far from appropriate (e.g. Bell and Roberts, 1991). In more extreme cases, such as in South Africa, the African population was subject to land alienation and enforced relocation, often concentrating the displaced population into environmentally marginal lands. All of these processes led to the erosion of indigenous systems of managing the natural environment, and the clearance of natural or semi-natural vegetation to make way for estate plantations or ranches. And where the environment is fragile – as in semi-arid lands – degradation has been the end result. Desertification should, therefore, be seen as a process of impoverishment, which includes not just plant or animal species and the soil but also indigenous human knowledge about how best to use them.

Industrialism and Resource Use

The last two centuries have witnessed a major shift in the human use of natural resources, associated with the development of industrial capitalism. The key to the development of industrial societies has been the switch from renewable to non-renewable resources, particularly in energy. In agricultural systems, energy inputs derive from photosynthesis (to achieve crop growth) and human or animal power, supplemented by water or wind power (typically to drive mills) and by wood fuel. All but the last of these are inherently renewable forms of energy. Even the forests can, in principle, be managed to achieve a sustainable yield either through replanting or by permitting sufficient natural regrowth of new trees. In practice, almost all countries had experienced a declining percentage of forest cover prior to the switch to non-renewable energy sources (Mather, 1992). The output from farming systems in the form of harvested grain, wool, and so forth will vary from one year to the next but is essentially predictable and sustainable, assuming no long-term decline in soil fertility.

The energy basis of modern industrial systems is fundamentally different. Its adoption marked a turning-point in human history comparable to the control of fire and the invention of agriculture (Simmons, 1989). The use of fuels such as coal, natural gas and oil has reduced the direct dependence of industrial society on biological energy fixation through green plants. In so doing, it has radically altered the structure of human–environmental relations. Hydrocarbon resources were themselves originally formed via photosynthetic fixation at various times during the last 500 million years of earth history but their energy content has been concentrated by subsequent geological processes. Consequently, the energy return from exploiting coal or oil is potentially far higher than can be achieved using wood fuel or wind power. Since 1800, these new energy sources have been exploited and allied to mechanical technology in order to produce material goods in ever greater numbers and of ever greater variety, as well as to allow them to be transported and distributed world-wide. These goods have, in turn, allowed the

Table 4.2 Human energy use at different stages in socio-economic development

	Plant and animal foods	*Home and commerce*	*Industry and agriculture*	*Transport*	*Total*
Technological society	10	66	91	63	230
Early industrial society	7	32	24	14	77
Advanced agriculturalists	6	12	7	1	26
Early agriculturalists	4	4	4	–	12
Hunters, fishers and gatherers	3	2	–	–	5

Note: Daily per capita energy consumption ($\times 10^3$ kcal).
Source: Data from Bennett (1976).

earth's surface to be transformed on a scale and a rate that was not possible under pre-industrial socio-economic conditions (table 4.2).

Industrialism was first developed in Britain and by the end of the nineteenth century it was firmly established in western Europe and eastern North America. This economic revolution was powered above all by coal and was associated with rapid urbanization and a shift of population away from the land. By the middle of the twentieth century, urban-industrial societies had become dominant in much of eastern Europe, especially in Russia and the Ukraine, and in Japan. Oil was becoming comparable in importance to coal as an energy source, and it also fed other industries such as petrochemicals. These industrial economies were not only the most productive; they also drew the rest of the world into a global economic network by exploiting raw materials from peripheral, non-industrial regions of the world, and by returning manufactured goods to them for sale. Some of these raw materials, such as cotton and rubber, derived from cash-crop agricultural production while other biological materials, like timber, were available naturally. Mineral resources, such as tin and copper, were an even more important target for economic exploitation; like hydrocarbons, these mineral resources are finite and effectively non-renewable on a human time scale.

To date, industrial economies have been based on increasing consumption of fossil fuels and other finite resources. This, in the long run, is unsustainable. It has led, therefore, to the suggestion that there may eventually be limits to economic growth (Meadows et al., 1972). To some thinkers, these are grounds for radically changing society and returning to self-sufficient, rural life-styles. To technological optimists, on the other hand, this represents a challenge to be overcome by efficient recycling of raw materials and by the development of new, alternative energy sources such as nuclear power (el-Hinnawi, 1982). Thus far, most commercial nuclear-power stations have used burner reactors whose fuel cycle consumes uranium and thus depletes uranium ore stocks in the same way that conventional power stations use up coal or oil reserves. However, nuclear power is potentially different in character from hydrocarbon fuels as an energy source in that it offers the prospect of an almost limitless energy supply, particularly if nuclear fusion were to become a technical reality. The grounds for concern over the development of nuclear power lie not with its exploitation of non-renewable resources but with the problems – so far unresolved – of dealing with the release of radionuclides into the environment either as radioactive waste or following a nuclear accident such as occurred at Chernobyl in 1986.

In the short and medium term, the exploitation of non-renewable minerals and fossil fuels has largely been determined by the relative cost of their extraction, transport and refining. Notwithstanding the oil crisis of the 1970s, the reality is that almost all of these resources have seen their prices fall in real terms on the world market during the last quarter of the twentieth century. New reserves have been found to counterbalance old ones which have been exhausted and greater mechanization has reduced other costs. The rapid economic growth of many East Asian countries, especially China with its huge population, seems set to lead to a surge in the global demand for energy and mineral resources which may yet reverse this trend within the next few decades. By contrast, there are already major and very urgent problems concerning the adequate supply of many of the world's renewable resources, notably fuel wood and other forest resources, water and agricultural land. An increasing number of rivers are now fully exploited for water supply and some, like the Colorado river, now dry up before they reach the sea. (Not all water resources are in any case renewable, in particular fossil groundwater, such as that being tapped by Libya's Great Man-Made River). Similarly, many tropical countries such as Burma, Nigeria and Brazil are experiencing a rapid and unsustainable rate of deforestation (Furley, 1994).

Environmental Pollution

Demand and supply economics has not, in any case, provided a satisfactory framework for resource utilization, even for non-renewable resources over the short to medium term. This is because there is an additional, and traditionally uncosted, environmental price to be paid for resource utilization in the form of pollution. Although pollutants include artificial elements and compounds, such as radioactive caesium or toxic DDT, pollution more usually involves naturally occurring compounds such as sulphur or nitrogen. This makes it difficult to establish whether the present state of an ecosystem has been modified by pollution or not. Lake Erie is much richer in nutrients than Lake Superior, but it is not evident whether this is because of sewage and industrial effluent from around Lake Erie's shores or because the geology of the two lake catchment areas is different anyway. A second problem that bedevils pollution studies is that pollutants move downstream or downwind from their point of origin to affect other localities. This makes establishing the source of pollution and consequently the cause of it very difficult.

In northern industrial regions, one of the most widespread forms of pollution impact on ecosystems has been the result of acid deposition. Increased atmospheric deposition of acids is believed to have had wide-ranging environmental consequences including tree death, accelerated weathering of building facades, damage to human health, and acidification of freshwater ecosystems. Acidification starts with the emission of oxides of sulphur (SO_2) and nitrogen (NO_x) into the atmosphere from car exhausts and from coal-fired power stations. It has long been known that atmospheric pollution in urban areas leads to dry fall-out of sulphur and that this poses an ecological threat, for example, to lichen. Precisely in order to avoid the problem of local dry fall-out of pollutants, high stacks were constructed from which emissions would be carried up into the atmosphere. Once in the atmosphere sulphur dioxide and nitrogen oxides undergo chemical transformation, involving combination with water vapour to produce dilute sulphuric and nitric acids. This acid rain is carried downwind by atmospheric circulation and is returned to earth as wet deposition in the form of rain or snow. Hydrological processes then lead the acidified waters into streams and lakes, with consequences for their plant and animal communities. There is a notable decline in both total biological activity and species diversity once waters move below pH c.6.0, with a characteristic decline in fish stocks. The precise ecological mechanisms associated with acidification of freshwaters are only partially understood, but one important aspect is the mobilization at low pH of aluminium which is toxic to fish and many other higher organisms.

However, acid precipitation clearly does not have a blanket effect over the landscape as might be expected from atmospheric deposition; catchment area conditions are also an integral part of the problem. In particular, the ability of catchment area soils and bedrock to neutralize acid precipitation critically influences the pH of the resulting runoff. Buffering capacities will be high in areas of basic bedrock such as limestone and low on granite and other acidic rocks. Vulnerability to acidification is therefore greatest where there is a combination of high-acid loadings and low-catchment area buffering capacities. Amongst the most vulnerable areas are the southern portions of Canada and Scandinavia. These areas have thin soils but innumerable lakes. Most of the latter in any case have relatively acidic waters. Chemical and fishery records in southern Scandinavia show a systematic downward trend since the 1940s (Almer et al., 1974). Palaeoecological analysis of lake sediment cores has provided another means of obtaining proxy records of water chemistry and biology which smooth out the random variations of individual years. They have shown that previously stable aquatic ecosystems in the vulnerable regions of Europe and North America have shown a marked decline in pH during the last hundred years or so, even in the most remote locations (Battarbee, 1994).

Many atmospheric pollutants have been recorded stratigraphically in environments ranging from peat profiles to ice cores. Even on the remote ice cap of Greenland, atmospheric deposition of lead has increased twenty times over the last two centuries (Murozumi, Chow and Patterson, 1969). Metal such as lead, zinc and copper are found at higher levels in acidified lakes (Renberg, Persson and Emteryd, 1994) and they have typically increased as pH has fallen. The advent of industrial capitalism has not only caused increased emission of pollutants into the atmosphere, it has also led to their direct release into soils and streams, frequently related to the intensification and industrialization of agriculture. One important consequence of this has been cultural eutrophication, or nutrient enrichment. Eutrophication can be caused by the application and subsequent runoff of chemical fertilizers to farmland or by the discharge of sewage or industrial waste into streams and rivers. In all of these cases, cultural eutrophication is an unintended, indirect consequence of land use, sanitation or industrial activity. Nutrients, particularly nitrate and phosphate, are washed away by runoff and carried downstream into rivers and lakes. In aquatic ecosystems, nitrogen and phosphorus are usually the limiting substances which hold in check greater primary productivity in the form of algal growth. Primary productivity may increase to the point where dead algae use up most of the oxygen as they decompose whilst sinking to the lake bed. Lake waters become de-oxygenated and higher organisms such as fish are restricted to the upper few metres of the lake where they are not starved of oxygen. Blue-green algae, which may be toxic, also increases more rapidly with eutrophication than other algal forms.

As with acidification, lake-sediment studies have uncovered cultural eutrophication throughout Europe and North America. Anomalously productive ecosystems in regions of otherwise oligotrophic lakes, such as Shagawa Lake, Minnesota, have been revealed as the product of human impact, in this case from mining and sewage-waste disposal (Bradbury and Waddington, 1973). The cultural causes of eutrophication are activities associated with industrial and urban society – piped sewage, chemical fertilizers and detergents, and industrial waste. Industrial society has modified the cycling of almost all natural elements to an extent that was impossible under previous modes of agricultural production. The global carbon cycle has been profoundly altered as have those of many other elements including key micro-nutrients such as phosphorus and toxins, such as lead. In the past, the costs of pollution, such as the loss of fish stocks or the dumping of toxic waste, have been met by adversely affected third parties or by no one save nature herself rather than by the producer or the consumer. A major challenge for the sustainable development of natural resources is, therefore, to establish mechanisms by which environmental costs can be included in economic calculations (e.g., carbon taxes, the polluter pays principle) which are both equitable and enforceable.

Conclusion

Human action has dramatically transformed the earth's surface and that transformation has accelerated towards the present day accompanied by phenomenal increases in population, and advances in technology and organizational capabilities. Each stage of socio-economic and technological evolution has provided a greater access to usable energy, leading to material improvement and an ability to transform further the natural environment. In many respects, our achievements in dominating nature and using it for our own ends have fed off our own success in the manner of a chain breeder re-action. For some resources at least, there are clear limits to this process. Deforestation, for example, has already transformed most of the deciduous forests of the northern mid-latitudes to agricultural land and is currently decimating many lowland tropical forests (Williams, 1990). Such changes to the earth's surface are critical to those affected directly but they also have significant impacts upon environmental change on a planetary scale. Global changes are in fact caused by an agglomeration of smaller-scale influences such as the alterations in surface albedo when land use is changed. The destruction of tropical moist forests, for example, makes a significant contribution to global climate change via the release of carbon into the atmosphere, as well as being a critical environmental issue in its own right through the loss of biodiversity (Brown and Pearce, 1994). In order to understand how earth-surface systems function and respond to human disturbance, a number of major international scientific research programmes have been initiated such as the International Geosphere-Biosphere Programme (IGBP) of the International Council of Scientific Unions (ICSU).

The United Nations has also been very active in calling attention to the deleterious nature of much of human impact on environmental resources. From the first UN Conference on the Human Environment in Stockholm in 1972, through the Brundtland Commission Report of 1987, to the Earth Summit at Rio de Janeiro, Brazil, in 1992, the nations of the world have been summoned to take definitive steps to ensure a more sustainable relationship with the environment. The Rio Conference was particularly significant because, apart from producing a number of guidelines and principles for fostering much better environmental management in most countries, it initiated two legally binding conventions for protecting the global environment. Moreover, the conference reinforced the role of non-governmental organizations in promoting a more concerned attitude and policy towards the environment. The establishment of a Commission on Sustainable Development in addition to the existing United Nations Environment Programme based in

Nairobi is also indicative of efforts to take more pro-active measures to protect the global environment.

Nonetheless, it is important to stress that human action has already seriously disturbed many earth-surface systems. Although the global environment has an intrinsic resilience to disturbance, the current rate of human-induced change cannot be sustained indefinitely without endangering the earth's potential capacity for recovery. A sobering example of the dangers of taking for granted the availability of natural resources, and one with which this chapter will conclude, comes from Easter Island in the eastern Pacific. This, the most isolated piece of inhabited land in the world, is renowned for numerous gigantic figures carved in stone and originally set in rows on stone platforms or *ahu*. They are the work of Polynesian settlers who first arrived there some 1500 years ago. Pollen cores from three crater lakes have shown that, at the time of the settlers arrival, Easter Island was largely covered by palm forest (Flenley and King, 1984). Although the forest had to be partially cleared to allow garden horticulture, timber continued to be a vital resource providing dug-out canoes on which fishing depended, wood for dwellings, palisades, and fuel.

It is almost certain that the stone statues could not have been moved and erected without the aid of supporting wooden timbers. The initial founding population grew so that between AD 1100 and 1650 the island possibly supported as many as 7000 people. This was the major period of monument construction and appears to have been associated with a society increasingly obsessed with rivalry and warfare. As demands on the finite timber resources of Easter Island increased, so trees were felled until eventually, as pollen records show, the palm forests were completely removed. On an island only 25 kilometres long, it would have been possible to see the last palm cut down knowing that there would be no more. Following this ecological crisis, the first European explorers encountered a barren, almost treeless island littered with toppled giant statues from a bygone age. The Easter Islanders had, it appears, destroyed the resources on which their society depended. Might the history of Easter Island be an appropriate lesson in microcosm for the human race in respect of the global ecosystem? Certainly, it is to be hoped that humanity as a whole will treat the common planetary home with greater care and foresight than did the Easter Islanders with theirs.

References

Adams, W.M. and Anderson, D.M. (1988) Irrigation before development: indigenous and induced change in agricultural water management in East Africa, *African Affairs*, 87:519–53.

Almer, B., Dickson, W., Ekstrom, C., Hornstrom, E. and Miller, U. (1974) Effects of acidification on Swedish lakes, *Ambio*, 3:30–6.

Battarbee, R.W. (1994) Surface water acidification, pp. 213–41 in N. Roberts (ed.) *The Changing Global Environment* (Oxford: Blackwell Publishers).

Bell, M. and Roberts, N. (1991) The political ecology of dambo soil and water resources in Zimbabwe, *Transactions of the Institute of British Geographers*, N.S. 16:301–18.

Boardman, J., Dearing, J.A. and Foster, I.D.L. (eds) (1990) *Soil Erosion on Agricultural Land* (Chichester: Wiley).

Bradbury, J.P. and Waddington J.C.B. (1973) The impact of European settlement on Shagawa lake, north-eastern Minnesota, U.S.A., pp. 289–307 in H.J.B. Birks and R.G. West (eds) *Quaternary Plant Ecology* (Oxford: Blackwell).

Brimblecombe, P. and Pfister, C. (eds) (1990) *The Silent Countdown: Essays in European Environmental History* (Berlin).

Brown, K. and Pearce, D.W. (eds) *The Causes of Tropical Deforestation* (London: University College Press).

Butlin, R.A. and Roberts, N. (eds) (1995) *Ecological Relations in Historical Times: Human Impact and Adaptation* (Oxford: Blackwell–IBG Special Publication).

Buttarbee, R.W. (1994) Surface water acidification, pp. 213–41 in N. Roberts (ed.) *The Changing Global Environment* (Oxford: Blackwell).

Butzer, K.W. (1992) The Americas before and after 1492: an introduction to current geographical research, *Annals of the Association of American Geographers*, 82:345–68.

Chang, T.T. (1976) The rice cultures, *Philosophical Transactions of the Royal Society of London*, series B, 275:143–55.

Crosby, A.W. (1986) *Ecological Imperialism: The Biological Expansion of Europe 900–1900* (Cambridge: Cambridge University Press).

Douglas, I. (1967) Man, vegetation and the sediment yields of rivers, *Nature*, 215:25–8.

Douglas, I. (1994) Land degradation in the humid tropics, pp. 332–50 in N. Roberts (ed.) *The Changing Global Environment* (Oxford: Blackwell Publishers).

el-Hinnawi, E.E. (ed.) (1982) *Nuclear Energy and the Environment* (Oxford: Pergamon).

Elwell, H.A. and Stocking, M.A. (1984) Estimating soil life-span for conservation planning, *Tropical Agriculture*, 61:148–50.

Flenley, J.J. and King, S.M. (1984) Late Quaternary pollen records from Easter Island, *Nature*, 307:47–50.

Foley, R. (1987) *Another Unique Species: Patterns in Human Evolutionary Ecology* (London: Longman).

Furley, P.A. (1994) Tropical moist forest: transformation or conservation?, pp. 304–31 in N. Roberts (ed.) *The Changing Global Environment* (Oxford: Blackwell Publishers).

Glacken, C.J. (1967) *Traces on the Rhodian Shore* (Berkeley: University of California Press).

Grove, R.H. (1995) *Green Imperialism: Colonial Expansion, Tropical Island Edens and the Origins of Environmentalism, 1600–1860* (Cambridge: Cambridge University Press).

Haigh, M.J. (1994) Deforestation in Himalaya, pp. 440–62 in N. Roberts (ed.) *The Changing Global Environment* (Oxford: Blackwell Publishers).

Haines-Young, R. (1994) Remote sensing of environmental change, pp. 22–43 in N. Roberts (ed.) *The Changing Global Environment* (Oxford: Blackwell Publishers).

Hobsbawn, E.J. (1964) *Karl Marx Pre-Capitalist Economic Formations* (London: Lawrence and Wishart).

Hooke, J.M. and Kain, R.J.P. (1982) *Historical Change in the Physical Environment* (London: Butterworth).

Kershaw, A.P. (1986) Climatic change and aboriginal burning in north-east Australia during the last two glacial/interglacial cycles, *Nature* 322:47–9.

Lee, R.B. and De Vore, I. (eds) (1968) *Man, the Hunter* (Chicago: Aldine).

Lewis, H.T. (1972) The role of fire in the domestication of plants and animals in southwest Asia: a hypothesis, *Man*, 7:195–222.

Martin, P.S. and Klein, R.G. (1984) *Quaternary Extinctions: A Prehistoric Revolution* (Tuscon: University of Arizona Press).

Martin, P.S. (1967) Prehistoric overkill, pp. 75–120 in P.S. Martin and H.E. Wright, Jr. (eds) *Pleistocene Extinctions* (New Haven: Yale University Press).

Mather, A.S. (1992) The forest transition, *Area*, 24:367–79.

McDonald, J.N. (1981) *North American Bison: Their Classification and Evolution* (Berkeley and San Francisco: University of California Press).

Meadows, D.H., Meadows, D.L., Randers, J. and Behrens, W.W. III (1972) *The Limits to Growth* (London: Earth Island).

Meyer, W.B. and Turner, B.L. II (eds) (1994) *Changes in Land Use and Land Cover: A Global Perspective* (Cambridge: Cambridge University Press).

Mortimore, M. (1989) *Adapting to Drought: Farmers, Famines and Desertification in West Africa* (Cambridge: Cambridge University Press).

Murozumi, M., Chow, T.J. and Patterson, C. (1969) Chemical concentrations of pollutant lead aerosols, terrestial dusts and sea salt in Greenland

and Antarctic snow strata, *Geochimica et Cosmochimica Acta*, 33:1247–94.

O'Hara, S.L., Street-Perrott, F.A. and Burt, T.P. (1993) Accelerated soil erosion around a Mexican highland lake caused by pre-Hispanic agriculture, *Nature*, 362:48–51.

Oldfield, F. (1983) Man's impact on the environment: some recent perspectives, *Geography*, 68:245–56.

Renberg, I., Persson, M.W. and Emteryd, O. (1994) Pre-industrial atmospheric lead contamination detected in Swedish lake sediments, *Nature*, 368:323–6.

Richards, P. (1985) *Indigenous Agricultural Revolution: Ecology and Food Production in West Africa* (London: Hutchinson).

Roberts, N. (1989) *The Holocene: An Environmental History* (Oxford: Blackwell).

Sauer, C.O. (1956) The agency of man on the earth, pp. 49–69 in W.L. Thomas Jr. (ed.) *Man's Role in Changing the Face of the Earth* (Chicago: University of Chicago Press).

Sheail, J. (1980) *Historical Ecology: The Documentary Evidence* (Cambridge: Institute of Terrestial Ecology).

Simmons, I.G. (1989) *Changing the Face of the Earth: Culture, Environment, History* (Oxford: Blackwell).

Vita-Finzi, C. (1969) *The Mediterranean Valleys* (Cambridge: Cambridge University Press).

Watts, M. (1983) *Silent Violence: Food Famine and the Peasantry in Northern Nigeria* (Berkeley: University of California Press).

Williams M. (1990) Forests, pp. 179–201 in B.L. Turner II et al. (eds), *The Earth as Transformed by Human Action* (Cambridge: Cambridge University Press).

Wittfogel, K.A. (1956) The hydraulic civilizations, pp. 152–64 in W.L. Thomas (ed.) *Man's Role in Changing the Face of the Earth* (Chicago: University of Chicago Press).

Wittfogel, K.A. (1957) *Oriental Despotism: A Comparative Study of Total Power* (New Haven: Yale University Press).

Worster, D. (ed.) (1988) *The Ends of the Earth: Perspectives on Modern Environmental History* (Cambridge: Cambridge University Press).

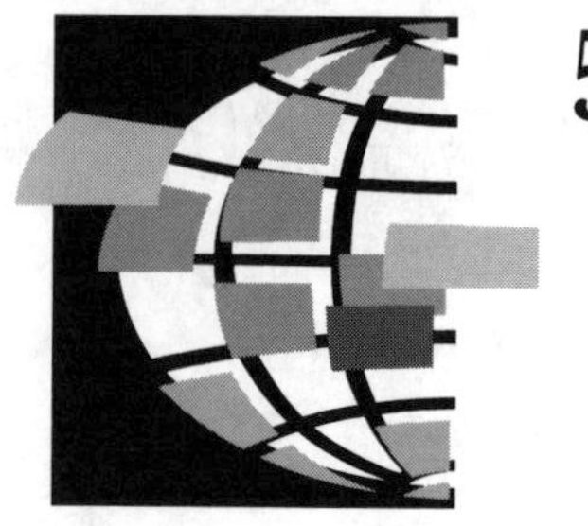

5 Climatic Change and the Future of the Human Environment

From time immemorial, people have been aware of the weather. Records of the weather exist, for instance, in the Bible and other ancient texts. Long-term values of weather parameters such as air pressure, air temperature, amount of precipitation, and recurrence of weather types determine the notion of climate. This, in turn, is one of the basic characteristics of natural conditions prevalent in a particular area.

For millennia, climate in different parts of the earth has been subject to cyclic changes of varying duration whose nature is not entirely clear. Minimum amplitude of climate variations typically lasts from 30 to 40 years. Average values for such length of time are used to characterize a climatic epoch. The climate of a place can thus be described as a statistical state of the atmosphere over a period of some 30 to 40 years. The primary factors determining climatic processes comprise the nature of the earth's surface and of the neighbouring world ocean, with solar radiation being the main source of energy.

Through the mechanism of general atmospheric circulation, climate fluctuations are experienced all over the earth's surface and over the ages. Whatever their causes or wherever their sources, the consequences of these fluctuations are manifest everywhere. They are imprinted on rocks and soils, on bottom sediments and coral reefs, on spores and pollens trapped and sealed in ancient deposits, and in the annual rings of trees. Only recently, ice cores extracted from the longholes of arctic ice sheets are proving to be another veritable source of evidence concerning climate changes that have occurred in the past.

This chapter is based on a contribution by Vladimir Kotlyakov.

Climate Changes over the Last Two Climatic Cycles

The ice core taken from a few longholes drilled in Greenland and the Antarctic regions has recently been analysed. Two holes were sunk at the top of the main ice dome in Greenland whose height stands at 3,235 metres above sea level. Here, at a height of 3,490 metres above mean sea level, average annual air temperature is minus 55.5°C, the annual amount of precipitation is hardly 23 mm. The thickness of the ice is close to 3,700 m, the entire body being made up of ice deposited in the course of several hundred thousand years. At present, the hole is 3,000 m deep. The core extracted has made it possible to analyse the last two climatic cycles (Genthon et al., 1987; Jouzel et al., 1989; Jouzel et al., 1993; Lorius et al., 1985).

The results of the analysis of the core from the hole at the Vostok station have been compared with the data of marine sediments. The Vostok station isotope profile provides information on the temperature conditions over the last 260,000 years and makes it possible to draw various conclusions on climatic fluctuations over the period. First, the peak of the first interglacial was approximately 2° warmer than during the Holocene; second, the final stages of the Dnieper (Riss) Ice Age were just as cold as the maximum of the Valdai (Wurm) Ice age; third, transition from the Dnieper Ice Age to the last (Riss-Wurm) interglacial was accompanied by a 12° warming, but the

Artist's impression of the city of Cologne in the year 2050, after a possible global warming due to the greenhouse effect has melted polar ice caps, raised the level of the oceans, and changed the climate of Western Europe. Source: Erik Viktor/ Science Photo Library/ Cosmos.

interglacial was short and was quickly superseded by the next ice age; fourth, in the last ice age, three periods of temperature minima can be easily distinguished with intervening temperatures being 4° and 6° higher respectively than during Late Pleistocene. By and large, the isotope-and-oxygen analysis of Antarctic and Greenland holes enables us to conclude that the cold epoch of Late Pleistocene ended 10,000–13,000 years ago, the height of the period occurring some 20,000 years ago, that is, simultaneously with the maximum of temperature drop on all other mainlands.

Besides, the isotope curve obtained at the Vostok station up to 110,000 years ago perfectly coincides with the appropriate isotope-and-oxygen profile of the deep sea deposits portraying the variation in the global volume of the ice. During the ice ages, there was accumulation of light (in terms of isotopic composition) ice on the continents. As a result, heavy isotopes of oxygen and hydrogen found their way into the oceans. This ensured that the findings of the Vostok station correlate well with the data of the changing volume of the entire continental ice on the earth's surface.

Calculations based on the depth core data indicate that during the ice ages there was twice as much snow deposited in the polar regions as there is at present. A conclusion such as this is quite natural. Various scholars had insisted as far back as early in this century that, in case of a global warming, evaporation from the surface of the ocean and the amount of falling precipitation will increase whereas, in case of cooling, the amount of precipitation will decline. Besides, during the ice ages, the masses of ice in the ocean kept growing, thereby distancing the sources of water vapour from the ice sheets.

The ice ages are characterized not only by a general cooling but also by an abrupt intensification of contrasts between different latitudes and between dryland and ocean. Consequently, the growing energy of all oceanic and atmospheric processes would also intensify. During the periods of global cooling the intensity of oceanic and atmospheric currents will increase and cyclonic processes on the boundary of ice sheets will become more active.

Proofs of a stronger atmospheric circulation during the ice ages were provided by the results of measuring the concentrations of continental and marine aerosols in the ice core taken from the long-holes. Typical indicators of the former are aluminium aerosols, and of the latter – sodium aerosols. Concentrations of both increase during the ice ages. At the Vostok station, the concentration of continental dust in Pleistocene ice is 30 times, and of marine aerosols five times as high as in Holocene ice. The principal reason for this is the more intensive winds due to the growing latitudinal contrasts. Of great significance was the general desertification of glacier areas. This extended to the continental shelves which had been drained of water as a result of the eustatic drop in sea level due to part of the water being used in forming the ice sheets.

Thus, all chemical parameters studied in the ice core lead to the conclusion that there was a sharp increase in the dust content of the atmosphere and heavier meridional circulation during the ice ages. This is usually associated with a growing difference of temperatures between the equatorial and the polar regions of the earth. However, dust content in the atmosphere by itself is a powerful climate-forming factor. The growing amount of dust and aerosols under conditions of cooling has the effect of further enhancing climate cooling.

When firn turns to ice, atmospheric air is encapsulated in the bubbles. By isolating the air from the core, it is possible to find out the composition of the atmosphere in the past and, in particular, to identify the content of green-house gases. Analysis of the core taken from the longholes showed that at the height of the Valdai Ice Age, the concentration of CO_2 was 25 per cent lower than during the Holocene (190–200 as against 260–280 ppmv, millionth parts, by volume). Obviously, the former level is typical of the ice ages, the latter, of the warm intervals.

The ice core taken from the Vostok station has made it possible to establish a good correlation between the changes in temperature and CO_2 content throughout both climatic cycles. Yet, while in transition from an ice age to an interglacial period, both the temperature and the content of CO_2 vary simultaneously. In the case of a reverse transition as happened, for instance, some 115,000 and 75,000 years ago, the concentration of CO_2 falls more slowly than the temperature does. It is obvious that the noted correlation is indicative of a cause-and-effect relationship. However, the data fail to identify the cause and the effect. Many experts regard the changing concentration of carbon dioxide as the cause but the aforesaid lag seems to underline the priority of temperature variations, followed by CO_2 changes which, in turn, enhance temperature fluctuations.

The real factor for such an enhancement may be the impact of the ocean and its changing circulation, the spread of ice and increased bioproductivity. In particular, during the cool spells a considerable part of atmospheric CO_2 could be absorbed by phytoplankton whose mass grew abruptly as a result of a more active upwelling and better nutrition of marine microflora. The analysis of the Vostok station cores also suggests that at different stages various mechanisms of temperature and carbon cycle interactions could play a crucial role.

The content of methane, another carbon compound, in ancient atmosphere is also closely associated with the march of palaeotemperatures. Sharp changes in methane concentration fall on both ice age–interglacial transitions: 150,000–135,000 and 18,000–19,000 years ago. During these periods, the concentration rose abruptly from 0.35 ppmv at the height of the Ice Ages to 0.6–0.7 ppmv during the interglacial optimums. The Valdai Ice Age is characterized by four maxima of CH_4 during relatively warm intervals, which is not as appreciable in the march of CO_2 changes.

The differences are most likely due to the origin of CO_2 and CH_4. While the content of CO_2 in the atmosphere is basically dependent on the processes in the ocean, the sources of the CH_4 are land-based: humid areas, deposits of hydrocarbons including gas hydrates, termite colonies and so on. In particular, the abrupt increase of the amount of atmospheric CH_4 after the termination of the ice ages could have been the result of a release of gas in the giant gas hydrate deposits sealed off by thick ice sheets within the polar continental shelves after the sheets had melted away.

Glaciological and oceanological data demonstrate a cyclic nature of climate fluctuations which portray astronomical factors analysed by Milankovic years ago. Changes with a period of 20,000 years are caused by the effect of the precession of the equinoxes; with a period of 40,000 years by the oscillations of the axis of rotation of the earth; and with a period of 100,000 years by the deviations of the eccentricity of the earth's orbit, pre-conditioned by a factor not quite clear yet.

The ice age–interglacial fluctuations are subject to the impact of fast feedbacks. These are due to the presence of water vapour in the atmosphere, to cloudiness, snow cover and sea ice as well as to slower feedbacks produced by slow changes of the structure and composition of the atmosphere. This, in a way, transfers the cool conditions of an ice age to the interglacial period. Against this background, it is worthwhile examining the kind of climate that obtained in the Northern Hemisphere during the past epochs of global warming.

Paleoclimates of the Global Warming Epochs

The current anxieties concerning possible global warming evoke interest in the climatic conditions over the interglacial period. These entail reviewing the climate during the optima of the Holocene era, during the last interglacial, and during the Pliocene era.

Throughout the Holocene, the warming in the temperate and polar latitudes of the Northern Hemisphere reached a maximum approximately 5,000 to 6,000 years ago. At that time, temperatures on the coast of the Arctic Ocean exceeded the average characteristic of the second half of the nineteenth century by 3° and more, and in the high latitudes of North America and Eurasia by approximately 2° (Andrews et al., 1981; Borzenkova and Zubakov, 1984; Khotinsky and Savina, 1985; Klimanov, 1982).

Summer temperatures in the middle latitudes as well as in Southeast Asia rose by 1° maximum compared to the value in the pre-industrial period. The warming reached two and more degrees in the maritime areas of China and Japan as the warm air was brought to the continent from the ocean due to the growing monsoonal activity. At the same time, in the interior tropical areas as well as in some regions of North America and Africa, a slight drop of temperature, roughly of about 1° was noted. Winter temperatures in Central Asia were somewhat higher than in the middle of the nineteenth century.

By and large, the temperature of the entire Northern Hemisphere during the Holocene optimum rose by 1°. Annual totals of precipitation were considerably higher in the tropical and subtropical areas such as the Sahara which had some 200 to 300 mm per annum. Humidification in the arid areas of Arabia and in the Thar Desert also increased. Farther to the north, the amount of precipitation increased, albeit at a much lower level. In the Middle East, on the Aral and Caspian Seas and in the many regions of China and Mongolia, it rose by approximately 50 mm.

Patchiness characterized the distribution of precipitation in the middle latitudes. In Northern Europe, humidification differed little from the present-day pattern. In the eastern areas of Europe's middle latitudes, precipitation was approximately 50 mm more than current annual totals. In some areas in the central part of the East European Plain and in the south west of Siberia, the annual totals of precipitation dropped slightly. In North America south of the 35° north latitude, humidification was higher than existing level, whilst farther to the north, precipitation decreased by more than 50 mm. An upward trend in precipitation is only now beginning to be felt in the high latitudes of North America and Eurasia.

Models of general atmospheric circulation in the middle Holocene indicate a rise in summer and a fall in winter air temperatures over most of North America. In Eurasia, the period was marked by a rise of summer temperatures whilst the interior continental regions of the Northern Hemisphere experienced a more intensive summer monsoon and aridization while the Arctic Basin saw some shrinkage in its ice cover (Anon, 1991).

The last interglacial (Eemian, Sangamon and Mikulimo), judging from the horizon in the columns of oceanic sediments and the isotope-and-oxygen scale, corresponds to the fifth stage and dates back to around 125,000 years, (Velichko et al., 1984). The strongest warming was observed in the high latitudes of the Northern Hemisphere, especially in the eastern part of Northern Eurasia. On Eurasia's arctic coast (Yamal and Taimyr Peninsulas), the temperature of July was 6–8° higher than that during the second half of the nineteenth century. In the month of January, the temperature here rose by 10–12° compared to those of the pre-industrial era. At present, severe cooling of these areas in winter is due to the dominance of the Siberian anti-cyclone. Much higher temperatures during the Last Interglacial indicate its considerable weakening and its offspur being forced out of Eastern Europe.

Scandinavia and the European Arctic regions were much warmer by 4–6°. Even during the current epoch, thanks to the Gulf Stream, these regions still feature a distinctly warm climate. Apparently, in the Last Interglacial, the arrival of the warm waters of the North Atlantic Stream was more powerful and penetrated much farther to the east than they do now. Air temperature fluctuations over the Atlantic and the Pacific oceans were not too great even in the high latitudes. North of the 35–40° north latitude, the temperature was only slightly higher than its present-day value. In the interior regions of Eurasia, winter temperatures were much higher. By contrast, the July temperatures did not deviate significantly from their present values. In Europe and in the south-west of Siberia they were 1–2° higher whilst in North-eastern Asia they were 2–3° higher. In the interior regions of Europe small positive deviations were also noted whilst in the south of Eastern Europe, in Kazakhstan and in the north of Central Asia these were negative but still small. By and large, tropical areas on the continents were characterized by both small negative and slight positive deviations of temperatures.

With regard to precipitation during the Last Interglacial, the high latitude areas of Eurasia exceeded their present-day value by 200–500 mm (Velichko et al., 1983) and in areas to the south, that is, areas between 65° to 45° north latitude by a smaller value. In the central regions of Western Europe, there fell some 300 to 500 mm more precipitation while the figure was only 50 mm more for most of the middle latitudes of Western Europe. At the same time, in the East European steppes, precipitation increased by 500–600 mm. The humidification of the southern part of Kazakhstan and the north of Central Asia was 100 mm higher than the present-day value. Unusually great were the positive deviations of precipitation in the Mediterranean Region and the north of Africa.

In short, during the Last Interglacial, average air temperature for the Northern Hemisphere was approximately 2° above the value noted in the second half of the nineteenth century. At that time, there was no tundra across most of Northern Eurasia. The northern forest boundary in Eastern Europe was 500–600 km further north. In the south-west of Siberia and in Kazakhstan, the forest-steppe boundary was shifting 200–300 km to the south and 500–600 km in Eastern Europe, thereby forcing out the steppe areas out of Europe (Grichuk, 1982). In the Mediterranean Region, semi-arid areas were afforested while savannas were found in super-arid regions. In the monsoon and tropical regions of North Africa, South and Southeast Asia, monsoon circulation intensified and latitudinal thermal exchange generally increased (Prell and Kutzbach, 1987).

Finally, throughout the Pliocene Optimum dating back some 4.3 to 3.3 thousand years, average annual temperature of the Northern Hemisphere was approximately 4° higher than in the second half of the nineteenth century. The most significant differences in the temperature regime occurred north of the 70° north latitude. In winter temperatures were 20° and in summer only 7–8° higher. The Arctic Basin was completely devoid of ice in summer while in winter sea ice covered a much smaller area compared with present-day situation. However, in some regions notably Kazakhstan, Central Asia, the Sahara and the Gobi Deserts, temperature deviations were negative which apparently was due to heavier precipitation in these regions.

Analysis of natural zonations during the Pliocene Optimum leads to the conclusion that continents in the Northern Hemisphere experienced higher precipitation totals. In the high and subtropical latitudes, the annual precipitation totals were 300 mm, in the middle latitudes, 100–150 mm higher. Higher levels of humidification were also noted in Central Asia, in the Middle East and in the Sahara. Here, the savanna vegetation that appeared to have prevailed during the Pliocene were eventually superseded by deserts (Anon, 1991).

For each of the above warming epochs, there was an appropriate rise in the content of carbon dioxide in the atmosphere indicating air temperature deviations from current values in the northern hemisphere. Such deviations were of the order of 1° for temperature and 270–280 ppmv for precipitation during the Holocene Optimum; 2° for temperature and 280–300 ppmv for precipitation during the Last Interglacial (Lorius et al., 1985); and 4° for temperature and 550–600 ppmv for precipitation during the Pliocene Optimum (Budyko, Ronov and Yanshin, 1985). Palaeoclimatic reconstructions and models of climate indicate a high sensitivity of the climate system to variation of greenhouse gases content in the atmosphere and to a change of orbital parameters. All of the warm epochs indicated above are characterized by a significant warming in the high latitudes and by relatively small temperature fluctuations in the low latitudes. Variations of annual precipitation totals are different in the tropical and middle latitudes and are still not entirely clear. All these peculiarities of climate changes must be considered when discussing the climate scenarios of the twenty-first century.

Changes of Climate, Glaciers and World Ocean Levels during the Twentieth Century

A characteristic feature of natural processes is their cyclic recurrence. The current ice age, which has dominated the earth for over 1 million years now, is distinguished by its alternation of glacial and interglacial conditions. The interglacial also includes the modern epoch, viz. the Holocene, even though the current area subjected to glacierization on the earth's surface is in excess of 16.3 million km^2. The warmest phase of the Holocene, that is, its optimum, is known to have been terminated some 6,000 to 5,500 years ago. At present, we are going through the second half of the interglacial interval characterized by a trend towards cooling. The value of this trend, however, is negligible, being around 0.001° per annum (Velichko, 1977). Far more appreciable are the century and the century-to-century fluctuations that determine the nature of our age.

The last major climate cooling on the earth, referred to as the Little Ice Age, occurred from the fifteenth to the end of the nineteenth century. Early in the twentieth century, the cooling gave way to a steady climate warming (figure 5.1). Yet, the temperature march in the course of this century has not been uniform. Temperature rise during the first half of the century reached its maximum early in the 1940s. Since then and up to the end of the 1970s, that is for nearly 40 years, the temperature levels had been unusually low. It was only in the 1980s that temperature began to rise again. This process has slowed down in the last few years.

Characteristically, the amplitude of temperature fluctuations in the twentieth century is directly related to geographical latitudes. Given the general temperature march, the annual fluctuations in the middle latitudes did not exceed + or – 0.4°C; farther to the north, it was + or – 1.0°C and in the high latitudes + or – 2.0°C. Thus, recent decades have been characterized by small global cooling despite a possible reverse impact of carbon dioxide gradually accumulated in the atmosphere as a result of human activity.

The temperature march indicated above was also reflected in the state of some environmental components found especially in the ice cover of the Arctic Basin. The maximum area of ice cover was noted in the mid-1910s. By the 1940s, the ice-covered area in the Eurasian Arctic Region declined by 1 million km^2. Early in the twentieth century, sea ice covered nearly 90 per cent of the Arctic Ocean. When glaciation was at its highest, that is, late in the 1930s and early in the 1940s, the area covered by ice shrank to 75 per cent. Climate cooling in the second half of the century led to expansion of the Arctic ice sheet by 0.8 million km^2 in the 1960s. However, by the mid-1970s, the ice-covered area had again declined to 0.4 million km^2 (Zakharov, 1981).

In accordance with the general trends over the last 100 years, mountain glaciation has also changed. The last advance of mountain glaciers, which marked the end of the Little Ice Age, occurred in the 1850s–1870s. Since then, glaciers have been retreating steadily although the process was sometimes interrupted by short advances. Such small advances were particularly frequent in the 1910s and 1920s when it was due to a certain rise of precipitation under

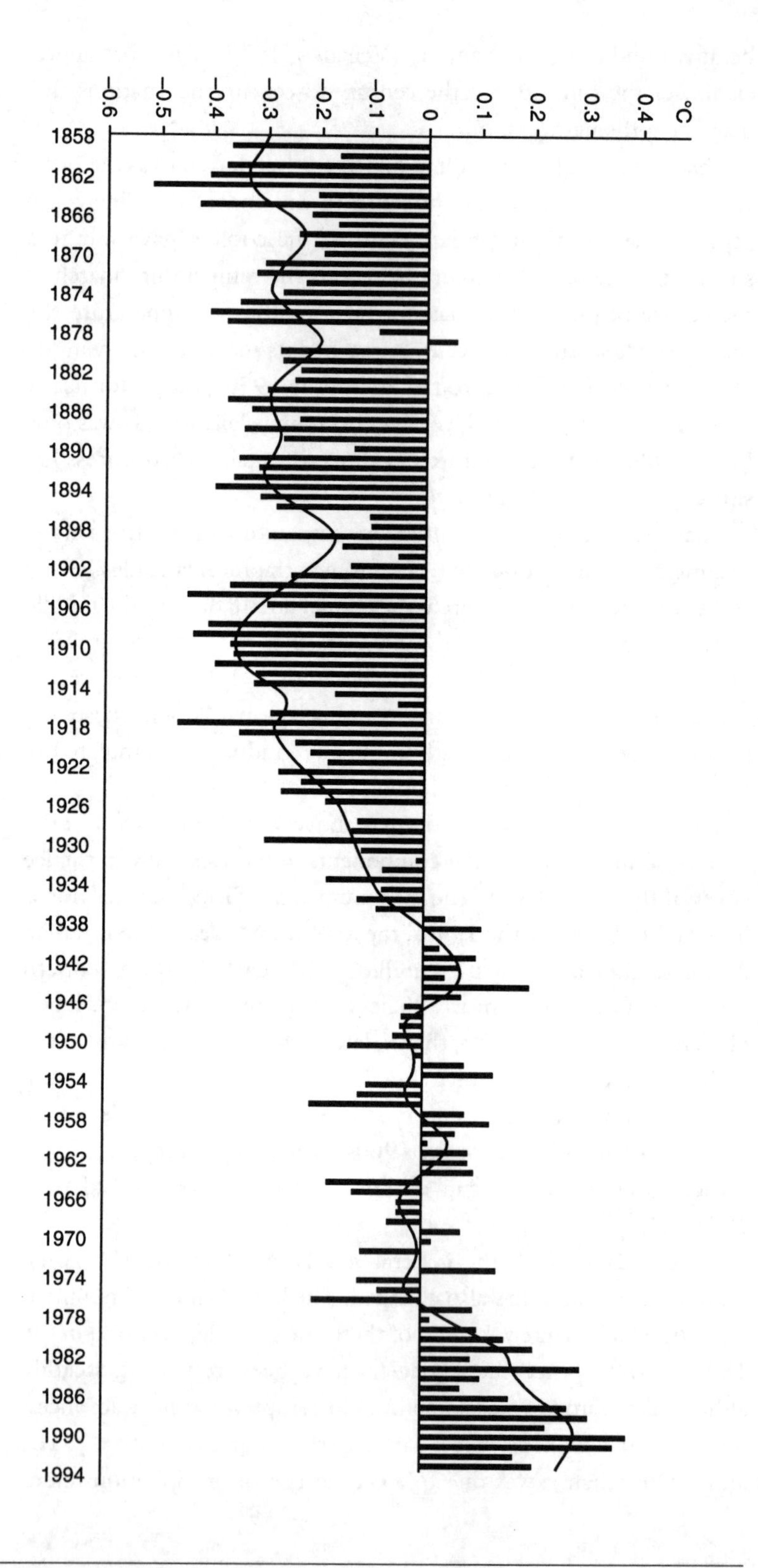

Figure 5.1 March of temperatures, 1858–1994

conditions of inadequate climate warming. A most intensive retreat of mountain glaciers was characteristic of the 1930s and especially of the 1940s when the temperature of summer months reached maximum values and precipitation somewhat declined.

The regime of mountain glaciation began to change appreciably from the end of the 1950s. While the number of advancing glaciers in the Alps in the 1940s was close to nil, in the mid-1960s, nearly 30 per cent of all glaciers surveyed in the Alps were advancing (in the mid-1870s, the figure had been nearly 55 per cent). By the 1960s, in the Caucasus, the velocities of glacier retreat dropped 2.5 times. On the other hand, the velocities of glacier movement grew 1.5–2.0 times and many glaciers began to advance. Among Scandinavian glaciers, there prevailed a positive mass exchange in the 1960s. The activity of glaciers in the North American Cordilleras and Alaska increased. The retreat of Central Asian glaciers slowed down while some glaciers there began to advance (Kotlyakov, 1994).

The changing regime of glaciers was reflected in fluctuations of snow-melt runoff from the mountain glacier regions. This grew by 10 per cent from the end of the nineteenth century to the 1930s and dropped by 5 per cent during the 1970s. The reversal of the mountain glaciers regime was similar to the manner in which the level of the Caspian Sea, the world's largest enclosed water body, changed. From the early 1880s, its level kept going down at short intervals. By 1970, it had dropped by 2.5m compared with 1930.

The ice sheets on the Arctic islands every year during the mid-twentieth century lose between 20 to 25 km^3 of ice. This represents around 1.5 per cent of their total volume over the last 50 years. Nonetheless, in the decades that followed, degradation of island glaciers in the Arctic regions slowed down. Apparently, climate warming during the 1930s and the 1940s had resulted in heavier precipitation in Central Greenland which, in conditions of negligible melting, produced an expansion of the ice sheet. A recurrence of climate warming in the 1980s and 1990s led to quick degradation of mountain glaciation.

Changes in the glaciation mass on the earth's surface constitute one of the crucial factors that determine the world ocean level. The data of over 500 stations indicate that, from the beginning of this century, the ocean level has, albeit not uniformly, risen at an average rate of 1.44 mm per annum. The rise over the last 100 years totals 16 cm. By and large, variations in the level of the ocean reflect temperature fluctuations with a lag of around 20 years. The ocean level rises largely as a result of water expansion due to temperature increases. The decreasing density of the upper 100-meter of the water layer when climate becomes warmer by 1° produces a rise of the ocean level of 3cm. The underlying 900-meter layer, in case of a similar climate warming, becomes 7–18 cm thicker whereas the depth thickness slowly expands by 3–6 cm. As a result of temperature rise by several tenths of a degree during the twentieth century, the world ocean level has, over a 100-year period, risen by 4–6 cm due to the expansion of the ocean water body which accounts for 40 per cent of the actual rise. The balance is most probably due to the current degradation of glaciation.

By and large, from 1894 to 1975, the body of the entire mountain glaciation shrank approximately by 5 per cent while the annual runoff from all mountain glaciers equalled 403 km^3. This amount exceeded by 23 km^3 the volume of precipitation that fell on these territories. The excess melt-water eventually went to the ocean and this contributed to the rise of the ocean level. Obviously, the melting of all mountain glaciers and the small polar ice domes led to the rise of the world ocean level by as much as a third or a half of the volume observed throughout this century, that is, by exactly the amount that can be put down as due to its thermal expansion (Meier, 1984).

The balance of Greenland ice-sheet mass, as deduceable from the calculation, is at present most probably close to equilibrium. Even though the area of the ice sheet is not shrinking, its volume is growing. As a result, the contribution of Greenland glaciation to the world ocean level is so insignificant nowadays. As far as the Antarctic ice sheet is concerned, its response to climate change tends on the whole to be slow. It is thus unlikely that the processes going on in the Antarctic ice sheet will cause any significant rise in the world ocean level at this stage. According to present estimate (Kotlyakov, 1994), annual total increment of ice in the Antarctic regions is equal to 2230 ± 280km^3 of water while the total ice discharge is equal to 2335 ± 430 km^3. This means that the effect of the current processes in the Antarctic regions is far from great. Thus, the current rise of the world ocean level is due to its thermal expansion and input of more water masses resulting from the melting of mountain glaciers with the major existing ice sheets having little significant impact. However, the situation may be reversed in case of a crucial climate change.

Environmental Changes in the Decades to Come

Greenhouse gases continue to be emitted into the atmosphere. This stimulates changes in the composition of the atmosphere which is conducive to climate warming. For the time being, however, the warming is within the frame of natural fluctuations that occurred in the past but the trend is already accelerating. There are reasons to suppose that the average air temperature early in the twenty-first century may rise by up to 1°C compared to the temperature of the pre-industrial era in the middle of the nineteenth century. By the middle of the next century, this difference may reach 2°C. Such serious climatic changes will certainly lead to changes in the environment.

It is absolutely clear that when the average global temperature rises, maximum warming occurs in the high latitudes. To the south (in the northern hemisphere), positive deviations of temperature will decrease. In some regions within a belt of between the 35° and 45° north latitude, negative deviations are also possible. Changes in the amount of precipitation as a result of the warming will give rise to a more complex picture. If global temperature rises by 1–1.5°, more precipitation will fall in the high latitudes. Some parts of the steppe and semi-desert area will become more arid. When temperature rises by 2° and more, the amount of precipitation increases in all latitudes.

In order to assess the response of the landscape mantle to the pending global warming, two basic methods have been used. The first is the paleographic analogs; the second is the nonequilibrium models. Both methods make it possible to correct paleographic analysis on the basis of the real time of the events (Velichko, 1977). The real issue is that, whereas in the case of a natural development of events, it will take the natural landscape at least a few hundred years to secure a correspondence with a new hydrothermal regime, with man-induced changes, climate warming of a similar scope will apparently occur in a matter of decades.

If we orient ourselves to the paleographic analogs alone, it turns out that a rise of global temperature of 1°C will cause the northern boundary of the forest zone in Eurasia to shift north by some 300 to 400 km. With a rise of 2°C, the shift will be of a magnitude of 500–600 km. In the latter case, the tundra zone on the mainland will disappear. Similarly, the subzone of broad-leaved forests will shift to the north. Apparently, the potential reserves of phytomass will, in case of a 1° general warming, increase by 20 to 25 per cent whereas in the case of a 2° general warming, the increase will be roughly of the order of 50 per cent. However, in real-time conditions, the shift of arboreal vegetation to the north at the very beginning of the twenty-first century will not exceed a few kilometres, rising possibly to a maximum of 10 km by the first third of the century (Velichko et al., 1991).

In case of global warming, the permafrost will be subject to profound changes. A warming of 1°C will result in a shrinkage of the permafrost area, its southern boundary everywhere retreating to the north. These processes assume a still bigger scale when the temperature of global warming rises to 2°. Judging by paleographic data, during the Holocene Optimum, the southern boundary of permafrost ran 200–300 km to the north of its present-day position while it went up to 500 km north during the Last Interglacial (Velichko and Nechaev, 1992). But considering the rapid man-induced temperature rise, it may turn out that by the early twenty-first century, the position of the southern boundary of the permafrost will change little in comparison with the present-day situation. It will only be at the end of the first third of the century that the boundary will shift to the north by approximately 100 km. At the same time, the depth of a seasonally melting layer will reach 0.8 m, that is, it will be thicker than now (0.2–0.4 m). Precipitation will also increase. All this will lead to greater water-logging and consequently to a heavier input of methane in the air which will affect the global carbon cycle. In Canada and Siberia, there will be more intensive processes of soil flow, thermokarst, thermal abrasion of the shorelines and the hazard of forest fires will increase.

Global warming will produce profound changes in semi-arid and arid regions of the world. Estimates indicate that during the first half of the twenty-first century, temperatures will rise in these areas, first by between 1° and 3°, and then by between 2° and 5°. Summer temperatures will remain unchanged, may even decline by 2° towards the middle of the century. Annual precipitation occasionally will increase by 200–300 mm. All this will augur well for agriculture and crop production as well as for agro-industrial development by the middle of the twenty-first century. Unfortunately, erosion processes and landslides will also become more active (Velichko, 1992).

In the mountains of the middle latitudes (the Alps, the Balkans, the Caucasus), there will be more active transfer and fall of precipitation, including winter precipitation. This will lead to more snow avalanches, mudflows and landslides. The avalanche hazard will increase in the mountains of the high and middle latitudes such as the Rocky Mountains of North America and others in Scandinavia and Siberia. Changes in the volume of mountain glaciers resulting from climate warming will, however, differ, depending on the region (Groswald and Kotlyakov, 1978). For example, according to the climate forecast, in the mid-latitude mountain countries of Eurasia during the first quarter of the twenty-first century summer temperature will remain at the existing level. Appreciable warming of winters will be offset by increasing precipitation with the result that glaciers will at first not experience any thermal effects. Little by little, however, the glaciers will begin to shrink fast everywhere. Due to this melting of the glaciers, the world ocean level will rise by 1–2 mm per annum. Subsequently, the rate of the ocean level rise will decrease as glaciation is reduced abruptly with an appropriate decline of glacier-derived runoff. At the same time in the Arctic region, where both the winter and summer temperatures will rise by nearly 10°, there will be an abrupt growth of glacier ablation and small arctic domes may disappear in a matter of decades. By and large, due to the melting of mountain glaciers and island glacier domes, the world ocean level will be likely, within the next 50 years, to rise by 15 to 25 cm.

As a result of the warming and despite a certain increase of precipitation, the Greenland ice sheet will lose 0.3 to 0.6 metres every year. The world ocean level will rise by 5 to 15 cm due to ice melting in Greenland during the period in question. It is believed that global climate warming will lead to the increase of precipitation in the Antarctic regions by at least 10 per cent and that this will add some 200 to 250 km^3 to annual accumulation there. However, rising temperature in the offshore regions will extend the ablation area causing gradual destruction of shelf glaciers. In general, the area covered by the ice-sheet mass in the Eastern Antarctic region may even increase.

Major changes are also likely in the area covered by the ice sheet of the Western Antarctic. Here, large ice shelves in the Ross and Filchner-Ronne region will begin to be destroyed rapidly. This will upset the dynamic equilibrium of the ice sheet the bed of which is below sea level. The ice sheet itself may be subject to catastrophic destruction and drop into the ocean. It is, of course, too early to say how this will happen. But, should temperature rise by 10° in the

Antarctic latitudes, such developments will be quite likely. Disintegration of the Western Antarctic ice sheet will lead to a 5 to 6 m rise of the ocean level which will be fraught with grave consequences for low-lying coastal areas. In case of a global temperature rise of 2° to 3° by the middle of the next century, the world ocean level may rise by 0.5 to 1.2 m. Out of this, some 0.3 to 0.8 will be due to thermal expansion of the water mass. If, as a result, the ice sheet of the Western Antarctic region begins to disintegrate, the ocean level will rise by a few more metres.

The above brief review indicates the anticipated consequences of climate warming which could transform the human environment by the middle of the next century. This will inevitably tell on the economic use of land and other natural resources and will call for serious adjustment of societies to the pending changes.

Is an Environmental Crisis Imminent?

So, what kind of a phenomenon is entailed in a future climate warming? Is this likely to precipitate a global environmental crisis or initiate a transition to new conditions that will include both positive and negative dimensions? It should be stressed from the outset that the natural landscapes respond to rapid climatic changes in a rather complex manner. The rate of response of particular landscape components such as vegetation, soils, and permafrost, is not the same and this will add to the instability of natural geographic systems in case of rapid climate changes.

During the first half of the twenty-first century, changes in the regime of precipitation and evaporation will have a strong impact on water resources, including water quality and soil moisture. Obviously, there will be problems of water availability in areas with limited water resources, especially in marginal agricultural regions. The rising snow line in the mountains will enhance the probability of winter and spring floods in mountain areas. The degrading mountain glaciation will reduce the regulatory role of glaciers. This has usually been favourable for agriculture on two scores. First, glaciers redistribute the runoff from the first to the second half of summer when the need for irrigation increases abruptly. Second, glaciers, in general, serve to concentrate the runoff during the summer period which is dry in most piedmont areas of the temperate zone.

The imminent climate warming, if it occurs, will result in the growing frequency of many extreme phenomena, especially of droughts. At the early stages of the warming, when the increase of precipitation will not be so obvious, there will be more dry winds and dust storms in some regions. In regions rich in peat deposits, there will be a strong likelihood of self-ignition. Forest fire hazards will be high. Fire smokes will pollute the atmosphere and add more carbon dioxide. A considerable rise of temperature in the North, say by 4–8° in summer and by 6–12° in winter, will produce extensive melting of the permafrost and in many places will result in severe water-logging. All this will change dramatically the living conditions of minority groups in the North, necessitating their adaptation to the rapidly changing conditions.

However, the changing conditions for human activity in the middle latitudes of the Northern Hemisphere as a result of climate warming will, in many respects, be positive. It is common knowledge that a twofold increase of CO_2 concentration in the atmosphere in conditions where moisture availability is adequate will increase crop yields by up to 30 per cent (Anon, 1991). The validity of this trend has been proven by the increase in wheat productivity in the USSR and the USA during the second half of the twentieth century (figure 5.2). True, crop yields have grown during this period due, in large measure, to advances in farming practices. Yet, there is no doubt that climate warming noted during the period has also accounted for no small share of this achievement. This becomes especially obvious when we compare actual productivity of crops in particular years with favourable weather conditions.

A special analysis carried out in Russia demonstrates good results from the future climate warming for the European part of Russia where nearly 80 per cent of the population and its economic potential are concentrated. As a result of a medium global warming of between 1–1.5°, precipitation totals in this region will grow by 100 mm per annum. On the other hand, in the northern and north-eastern parts of European Russia, this could mean approximately a 400 per cent increase for the annual total of active temperatures and a substantial augmentation of agricultural productivity. Consequently, harvest of grain may increase by some 15 to 20 per cent. This will more than compensate for a possible decrease in grain production in the Lower Volga area where some increase of arid conditions can be expected.

Climate warming in the middle latitudes will bring about a

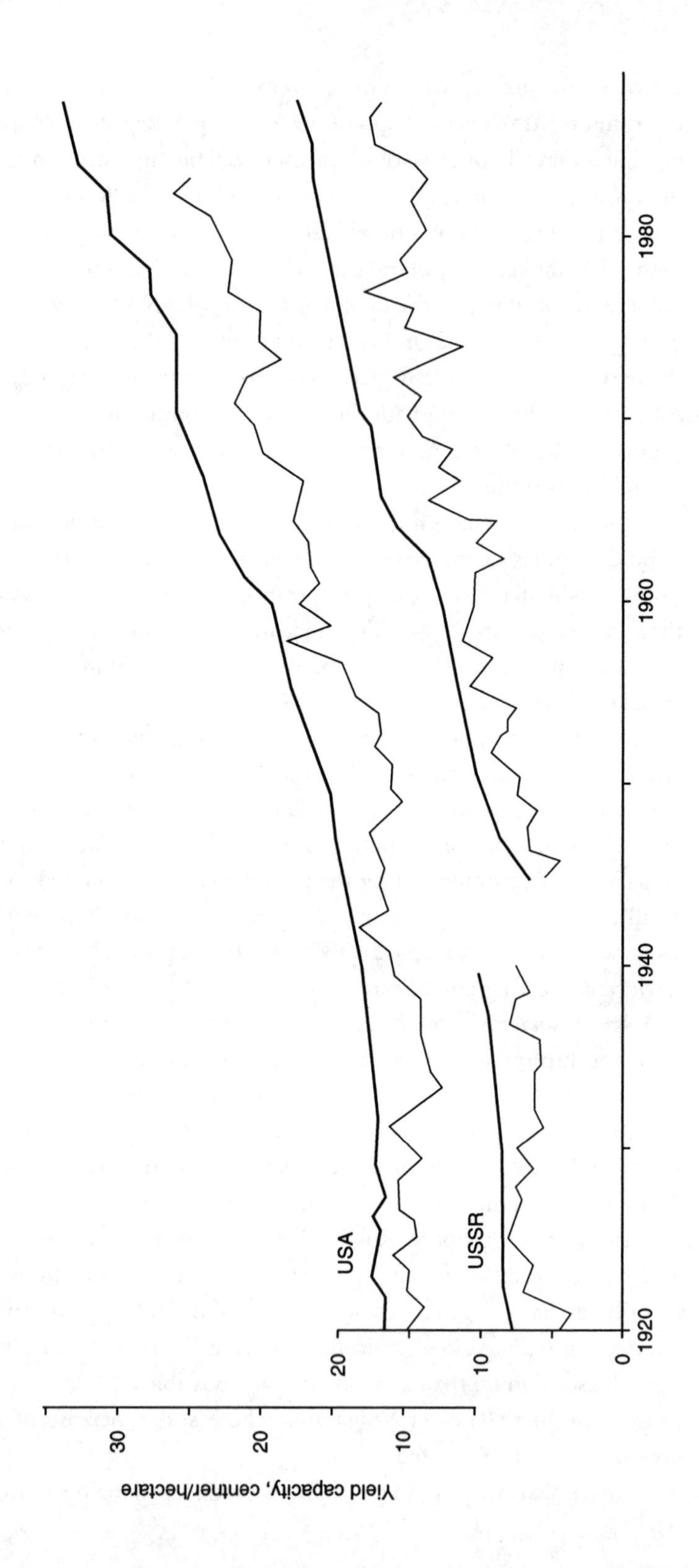

Figure 5.2
Increase in wheat productivity in the USSR and USA

considerable increase in vegetal resources. In the near future, an increment in biomass of some 30 per cent can be expected in the taiga, with new forest ranges emerging in the tundra region. Grasslands will supersede some marshes. Medico-geographical conditions will also improve. With a rise of temperature of 2°, general morbidity rates will drop by about 4 per cent and, in the case of particular diseases, by as much as 10 per cent.

However, destructive processes caused by human activity will likely be superimposed on these environmental changes. Processes such as deforestation, the spread of savannahs and desertification, will assume large scale proportions in developing countries (Mashbitz, 1988). By and large, areas under forests in these countries have, for the last 40 years, decreased from 16 per cent to 7 per cent. Eight million km^2 out of the 14.5 million km^2 of what used to be forest areas are now under crops, 3.5 million km^2 are rangelands and 3 million km^2 are under slash-and-burn farming (Malingrean and Tucker, 1988).

In the nineteenth century, on the basis of expeditionary research in Latin America, the celebrated geographers, Alexander von Humboldt and A. I. Voyeikov established a correlation between uncontrolled cutting of forests in the tropics and the aridization of the climate. The tragic experience of several generations of the people of these countries confirms these conclusions. Uncontrolled cutting of tropical forests sharply diminishes the biomass potential and is detrimental to the natural carbon cycle.

Conclusion

According to the United Nations Environmental Programme (UNEP), anthropogenic desertification threatens 35 per cent of the dryland areas of the world. The two processes, namely deforestation and desertification, constitute an integral part of the environmental crisis that has spread all over the world. The impact of these and other man-induced processes on the environment is closely connected with the changes resulting from anthropogenic climate warming. In a bid to better understand the future of the human environment, geography is faced with the question of how far human beings can go on the way they are doing and what management strategy can be used to secure joint development of nature and human society through sensible territorial organization?

The basis for preserving the quality of the environment is directly

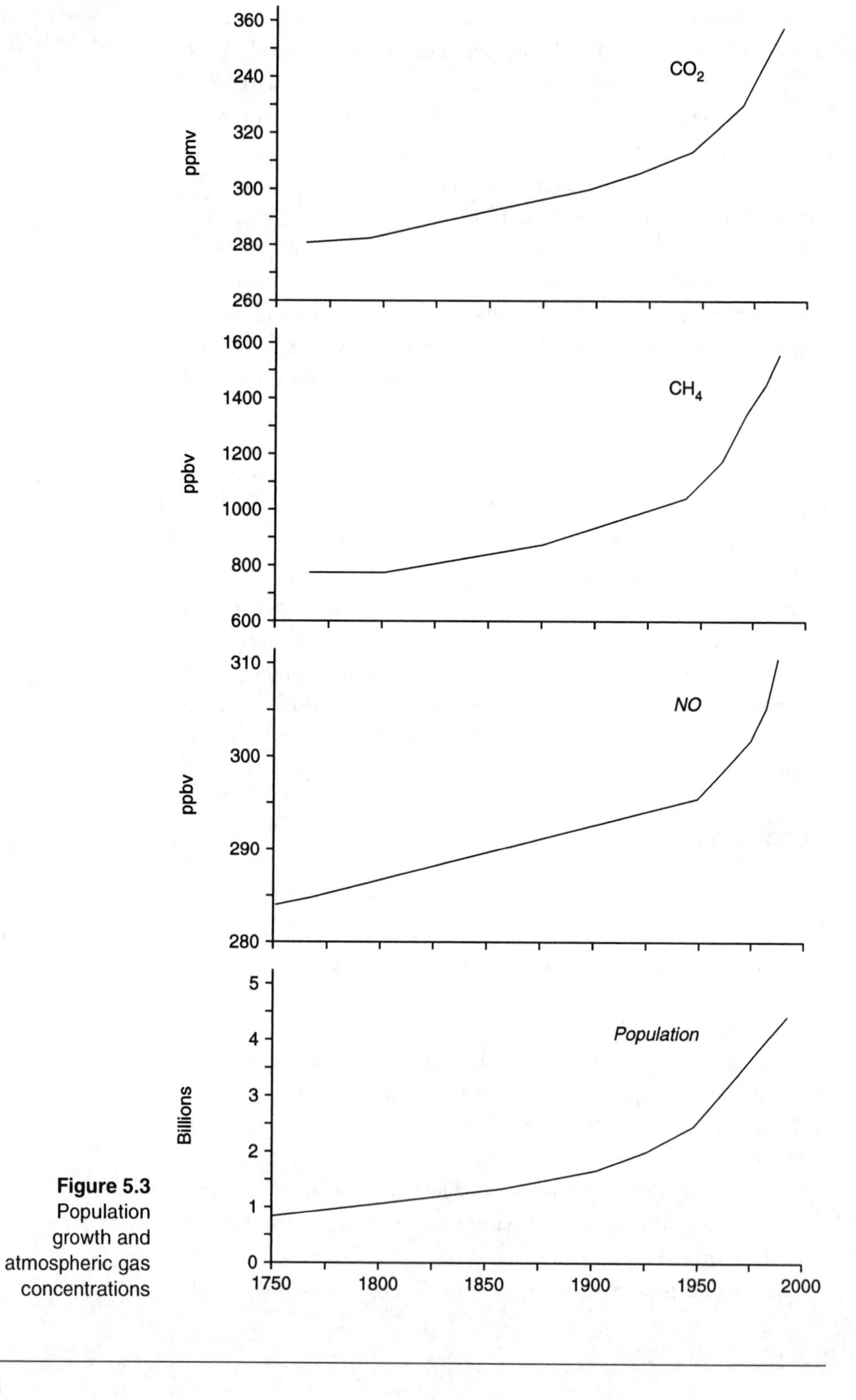

Figure 5.3 Population growth and atmospheric gas concentrations

related to the value of primary biological production obtained through photosynthesis and consumed by human beings. Prior to the beginning of the twentieth century, human beings consumed only one per cent of biological production in nature. This did not cause any significant changes in the environment. At present, they consume a significant percentage of the production of the biosphere and in the process actually destroy the natural geographical systems (Danilov-Danilyan and Kotlyakov, 1993)

An appreciable build-up of greenhouse gases in the atmosphere began some 200 years ago. It is significant that the exponential rate of concentration of these gases and the rapid rate of population growth on the earth's surface follow a similar trend pattern (figure 5.3). It is a matter of general knowledge that in the 1950s world population grew by 50 million per annum. The forecast for the 1990s gives the figure as 96–98 million per annum. At this rate, nobody can tell if the earth will be able to endure such an anthropogenic pressure. One thing is clear. Rapid population growth constitutes one of the chief factors responsible for the deteriorating environment already causing great stress in many regions of the world.

Geography can play a substantial role in defusing environmental tension. Unless geographic problems are resolved, it is impossible to produce a concept of environmental security and to evolve a strategy of sustainable development. How should humankind develop to preserve the environment in a state that would allow people to live on the planet almost indefinitely? This is a crucial question to which different answers may be given. At present, both scholars and politicians are carried away by a resource, or a technogenic concept of development. This proceeds from the assumption that humankind is capable of solving all environmental problems and can deal with issues of environmental security mostly by technological means, that is, by reforming and adjusting the economy on the basis of new technologies, without restricting the rate of economic growth.

However, in the context of present-day knowledge of the environment, it is necessary to begin to think in terms of a biosphere concept of development. This entails the notion of an environmental imperative. This notion hinges on a realization of the danger of destroying the biospheric mechanisms that stabilize the environment. The biosphere concept of development thus includes nature-conservation activity which it sees as an integral part of development planning even at the local level. To provide for global and regional environmental security and sustainable socio-economic development, effort must be made to conserve and minimally disturb the biosphere. Anthropogenic components of global changes must be 'isolated' and studied in the context of the interaction between natural and human-induced factors in the course of societal evolution. Such studies must also investigate what happens when processes involving different time scales (geological, historical and political) impact on spatial landscape structures.

Humankind is faced with the challenge of numerous induced modifications to the environment. Most of these changes are predisposing to future climate warming and other anthropogenic climatic processes. Human society is about to enter a period requiring serious adjustment to rapidly changing environmental and economic conditions. The task of geography in such a situation is to provide the basis for a better understanding of the spatial peculiarities of imminent changes and for realistically confronting whatever problems may arise in the process of securing a stable and sustainable development of nature and society.

References

Andrews, J.T. et al. (1981) Relative departures in July temperatures in northern Canada for the past 6,000 years, *Nature*, 289:164–7.

Anon, (1991), *Predstoyaschchie Izmenenia Klimata (Pending Climatic Changes)* (Leningrad: Gidrometeoizdat).

Borzenkova, I.I. and Zubakov, V.A. (1984) Klimaticheskii optimum golotsena kak model globalnogo klimata nachala 21 veka (Climatic optimum of Holocene as a global climate model for the beginning of the 21st century), *Meteorologia i Gidrologia*, 8:69–77.

Budyko, M.I. Ronov, A.B. and Yanshin, A.L. (1985) *Istoria Atmosfery (History of the Atmosphere)* (Leningrad: Gidrometeoizdat).

Danilov-Danilyan, V.I. and Kotlyakov, V.M. (eds) (1993) *Problemy Ekologii Rossii (Problems of Russia's Ecology)* (Moscow).

Genthon, C. et al (1987) Vostok ice core: climatic response to CO_2 and orbital forcing changes over the last climatic cycle, *Nature*, 329 (6138):414–18.

Grichuk, V.P., (1982), Paleoekologischeskie rekonstruktsii, osnovannye na paleobotanicheskikh dannykh (Palaeoecological reconstructions based on paleobotanical data), pp. 120–1 in *Paleogeografia Yevropy za poslednie 100 tysyach let (Palaeogeography of Europe for the Last 100,000 Years)*, (Moscow: Nauka).

Groswald, M.G. and Kotlyakov, V.M. (1978) Predstoyashchie izmenenia klimata i sud ba lednikov (Anticipated change of climate and the future of glaciers), *Izv. An SSSR. Ser.geogr.*, 6:14–26.

Jouzel et al. (1989) The Antarctic climate over the last glacial period, *Quaternary Research*, 31 (2):135–50.

Jouzel et al. (1993) Vostok ice cores: extending the climate records over the penultimate glacial period, *Nature*, 364(6436):407–12.

Khotinsky, N.A. and Savina, S.s. (1985) Paleoklimaticheskie skhemy territorii SSSR v borealnom, atlanticheskom i subborealnom periodakh golotsena (Paleoclimatic schemes of USSR territory during the Boreal, Atlantic and Subboreal Periods of the Holocene), *Izv. AN SSSR ser.geogr.*, 4:18–34.

Klimanov, V.a. (1982) Klimat Vostochnoi Yevropy v klimaticheskii optimum golotsena-po dannym palynologii (Climate of Eastern Europe during the climatic optimum of the Holocene – based on palynological data), pp. 251–8 in *Razvitie prirody territorii SSSR v pozdnem pleistotsene i golotsene (Natural Development in the USSR Territory in Late Pleistocene and Holocene)*, (Moscow:Nauka).

Kotlyakov, V.M. (1994) *Mir Snega i L da (The World of Snow and Ice)* (Moscow: Nauka).

Lorius, C. et al. (1985) A 150,000 year climate record from Antarctica, *Nature*, 316(6029):591–6.

Malingrean, J-P. and Tucker, C.J. (1988) Large-scale deforestation in the Southeast Amazon Basin of Brazil, *Ambio*, 1:49–55.

Mashbitz, Ya.G. (1988) Razvivayushchiesya strany vc sisteme globalynkh problem chelovechestva (Developing countries in the system of mankind's global problems), *Izv. AN SSR, ser.geogr.*, 6:14–26.

Meier, M.F. (1984) Contribution of small glaciers to global sea level, *Science*, 226(4681):1418–21.

Prell, W.L. and Kutzbach, J.E. (1987) Monsoon variability over the past 150,000 years, *Journal of Geophysical Research*, 92:D7.

Velichko, A.A. (1977) Globalnye izmenenia klimata i reaktsia landshaftnoi obolochki (Global climatic changes and a response of the landscape mantle), *Izv. AN SSSR, ser.geogr.*, 5:5–22.

Velichko, A.A. (1992) Zonalnye i makroregionalnye izmenenia landshaftno-klimaticheskikh uslovii vyzvannykh parnikovym effektom (Zonal and macro-regional changes in the landscape-and-climatic conditions produced by the green house effect),*Izv. AN SSSR, ser.geogr.*, 2:89–102.

Velichko, A.A. and Nechaev, V.P. (1992) K otsenke dinamiki zony mnogoletnei merzloty v Severnoi Yevrazii pri globalnom poteplenii klimata (On evaluating the dynamics of the permafrost zone in Northern Eurasia under global climate warming), *Doklady Akad. Nauk*, 324(3):667–71.

Velichko, A.A. et al. (1983) Paleoklimat territorii SSSR v optimum poslednego – mikulinskogo – mezhlednikovya (Palaeoclimate of USSR Territory during the optimum of last – Mikulino – Interglacia), *Izv. AN SSSR, ser.geogr.*, 6:30–45.

Velichko, A.A. et al. (1984) Klimat severnogo polusharia v epokhu poslednego, mikulinskogo mezhlednikovya (Climate of the Northern Hemisphere during the epoch of the last – Mikulino – Interglacia), *Izv. An SSSR, ser.geogr.*, 1:5–18.

Velichko, A.A. et al. (1991) Kotsenke dinamiki yestestvennykh geosistem lesnoi i turndrovvoi zon pri antropogennykh izmeneniyakh klimata (On evaluating the dynamics of natural geosystems of the forest and tundra zones under man-induced climatic changes), *Vestnik AN SSSR*, 3.

Zakharov, V.F., (1981), *L dy Arktiki i sovremennye Prirodyne Protsessy (Arctic Ice and Current Natural Processes)* (Leningrad: Gidrometeoizdat).

6 Societal Responses to Environmental Hazards

The concern with climate change sets the scene with respect to a human environment that is in a continuous state of flux. Even on a long-term and global scale, the human environment is constantly experiencing change. Because of the fluidity and connectedness provided by the blanket of the atmosphere and the surrounding oceans, these changes have the potential of affecting all parts of the world. Some of these changes are nature-induced but these take place over a sufficiently long time for the earth to adjust gradually to them. Not so the changes caused by human beings. These anthropogenic changes have intensified in the last half century such that no one is certain how well the earth is adjusting and how close we are to real disaster. Global warming induced by greenhouse gas emissions, the potential sea-level rise that could result from such warming, and the implications of the continuing depletion of the stratospheric ozone by chlorofluorocarbon releases – all indicate the uncertainties under which the future of the human occupance of the earth's surface must labour.

But environmental changes are not all long term in their initiation nor global in their impact. Virtually in all parts of the world the environment at one time or the other produces extreme natural events which constitute immediate hazards and sometimes disasters and catastrophes to human beings. These extreme events include, among others, tropical cyclones, hurricanes, floods, windstorms, drought, coastal erosion, frost, snow storms and avalanches, volcanic eruptions and earthquakes. Indeed, it is to the extent that the frequency, magnitude and timing of these extreme events are based on imperfect knowledge that they pose a serious menace to local or regional occupance of portions of the earth's surface. In these circum-

This chapter is based on a contribution by Susan Cutter.

stances, these extreme events can only be foreseen as probabilities whose time of occurrence is unknown and to which, therefore, human societies have learnt to respond and to adjust over the centuries.

On 1 January 1990, the United Nations General Assembly proclaimed the start of the International Decade of Natural Disaster Reduction (IDNDR) and, by all accounts, not a minute too soon. The first half of the decade has been riddled with unprecedented environmental disasters: earthquakes in Zanjan, Iran (1990), Northbridge, California, USA (1994), and Kobe, Japan (1995); tropical cyclones and flooding in Bangladesh (1991); volcanic eruptions from Mount Pinatubo (1991); flooding on the Mississippi River, USA (1993); and the most costly disaster to the United States, Hurricane Andrew (1992), to name just a few.

The IDNDR is a multinational effort designed to shift social and political attention from post-disaster recovery to pre-disaster planning, preparedness and prevention. Through cooperative arrangements between national and international scientific committees, improved capacity to predict such potential disasters is being developed. The IDNDR also promotes improvements in warning systems, community awareness and disaster preparedness at both local and national levels. Disaster preparedness and mitigation are key elements in the IDNDR strategies which are designed to lessen the impacts of disasters on society. The importance of natural disasters in the deterioration and decline of human social conditions is an implicit assumption behind the IDNDR's activities. Geographical contributions to these activities are prevalent since hazards research has gained in importance within the discipline in recent years and has always emphasized a desire to use enhanced knowledge to reduce human suffering. This chapter thus provides an overview of contemporary hazards research and its importance in understanding societal responses to environmental hazards.

Definitional Clarification

Historically, the terms hazard, risk and disaster were used interchangeably although each term has a very precise meaning. *Hazard* is the broadest term and reflects a source of danger or the potential for harm. *Risk* is the likelihood or probability of an event occurring. Hazards include risk (e.g., probability), impact (or magnitude) and

Inhabitants of Banbam Village fleeing an eruption of the volcano Mount Pinatubo in the Philippines. *Source:* Van Cappellen/REA.

contextual (socio-political) elements. In other words, hazards are threats to people and the things they value (Cutter, 1993). They are, therefore, socially constructed, in the sense that people can contribute to, exacerbate and modify hazards. Hazards can vary by culture, gender, race, socio-economic status and political structure. *Disaster*, on the other hand, is a singular hazard event, unlike those mentioned above, that has a profound impact on local people or places either in terms of loss of life or injuries, property damages or environmental impact. While geographers traditionally studied hazards, the newer generation of scholars is branching out to examine the spatial dimensions of risks and disasters as well.

Nonetheless, despite 50 years of geographical research on hazards, a series of unanswered questions remains perplexing. These questions provide the foci for contemporary hazards research. Implicit in each of them is the need to understand the physical and social processes that give rise to hazards as well as the spatial and temporal variations in the phenomena and in their outcomes. Some of these unanswered questions include the following:

1. Are societies becoming more vulnerable to environmental hazards and disasters?
2. What social/physical factors influence changes in human occupance of hazard zones?
3. How do societies perceive and estimate the occurrence of environmental risks and hazards?

4 How do people respond to environmental hazards and what accounts for differential adjustments (in the short term) and adaptation (in the longer term)?
5 How do societies mitigate the risk of environmental hazards and prepare for future disasters?
6 How do local risks and hazards become the driving forces behind global environmental changes?

The Nature of Environmental Hazards

Although hazards derive from extreme events in nature, it is important to stress that not all extreme events constitute hazards to human beings. Moreover, an extreme event in nature may at the same time be both a hazard and productive resource. Thus, a flood which may destroy crops and farmsteads may at the same time provide rich fertilizer for agricultural land. Similarly, a lightning stroke, which may kill a human standing in a forest, may at the same time start a fire essential to the maintenance of a robust forest ecosystem. The hazard then is not the flood itself or the lightning stroke. It is the risk involved in occupying a place liable to such occurrence. In other words, hazard results from the interaction between natural and social systems. Natural events are neither benevolent nor malevolent to human societies. They are in fact neutral in so far as they neither determine nor constrain what can be done with them. Consequently, it is human beings and society at large that transform elements of the environment either into resources or hazards by using them for economic, social or aesthetic purposes.

There are many ways of categorizing extreme natural events. A simple schema, for instance, would recognize the domain of their occurrence as either geophysical or biological (Burton, Kates and White, 1993). In the case of the former, a distinction is made between meteorological and geomorphic occurrences; in the case of the latter, between floral and faunal occurrences. Meteorological extreme events, for instance, would include such incidents as blizzard, drought, flood, frost, tornado, hurricane and windstorm. Geomorphic extreme events cover such occurrences as avalanche, earthquake, erosion, landslide, shifting sand, tsunami and volcanic eruption. Of floral extreme events major infestations of unwanted vegetational elements such as water hyacinth, weeds, fungal diseases, dutch elm and blister rust can be mentioned; whilst for the faunal there are outbreaks such as locusts, grasshoppers, rabies, bubonic plague, typhus and various bacterial, viral and protozoal diseases.

However, as already emphasized, when these occurrences have risen to the status of being regarded as hazards, it is clear that quite different characteristics in their incidence become of paramount importance. Thus, for example, when meteorologists describe a snowfall as heavy, they are referring to characteristics such as the depth of snow accumulation, the water equivalent of snow as well as the accompanying conditions of wind and temperature. For such an event to become hazardous, the frequency and time of day of its occurrence, its disruptive capacity for economic and social life in the community and the period elapsing between the onset and peak fall become critical in societal evaluation (Baumann and Russell, 1971). Even with this, the relationship between the occurrence and its social impact is not a simple linear function. It depends on the ways in which the people of the area concerned normally cope with such events. Below a certain critical threshold, such an occurrence may not cause any significant damage or disruption to the point of being considered a hazard. Once this threshold is crossed, considerable damage and dislocation may be expected.

It is the specification of these relations and the definition of the threshold levels for a given place or society that pose significant problems for geographical research in this field. In human terms, at least seven dimensions of hazardous events have been recognized. These include magnitude, frequency, duration, areal extent, speed of onset, spatial dispersion and temporal spacing (Burton et al., 1993). The first five concern the aggregate of separate events; the last two refer to the distribution of a population of such events over space and time. The search for measurement of these various dimensions appropriate to the analysis of societal responses remains challenging. Magnitude, frequency and areal extent, for instance, describe the strength or force of the event, how often it can be expected to occur and over what area. In general, the greater or more powerful is the hazard event, the less adequate the available technology to control or mitigate its impact. On the other hand, the more frequently a hazard occurs, the greater the need to take steps to respond to or accommodate it. The larger the area affected, the broader the segment of society likely to be subjected to loss or disruption. The significance of the dimensions of speed of onset and of duration is chiefly in terms of emergency preparations and of the physical and societal capacity to establish and operate a warning system. Where a hazardous event strikes rapidly or usually has a short duration, little can be done. Where the duration is such that there is a long period of time between the onset and the peak of occurrence, the range of possible responses is also usually greater and a capacity developed to be able to predict such events on the basis of precursory phenomena such as pattern of air mass circulation or of tectonic stress.

These various dimensions of environmental hazards present complex problems of measurement and each profoundly affects the kinds of societal responses that are appropriate. For example, the magnitude of a hazard event is often particularly difficult to assess because the instrument of measurement – persons, their possessions and their activities – is subject to significant and rapid changes. Thus, two droughts may be of similar magnitude in terms of moisture deficiency, duration and areal extent, and yet have a very different magnitude in terms of their social impact or societal response because of differences in density of population, kinds of crops being grown, level of agricultural technology, capacity for rapid cooperative action and so on. As a result, the ways in which extreme events are measured are very much a matter of cultural convention, scientific knowledge, data availability and functional needs. For example, the ideal measure for the size of an occurrence would be to estimate the damage potential posed by the event. This ideal, however, cannot be met by current seismological data, despite widespread usage of both physical and human scales and the relatively large effort expended by the world's seismological researchers to measure the magnitude and intensity of each occurrence.

It is, however, possible to rank or group hazard events on a number of bases. One, for instance, would be in terms of the energy release per unit of area or per unit of time during the duration of the extreme event. Earthquakes and tornadoes are examples of events with high energy release per unit of area and time; whilst drought or erosion are illustrative of events with low energy release. On the basis of this energy release characterization, it is thus possible to appreciate differences in societal responses to both types of hazard because of the magnitude and immediacy of the disturbance.

Another way of categorizing hazards could be on the basis of the seven dimensions identified above. At one extreme would be a major earthquake. This is an infrequent extreme event, of short duration and relatively concentrated in space. Its speed of onset is usually high and it occurs in a more or less random fashion. At the other extreme is a severe drought event. This is usually of greater frequency, much longer duration, more widespread over the globe, slower in onset, more diffuse and somewhat more random. Between these two extremes one can group most other hazard events.

Whatever the categorization, one can also distinguish between hazard events that are pervasive and those that are intensive. A drought is the best example of the former; a tornado is the archetype of the latter. Both types have differential impact on places and the typology would appear to represent a continuum. Drought, for instance, defined as a reduction in the natural availability of water for human use, whether in rural or urban areas, is a pervasive phenomenon which may affect large areas of a country and a significant proportion of the population. In Kenya, for instance, national drought may affect most ecological zones in the country and significantly reduce the production of as much as 10 per cent of the population (Wisner and Mbithi, 1974:90). It may last for two or more growing seasons and generally involve serious loss of production in most ecological zones involving usually two or more provinces. This type of drought seems to occur about once a decade and may sometimes give rise to heavy livestock losses, occasionally amounting to some 40–50 per cent or more of herds. Drought, as a pervasive hazard, is frequent, low in energy output, slow to develop and has stimulated a social response of high level of expenditure on prevention of effects. Tornadoes, by contrast, are intensive hazards that threaten the lives of many people living within their zone of occurrence. In the United States, for instance, some 40 million people living in the Midwest, the Great Plains and the Gulf States are believed to be at risk from high incidence of tornadoes. Tornadoes are, however, comparatively rare events with high energy outputs, highly localized and with a very rapid onset. Moreover, although between six and seven hundred tornadoes occur every year, their average path is quite small. There is therefore little incentive to invest in protective measures since the probability of their being needed in any one place is small, and given the force of a tornado, such measures are often not effective. The emphasis, thus, is to try to improve warning systems and to construct tornado shelters or storm cellars.

The pervasive–intensive differences are not mutually exclusive categories but rather a continuum along which different types of event or particular variations of a given type may range. Extreme events at the pervasive end of the continuum include, apart from drought, fog, heat wave, excessive precipitation and snow. Towards the intensive end, there are, besides tornadoes, earthquakes, landslides, hail, volcanoes and avalanches. Other hazards are not so easy to place on the scale because, depending on a particular situation, they can be found at both ends of the scale. A good example is the case of floods. Flash floods in small, mountainous watersheds are localized, high-energy output phenomena close to the intensive end of the scale whereas floods on the mainstream of a great river tend to be more towards the pervasive end.

Modes of Societal Response to Hazards

Societal responses to hazards are related both to perception of the phenomena themselves and to awareness of opportunities to make adjustments. It is unusual for a society to be unaware of the existence of possible hazards in its environment. Yet, the perception and definition of the threat constituted by such hazards may differ markedly between and within societies. A wide range of responses thus characterizes the manner in which a society may act to minimize the impact of a hazard. These extend from immediate actions in the face of danger to long-term programmes of accommodation or adjustment. The former includes activities such as the setting up of warning systems, flood fighting with sand bags or emergency evacuation. The latter includes the planting of crops less susceptible to drought, the construction of levees or the construction of buildings designed to withstand earthquakes. Besides these, many normal actions taken in areas susceptible to particular hazards often have the incidental effect of reducing vulnerability to hazards. Building a dam to store water for irrigation purposes may also have the additional advantage of reducing the impact of drought conditions.

These different ways of coping with natural hazards are described as *adaptation* when they entail long-term cultural and biological changes and as *adjustment* when they refer to purposeful or incidental actions taken by human beings. Recognizing these two broad types of societal responses to environmental hazard, Burton, Kates and White (1993:224) suggest that most societies seem to have developed four major modes of coping with natural hazards. These are: loss absorption, loss acceptance, loss reduction and change of use and livelihood. Moving from one mode to another involves crossing a significant threshold of awareness, action or intolerance. The first mode – *loss absorption* – is recognizable in a situation in which a society has been absorbing the impact of an environmental hazard for a long time without being aware that it is doing so. This form of coping is, on one hand, characteristic of biological and cultural adaptation whilst, on the other hand, it is what is entailed in incidental adjustment.

The first mode is separated from the second – *loss acceptance* – by the threshold of awareness. In this second mode, the society has become aware of the existence of the hazard but chooses to bear the loss through various institutional or organizational mechanisms, often involving sharing the loss with a wider group than those directly affected. The mode of loss acceptance yields to the third mode of *loss reduction* when another threshold is crossed, namely that of a determination to take positive actions to reduce the loss. Finally, a fourth mode is reached when loss reduction gives way to *radical change* in either location or resource use. This occurs when a third threshold has been crossed, that of intolerance, when society decides that the loss is no longer tolerable or acceptable.

Although the four modes of coping appear universal, a steady progression through the modes is not invariably found. A society may be coping in one mode with respect to flood hazards and in another mode in dealing with drought. The thresholds themselves are not fixed levels, and movements in relation to them may be occasioned by a shift in one of several variables resulting from changes in cumulative individual, collective and national choices.

Loss absorption capacity is a function both of natural events and of social systems. Society and environment interact to form a unique pattern of hazard exposure and of resilience to extreme events. Two societies occupying the same environment may, by virtue of their different adaptations, both cultural and biological, and their incidental adjustments, react differently to hazards. For example, small fluctuations in moisture availability may be hardly noticed on a farm operating within a traditional agricultural system but may cause tremendous losses to a commercial plantation. The extent to which a society remains unaffected by such fluctuations in nature, fluctuations which for others would be regarded as of hazardous proportions, is dependent upon its absorbent capacity. This capacity reflects the sum of customary practices and devices such as land use which have become embedded in everyday livelihood such that the society involved can better cope with extreme events from its environment without being made aware of any significant harm.

The built-in absorbent capacity of a society may itself involve costs. A large number of hazard-related actions unconsciously practised by different societies are not achieved and maintained without allocation of resources. The social price of high level absorbent capacity may be the payment of a continual environmental rent not consciously recognized as such. Thus, it is possible that the intercropping practised in tropical agriculture provides protection against drought and small moisture deficits without the societies being consciously aware that it is performing that function. It is when attempts are made to replace the cropping pattern to a monocultural system in the interests of greater efficiency that the nature of this rent is perceived in the erosional instability that accompanies the change.

It is, however, impractical to attempt to assess the whole of the social system in terms of its adaptations to potential damages from natural hazards. Nonetheless, the capacity for incidental adjustments leading to cultural adaptation are of basic importance insofar as they indicate a society's ability to absorb the effects of extreme environmental fluctuations without being conscious of doing so. In agricultural societies, over-capacity in production and the development of storage facilities to contain the surpluses may not be seen as the in-built safety margin to deal with environmental hazards of drought. The situation, for instance, could vary greatly in a society where at one time much of the agricultural production was by tenant farmers in a landlord-dominated economy and at another when it is a joint activity of members of a commune.

Where an extreme event just exceeds the awareness threshold of most of its members, a society may respond passively to an environmental hazard through loss acceptance. People prefer to bear known ills or hazards rather than take action, the outcome of which may be uncertain. Periodic flooding of a farmer's fields or the mild tremor of a barely perceptible earthquake, or damage resulting from a tidal surge along the shore may be passed off with a shrug as inconsequential. Provided the impact or the damage is not great, many societies, indeed, exhibit a tremendous capacity to learn to live with hazard events rather than do anything about them. Such loss acceptance, however, does not rule out various actions to provide for the more vulnerable members of society such as the poor, the weak and the aged. These actions can take various forms including gifts, fiestas, transfers of wealth and resources, compensation payments, insurance, tax concessions and so on. The more evenly the losses are spread and the smaller the amount that each household or group has to bear, the more tolerable and acceptable the situation.

The decision to move to the more dynamic mode of loss reduction occurs when the action threshold is crossed, that is when individuals or societies are galvanized into a search for effective adjustments to minimize the damaging impact of an environmental hazard. In many instances, the crossing of this threshold arises in relation to a specific hazard event, usually a very disastrous or catastrophic one. Swift emergency action follows, to be succeeded by a resolve to learn from the experience and, if possible, to prevent a recurrence. Efforts are then directed either to controlling the hazard event itself or to reducing the vulnerability of individuals and groups in the society.

Adjustments to environmental hazards with a view to loss reduction may be separated into those that are purposefully adopted and those that are incidental, in the sense that they are not primarily hazard related but have the effect of reducing potential losses. The latter type of responses include many cultural practices and technological developments. For example, improvement in the quality and strength of building materials in the United States, particularly the change from wood to brick, has considerably reduced vulnerability to tornado damage, although this has not been the purpose of the change. Similarly, advances in communication systems, including the use of radar, satellite observation, and radio and telephone links, have provided previously unavailable opportunities for hazard response that are incidental to their main purposes. Transportation development in many developing countries such as India, providing increased capacity to move large quantities of food materials over considerable distances, has reduced vulnerability to low-energy or pervasive hazards such as drought and famine.

With regard to purposeful adjustment strategies, there is often a preference for adjustments that seek to control the natural events themselves over those that require behavioural, social or institutional changes in society. The preference for learning to live with it, characteristic of the acceptance mode, now translates into learning to prevent it. In both cases, changes in the societal side of the human–environment equation is avoided or put off for as long as possible. The initial concentration on modifying the hazard event is often followed by a more mixed strategy in which ways of preventing the effects of the hazards are developed and adopted.

There is, however, considerable overlap between such purposeful adjustments and cultural adaptations. Indeed, a response initially adopted as an adjustment, either purposefully or incidentally, may gradually become transformed into a cultural adaptation. The critical point at which a change in the pattern of land use ceases to be considered an adjustment to hazardous conditions and becomes accepted as the norm is not itself as important as the fact that it determines a society's absorbent capacity to cope with and accommodate natural hazards.

Finally, a point is reached, although very rarely, when the pressure of a hazard event becomes intolerable and forces a society to undertake a wholesale reappraisal of the situation. When such an event occurs, the threshold of loss intolerance is crossed. The society involved is then in a position to consider choices that otherwise would have been thought impossible. At this point, three lines of radical action are open to such societies: they can make a substantial change in the way they use the environmental resource; they can change their location; or they can combine both lines of action. Usually, changes of location or resource use are associated with institutional and social changes. Thus, resource use may become more extensive if new lands are occupied at a lower population density than previously or with lower capital investment per unit area. Changes from more to less intensive use and the reverse may be undertaken in response to hazards.

A change in hazard exposure often accompanies a change in location. For most of the world's societies, attachment to place, especially native place, is extremely strong. While in some sense a change in resource use is interchangeable with relocation, there seems nevertheless a desire to stay in the traditional location rather than to move. For many cultures, place and use are so intertwined that a drastic move may require a change of livelihood that is difficult for many to contemplate even under pressure of the most extreme events. Indeed, it would appear that once human settlement has been achieved and a commitment to a location has been made in terms of investment of capital and sense of affinity or identity with that place, complete abandonment rarely occurs. Even in relatively mobile societies such as in the United States, the stability of houses, infrastructures and industry make changing locations the least attractive of options. Organized attempts to remove people and property from floodplains usually flounder, and efforts to relocate the centre of a city into a less earthquake-prone site often encounter enormous resistance from the citizens even though the new site may be only a few hundred metres away. In short, it is usually more feasible to reduce hazard loss potential by changing use than by changing location.

Disaster Trends

There is no simple answer to the question: are societies becoming more vulnerable to environmental hazards? The frequency and magnitude of natural disasters have been steadily increasing for the last 30 years with a noticeable peak in 1991, the worst year for disasters in decades (figure 6.1). The world's less developed countries suffered approximately 97 per cent of these disasters and accounted for approximately 99 per cent of the deaths attributed to natural disasters (UNEP, 1993). While numerical estimates of mortality and injury are often questionable, the loss of life from natural disasters is enormous (table 6.1). Tropical cyclones and earthquakes are the greatest sources of fatalities.

Economic losses from natural disasters are greatest in the developed world and have tripled during the last 30 years. During the 1960s, for example, disaster losses were estimated at US$40 billion and by the 1980s these losses had risen to $120 billion. In the first half of the 1990s, cumulative losses were already beyond $160 billion (figure 6.2). In the United States, losses from Hurricane Andrew ($30 billion) and the Northridge earthquake ($30 billion) made these the most disastrous events to affect the nation. In Japan, losses from the Great Hanshin-Awaji (Kobe) earthquake are running at $50 billion (Domeisen, 1995). It is paradoxical to note that economic losses from two of the top ten natural disasters since 1945 occurred at the beginning of the 1990s, the start of the Decade of Natural Hazard Reduction (table 6.2).

Industrial accidents are a by-product of economic development and often have a much smaller catastrophic potential than natural hazards. Often they do not result in immediate fatalities, but pose longer term threats to human health and ecosystem stability. Many industrial accidents are associated with energy production and

Table 6.1 Top natural disasters, 1945–90

Year	*Location*	*Type*	*No. of Deaths*
1970	Bangladesh	Tropical cyclone	300,000
1976	China	Earthquake	242,000
1991	Bangladesh	Tropical cyclone	132,000
1948	Soviet Union	Earthquake	110,000
1970	Peru	Earthquake	67,000
1949	China	Flood	57,000
1990	Iran	Earthquake	40,000
1965	Bangladesh	Tropical cyclone	36,000
1954	China	Flood	30,000
1965	Bangladesh	Tropical cyclone	30,000
1968	Iran	Earthquake	30,000
1971	India	Tropical cyclone	30,000

Note: Table is based on estimated number of fatalities.
Source: Cutter (1994); UNEP (1993); Tolba et al. (1992).

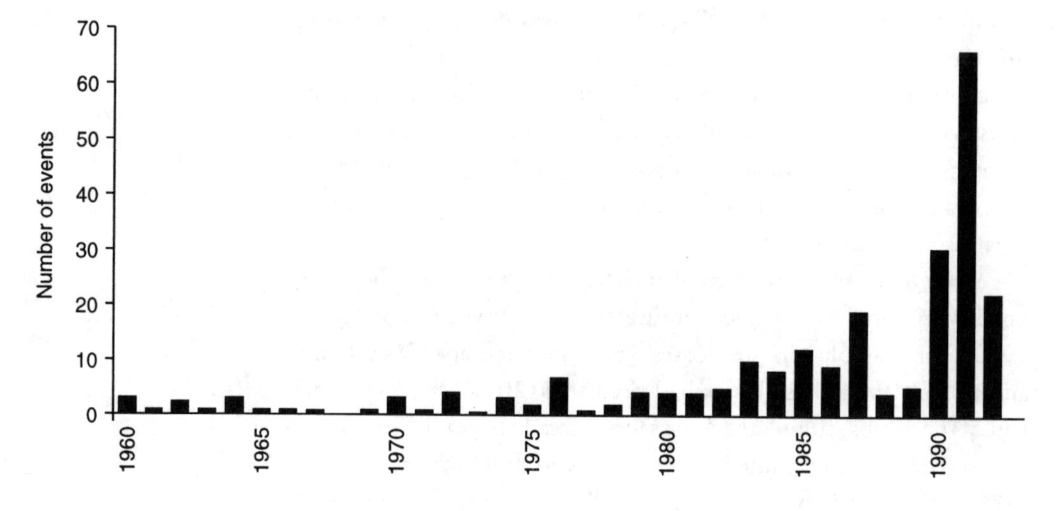

Figure 6.1 Frequency of major natural disasters, 1960–93 (adapted from Tolba et al., 1992, with additional data from UNEP, 1993)

distribution such as oil tanker accidents (Exxon Valdez, Aegean Sea and Braer) and intentional spills (Persian Gulf conflict 1991)(table 6.3). Chemical disasters have steadily increased since the 1960s (figure 6.3) with a decline in industrial accidents during the 1990s. As was the case with natural disasters, two of the top industrial disasters occurred in 1992. The first was the mine explosion and gas leak in Zonguldak, Turkey, and the second was the sewer gas explosion in Guedalajara, Mexico which killed some 210 people.

Regionally, natural disasters are more prevalent in less developed countries where increasing urbanization and environmental degrada-

Table 6.2 Losses from natural disasters, 1985–95

Year	*Location*	*Event*	*Economic losses (US$bn)*
1995	Kobe, Japan	Great Hanshin earthquake	50.00
1992	Florida, USA	Hurricane Andrew	30.00
1994	California, USA	Northridge earthquake	30.00
1993	Midwest, USA	Mississippi floods	12.00
1989	Caribbean, USA	Hurricane Hugo	9.00
1990	Europe	Winter storm, Daria	6.80
1989	California, USA	Loma Prieta earthquake	6.00
1991	Japan	Typhoon Mireille	6.00
1993	Northeast, USA	Blizzard	5.00
1987	Western Europe	Winter gale	3.70
1990	Europe	Winter storm, Vivian	3.25
1992	Hawaii	Hurricane Iniki	3.00
1995	Florida, USA	Hurricane Opal	2.80
1990	Europe	Winter storm, Wiebke	2.25
1991	USA	Forest fire	2.00
1990	Europe	Winter storm, Herta	1.90
1991	California, USA	Berkeley–Oakland Hills fire	1.60

Source: Domeisen (1995); Showalter et al. (1993); *New York Times* (1991, 1995).

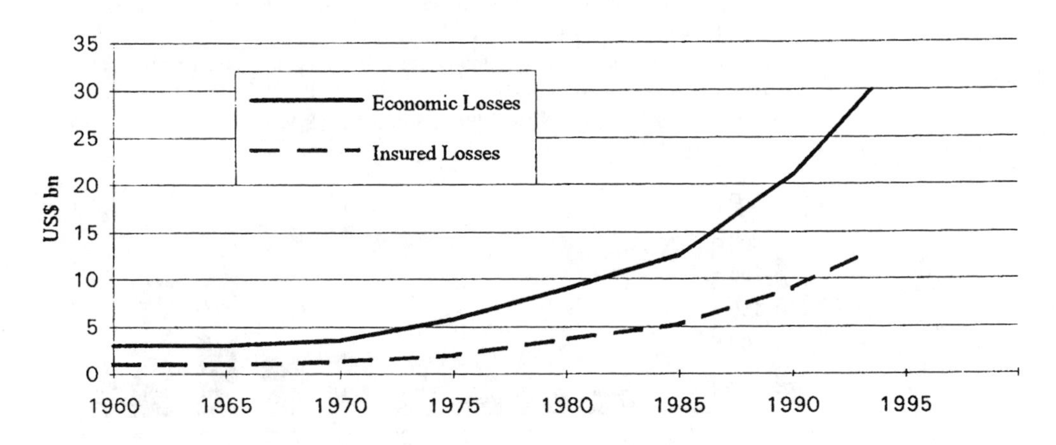

Figure 6.2 Losses from natural disasters, 1960–93 (adapted from Domeisen, 1995)

tion make these areas more vulnerable to the impacts of natural events. Industrial disasters, on the other hand, are more common in the developed nations. Regionally, southern and eastern Asia have the greatest fatality rate from natural disasters, Bangladesh being the most extreme case.

Table 6.3 Top industrial disasters, 1945–90

Year	*Location*	*Type*	*Number of Deaths*[a]
1984	Bhopal, India	Toxic vapour/ methyl isocyanate	2,750–3,849
1982	Salang Pass, Afghanistan	Toxic vapour/ carbon monoxide	1,500–2,700
1956	Cali, Colombia	Explosion/ ammunitions	1,200
1958	Kyshtym, Russia	Radioactive leak	1,118[b]
1947	Texas City, TX	Explosion/ ammonium nitrate	576
1989	Acha Ufa, Russia	Explosion/natural gas	500–575
1984	Cubatao, Brazil	Explosion/gasoline	508
1984	St. Juan Ixhautepec, Mexico	Explosion/natural gas	478–503
1992	Zonguldak, Turkey	Mine explosion/gas	388
1983	Nile River, Egypt	Explosion/natural gas	317
1992	Guadalajara, Mexico	Sewer explosion/gas	210
1986	Chernobyl, Ukraine	Explosion/radioactivity	31–300[b]

This table is based on estimated fatalities.

[a] Estimates vary widely depending on the source(s) used, therefore ranges are provided where discrepancies exist.

[b] Reported fatality figures reflect immediate deaths only, not longer term fatalities associated with the exposures.

Sources: Cutter (1994), UNEP (1993), Tolba et al. (1992).

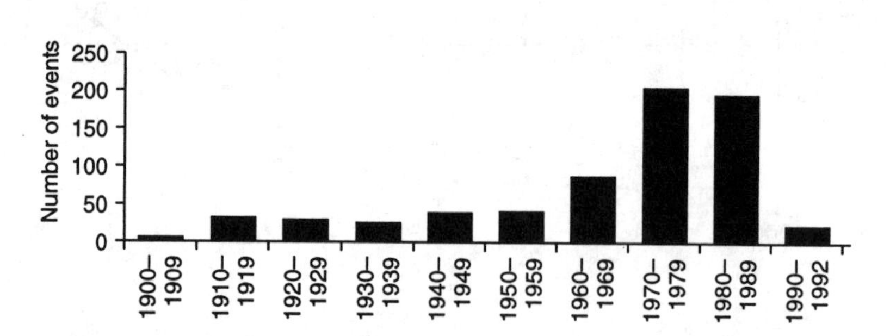

Figure 6.3 Chemical accidents, 1990–92 (data from Cutter, 1993)

Disaster Proneness: Towards Measuring Vulnerability

In 1990, UNDRO produced its first assessment of the vulnerability of nations to natural disasters. Working within the framework of economic impacts caused by natural disasters, UNDRO created its disaster proneness index for individual countries. The index provides a measure of the total economic effect of disasters over a 20-year period as a percentage of the total annual GNP. Only significant disasters, defined as those causing financial damages assessed at more than one percent of the country's annual GDP were included (UNEP, 1993). While preliminary in nature and fraught with all types of assumptions, the disaster-proneness index does provide some global comparative statistics on the vulnerability of countries to disasters.

Not surprisingly, some of the most disaster-prone countries are those with hazards (such as tropical cyclones) with repeated frequencies and hits during the last 20 years. Thus, Caribbean countries such as Monserrat, Dominica, St Lucia and the Pacific island nations of Vanuatu, and Cook Islands rank among the top ten. Other countries had one disastrous event during the last 20 years which inflated their ranking on the index. Figure 6.4 maps the disaster proneness of countries. In addition to the island nations mentioned above, Central American nations such as El Salvador, Honduras and Nicaragua, Sahelian countries including Burkina Faso, Ethiopia and Mauritania, and Asian countries such as Bangladesh, are the most disaster prone. The inverse relationship with national wealth comes as no surprise. Not only are these nations the most disaster prone, they are among the least able to respond to the aftermath of a disaster and mitigate the impacts of future ones.

While provocative, the disaster-proneness index does not measure those factors that govern the increasing vulnerability of countries to hazards. It is recognized, for instance, that urbanization, industrialization and technology, all influence the impact of hazards on places, often making local residents more vulnerable generally. Population pressure, poverty and gender relations influence vulnerability by making certain segments of the population more susceptible to the impact of disasters once they occur. These factors are not included in the disaster-proneness index, yet they are vital in understanding why some countries and certain populations within those countries are disproportionately affected by hazards. There are both spatial and temporal dimensions to this biophysical and social vulnerability that are not fully understood nor incorporated into this index.

There are a number of confounding concerns that prohibit simple answers being proferred to the questions posed earlier in this chapter.

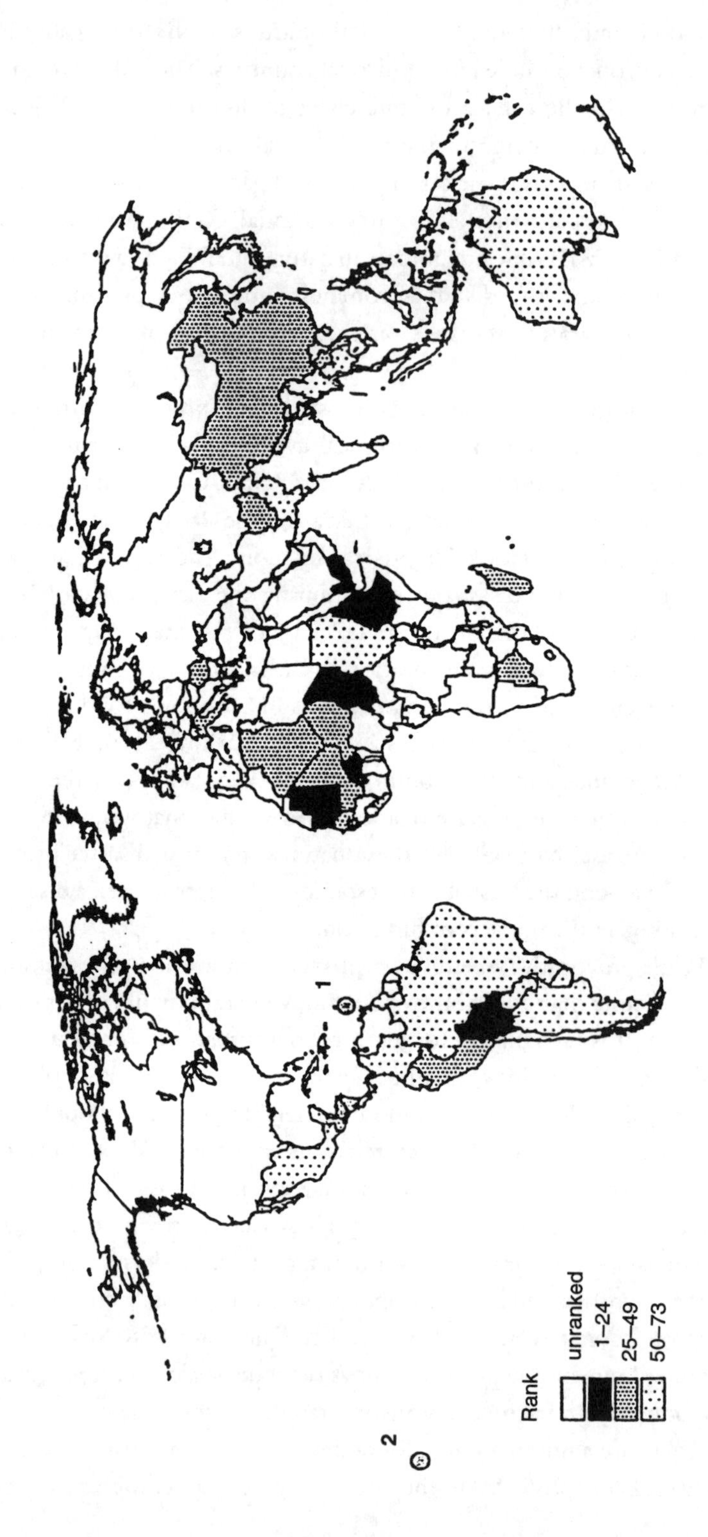

Figure 6.4
UNDRO's disaster-proneness index.
Note: Data are based on significant disasters, 1970–89, and are ranked according to the total index variable. The higher the rank, the more disaster-prone the country
Source: UNEP, 1993

First and foremost, the very nature of hazards has changed from singular natural events, sometimes referred to as acts of God, to more complex phenomena involving the interaction between natural, social and technological systems. The development of typologies based on the etiology of events no longer works nor does the distinction between natural and technological hazards. Second, responses to hazards are now viewed as embedded within larger social and environmental constructs where it is increasingly difficult to separate the impacts of specific disasters or hazards from broader based social or environmental contexts. A direct result of the present situation is the increased complexity of hazard-management systems. The expanded range of management alternatives has gone beyond the geotechnical, especially as risks and hazards have become more politicized and decisions on them are made on the basis of social choices rather than just technical prowess (Kates, 1985; Mitchell, 1990.).

Within this framework, geography plays a pivotal role. Geographical scale is crucial to understanding the distribution, impact and reduction of hazards (Cutter, 1994). Scale is an important parameter in detecting and monitoring the impacts and consequences of environmental hazards. The discovery of new hazards and the rediscovery of older ones with more dispersed and cumulative impacts necessitate the globalization of risk- and hazard-management systems. Yet, most hazard studies continue to be localized case studies. The articulation of global processes and local impacts will continue to challenge the scientific understanding of hazards.

Geography provides linkages between physical processes and human contexts and helps to define the areal or spatial extent of a hazard. There are a number of geometric manifestations of hazards (Turner et al. 1990; Zeigler, Johnson and Brunn, 1983) but these have not been analysed in any systematic way. The application of geographic information-processing techniques such as remote sensing and GIS can help delineate hazard zones but they are at present not widely used.

Data difficulties inhibit better understanding of the broad patterns of hazards distributions as well as societal responses. While international comparative statistics are available, their reliability is often questionable due to the inaccuracies and inconsistencies in reporting and record keeping. The most basic of data on a disaster such as its location, magnitude and duration are often missing, incomplete or withheld for national security purposes. Measuring the impacts of disasters is more problematic. Most databases concentrate on three main criteria: mortality, number of people affected and damage estimates (usually in US dollars), each with its own inherent biases in data collection. For example, most of the damage estimates are made in local currency and then adjusted to the US dollar standard. Fluctuations in exchange rates and inflation from year to year often render these estimates meaningless.

For example, the United Nations Environmental Programme (UNEP) maintains a disaster database but only reports them when at least 30 immediate fatalities occur. In addition to the UNEP, the Centre for Research on the Epidemiology of Disasters (CRED) in Brussels, Belgium, and the United States Office of Foreign Disaster Assistance also maintain global databases on natural disasters (International Federation of Red Cross and Red Crescent Societies, 1995). As should also have become patently clear, all these efforts focus on disasters arising from extreme natural events. Rarely are multiple origins considered such as combined cyclones and flooding, nor are hazards arising from more chronic conditions such as drought (which could facilitate a famine disaster or forest fires).

Human-induced hazards are increasing in importance, yet relevant global data are rare. Oil spills, chronic toxic contamination and pollution are good examples. Industrial-accidents data are collected, with the OECD databases being among the best. Statistics are also collected on oil spills by such agencies as the International Tanker Owners Pollution Federation Limited and the Oil Spills Intelligence Report. Nuclear accidents are monitored by the International Atomic Energy Agency (IAEA). Data on the transboundary movement of hazardous waste are difficult to collect because of the lack of international agreement on what constitutes hazardous waste. The Basel Convention comes closest to a universal definition by providing a list of regulated categories of hazardous waste. Other sources of data include the UNEP's International Register of Potentially Toxic Chemicals (IRPTC) and the APELL (Awareness and Preparedness for Emergencies at Local Level) Programme (Tolba et al., 1992).

Unfortunately, basic data on the range and extent of hazards has not kept pace with needs. Detailed information on the human occupance of hazard zones and adjustments is generally only available at a very localized level. It is possible, for instance, to monitor and even model the physical systems response to hazards and ultimately to assess the biophysical impacts at both global and local levels.

However, there are few global databases on human occupance and societal adjustments to environmental hazards. Social data are unreliable in many world regions, further hampering efforts at assessing the social consequences of environmental hazards.

Just as basic data are lacking for determining the range, extent and response to environmental hazards, so is there an absence of serious theoretical constructs for helping with the understanding of the processes by which hazards are produced and the types of options that are available for mitigation and recovery. Hazards research is a very active component of the nature–society emphasis within geography and has been so for more than half a century. Most of the early research was practice driven and was concerned with why people occupy hazardous environments and what policy outcomes reduce the impact of such locational decisions. In recent years, more emphasis has been placed on theory development – hazards in context (Kirby, 1990; Palm, 1990), social theories of risk (Johnson and Covello, 1987; Krimsky and Golding, 1992), social amplification of risks (Kasperson et al., 1988; Kasperson, 1992) and vulnerability (Blaikie et al., 1994; Liverman, 1990; O'Riordan, 1986). Despite these attempts there is still some frustration at the lack of an integrated theory on how people respond to and interact with the environment. For example, there is still a need to understand the linkages between macro-level processes and micro-level impacts, as well as the connections between dynamic systems and static phenomena. Within the geographical community as well, there is much still to be done with regard to longitudinal analyses of localized hazardous environments as well as temporal changes in residents' responses to single events or multiple hazards. While advancing understanding of nature–society interactions, hazards research still requires additional theoretical development to ascertain what makes places and people vulnerable to environmental hazards.

Reducing Vulnerability: Short-term Adjustments

When a disaster strikes, the immediate societal adjustments are to rescue the survivors and re-establish the lifelines of the ravaged community. These relief operations include medical supplies, food, shelter, water and power. In some instances, the rescue and relief operations are within the capabilities of the affected country. Often, however, the disaster is of such magnitude that the individual country cannot cope and international relief efforts have to be mobilized through international relief organizations such as the Red Cross/Red Crescent, as well as through cooperative arrangements within the United Nations Disaster Relief Organization (UNDRO). Once the lifelines have been re-established and the crisis period is over, the recovery phase begins. Recovery adjustments are temporary in nature and provide for the beginning of normalcy after an event. The use of temporary shelters during the recovery period gives way to building permanent structures during the reconstruction phase. Throughout the recovery and reconstruction phases, hazard mitigation continues. Some of the mitigation options are structural in nature, such as the use of steel-reinforced construction materials in seismic areas, or the use of elevated pilings in flood-prone areas. Other mitigation strategies are non-structural in their emphasis and involve land-use planning and management, insurance, and pre-event preparedness including warning systems.

While the range of potential adjustments to hazards has increased over time, individual access to adjustments is more restricted now than in the past and has become a function of social class, income and life circumstances. Increasingly, nations and societies are becoming polarized between rich and poor, powerful and powerless, and divided by ethnic divisions or subcultures. The chasm between the haves and the have nots is widening within societies and between nations. Ultimately, the ability to respond to environmental hazards is constrained by these divisions as we have seen time and time again.

Poverty and environmental degradation are linked and create an impoverishment–degradation spiral (Kates and Haarmann, 1992; Mellor, 1988). The driving forces behind environmental degradation are development/commercialization along with population growth and poverty. Natural hazards accelerate the process and further restrict the use of those remaining natural resources, many severely degraded already.

Wealth plays an enormous role in fostering opportunities for adjustments to these deteriorating conditions and recovery from hazards. Poverty restricts one's ability to maintain the simplest of adjustments, such as protective work, because of the lack of skill and labour to undertake the improvement, lack of needed inputs for rebuilding, or lack of access to education and thus knowledge of public programmes for recovery. Without capital or power, poor people who often live on marginal lands to begin with, eventually get displaced from them, and begin a migratory odyssey as environmental refugees. In many countries, these refugees – mostly women and children – are the forgotten casualties, a subgroup who are often least able to adjust to environmental hazards (Cutter, 1995).

Urbanization is one of the key processes that influences vulnerability to environmental hazards. Not only are the world's mega-cities becoming more populated, they are also situated in some of the most natural hazard-prone areas of the world, namely along coastlines, and in active seismic areas. Air pollution, toxic chemical contamination, and poor water quality plague the world's mega-cities as well. The elderly and the children are most susceptible to air pollution episodes, be they in developing or developed world cities. Los Angeles, Mexico City, Beijing, Seoul and Cairo fail to meet more than half of the World Health Organization's standards for air quality. Lead contamination (which contributes to learning disabilities in children) is on the rise in cities in developing countries as the use of motor vehicles using lead-based fuel rises (lead-free fuel is more expensive). There are many other societal trends affecting the hazardousness of places and the response of people that require further exploration by hazard geographers (table 6.4).

Conclusion: Making the Tent Larger – Expanding the Hazards Reduction Decade

Hazard reduction will only come about with profound changes in society. Geotechnical solutions such as levees or seismic-proof buildings will provide short-term relief but ultimately will exacerbate hazards in the future. Nations must address why people live in hazardous environments in the first place, how they respond and adjust to environmental hazards, and what types of mitigation programmes are appropriate at the local and national level. Hazard-reduction strategies will vary from region to region depending on the range of hazards that affect local places. Geographers are key players and have much to contribute to these ongoing discussions. An understanding of physical processes and of societal responses provides a unique perspective for examining the relationship between society and nature and for working towards improving the human condition.

While disasters capture immediate attention when they happen, it is necessary to be cautious and to try to understand that hazards are part of daily life. It is not only the extreme natural event such as an

Table 6.4 Social trends affecting environmental risks and hazards

Lessening the Impacts
improved building technology
better detection and warning systems
improved health care
improved environmental regulations
environmentally-sound development
better understanding of risks and hazards
improved educational opportunities
Aggravating the Impacts
occupancy of hazard zones
ageing populations
ageing infrastructure
increasing populations
urbanization
migration
industrialization
resource exploitation
increasing poverty
reliance on complex technological systems

Source: Showalter et al. (1993).

earthquake or a hurricane that needs to be planned for. Precautions must be taken for those hazards that are experienced on a daily basis concerning the air human beings breathe, the water they drink and food they eat. Poor water quality and sanitation kill more people in developing countries than all natural disasters combined. Clearly, the International Decade of Natural Hazard Reduction is not focusing on these chronic everyday hazards, which may be more costly to societies in the long run in terms of loss of life, injury and reduced productivity than the occasional natural event. Now is the time to expand the mission of the International Decade to include an all-hazards approach to understanding the vulnerability of societies to environmental risks and hazards. In so doing, improvements in the human condition will occur and the major goal of the decade will have been achieved.

References

Baumann, D.D. and Russell, C. (eds) (1971) *Urban Snow Hazard: Economic and Social Implications* (Urbana, IL, University of Illinois, Water Resources Center Research Report No. 37).

Blaikie, P., Cannon, T., Davis, I. and Wisner, B. (1994) *At Risk: Natural Hazards, People's Vulnerability and Disasters* (London: Routledge).

Burton, I., Kates, R.W. and White, G.F. (1993) *The Environment as Hazard*, 2nd edition (New York: The Guilford Press).

Cutter, S.L. (1993) *Living with Risk: The Geography of Technological Hazards* (London: Edward Arnold).

Cutter, S.L. (1994) *Environmental Risks and Hazards* (Englewood Cliffs, NJ: Prentice-Hall).

Cutter, S.L. (1995) The forgotten casualties: women, children, and environmental change, *Global Environmental Change*, 5, 3:1818–94

Domeisen, N. (1995) Disasters: threat to social development, *Stop Disasters*, 23, Winter: 7–9.

International Federation of Red Cross and Red Crescent Societies (1995) *World Disaster Report* (Dordrecht: Nijhoff).

Johnson, B.B. and Covello, V.T. (eds) (1987) *The Social and Cultural Construction of Risk: Technology, Risk, Society* (Dordrecht: D. Reidel).

Kasperson, R.E. (1992) The social amplification of risk: progress in developing an integrative framework, pp. 153–78 in S. Krimsky and D. Golding (eds) *Social Theories of Risk* (Westport, CT: Praeger)

Kasperson, R.E., Renn, O., Slovic, P., Brown, H.S., Emel, J., Goble, R., Kasperson, J.X. and Ratick, S. (1988) The social amplification of risk: a conceptual framework, *Risk Analysis*, 8, 2:177–87.

Kates, R.W. (1985) Success, strain, and surprise, *Issues in Science and Technology*, 2, 1:46–8.

Kates, R.W. and Haarmann, V. (1992) Where the poor live, are the assumptions correct? *Environment*, 34, 4:4–11, 25–8.

Kirby, A. (ed.) (1990) *Nothing to Fear: Risks and Hazards in American Society* (Tuscon: University of Arizona Press).

Krimsky, S. and Golding, D. (eds) (1992) *Social Theories of Risk* (Westport, CT: Praeger).

Liverman, D. (1990) Vulnerability to global environmental change, pp. 27–44 in R.E. Kasperson, K. Dow, D. Golding and J.X. Kasperson (eds) *Understanding Global Environmental Change: The Contributions of Risk Analysis and Management* (Worcester, MA: Clark University, The Earth Transformed Program).

Mellor, J.W. (1988) The intertwining of environmental problems and poverty, *Environment*, 30, 9:8–13, 28–30.

Mitchell, J.K. (1990) Human dimensions of environmental hazards:

complexity, disparity, and the search for guidance, pp. 131–75 in A. Kirby (ed.) *Nothing to Fear: Risks and Hazards in American Society* (Tuscon: University of Arizona Press).

O'Riordan, T. (1986) Coping with environmental hazards, pp. 272–309 in R.W. Kates and I. Burton (eds) *Geography, Resources and Environment, vol. II – Themes from the Work of Gilbert F. White* (Chicago: University of Chicago Press).

Palm, R.I. (1990) *Natural Hazards: An Integrative Framework for Research and Planning* (Baltimore: Johns Hopkins University Press).

Tolba, M.J., el-Kholy, O.A., El-Hinnawi, E., Holdgate, M.W., McMichael, D.F. and Munn, R.E. (1992) *The World Environment 1972–1992* (London: Chapman and Hall).

Turner B.L.II, Kasperson, R.E., Meyer, W.B., Dow, K.M., Golding, D., Kasperson, J.X., Mitchell, R.C. and Ratick, S.J. (1990) Two types of global environmental change: definitional and spatial-scale issues in their human dimensions, *Global Environmental Change*, 1, 1:14–22.

United Nations Environment Programme (UNEP) (1993) *Environmental Data Report 1993–94* (Cambridge, MA: Blackwell).

Wisner, B. and Mbithi, P.M. (1974) Drought in Eastern Kenya: nutritional status and farmer activity, pp. 87–97 in G.F. White (ed.) *Natural Hazards: Local, National, Global* (New York: Oxford University Press).

Zeigler, D.J., Johnson J.H.Jr. and Brunn, S.D. (1983) *Technological Hazards* (Washington, D.C..: Association of American Geographers).

Part III

Spatial Structures and Societal Processes

7 Market Forces, Cultural Factors and Locational Processes

An important issue in the human occupation of the Earth's surface concerns the spatial structures created by different kinds of societies. Traditionally, geographers sought to explain locational patterns by reference to the local or regional specifics of the interaction of people and environment. This was particularly true of the ideographic approach prevailing until the 1950s, which stressed the uniqueness of places. Culture was viewed as both influenced by and contributing to the particular way in which human landscapes were formed in different parts of the world. However, the emergence of the nomothetic approach associated with the so-called quantitative revolution and reformation of human geography as locational analysis or spatial science, which dominated the 1960s and 1970s, turned attention towards more universalistic interpretations in which the contribution of cultural variability tended to be subdued if not entirely ignored. The resurgence of cultural geography in the 1980s, in the context of the postmodern preoccupation with differences and the disdain for grand theories and meta-narrative, focused human geography once again on the local and the particular, with critical implications for the supposedly universal theories which constituted the foundation of locational analysis for a quarter of a century.

Notwithstanding the role of the market in determining the character of various spatial structures, locational processes can hardly be ignored in human geography. As a major mechanism for the allocation of land resources to particular uses, the market is a fact of daily life in modern societies. Its overall impact on locational decision-making by households and firms has been important in shaping many aspects of the configurations of human settlements. However, there is no gainsaying the fact that cultural factors do play an equally impor-

This chapter is based on a contribution by David M. Smith.

tant part in decision-making processes. In this chapter, an attempt is made first to understand the nature of the market as an economic institution; second, based on the importance of market forces, I shall review the explanatory value of the principal models of locational processes developed on the foundation of conventional neo-classical economic theory; and third, I shall examine the way in which so-called behavioural factors were introduced into these models, suggesting that this comprised a rather restricted reading of those influences which might be subsumed under the heading of culture. A broader interpretation of the role of culture provides a reminder that it has always been an active agent in spatial economic processes. This leads on to the argument that the kind of economy assumed in location theory, regulated by competitive market forces, is itself the expression of a particular culture which is historically and spatially specific, and which has undergone significant changes in recent years. Claims as to the universality of approaches to locational analysis grounded in such an economy must therefore be modified. More culturally sensitive interpretations are required although models built on neoclassical economic foundations may still serve the more limited role of technical devices for optimal resource allocation.

The Market as an Economic Institution

Although the market is a dominant feature of life in virtually all societies, its role in the articulation of social and economic processes varies considerably from one society to the other and particularly between historical times. This has led Polanyi (1968) to distinguish between three types of market exchange. The first is a market exchange involving no more than the locational movement of products among people, a situation no different from barter. The second type of market exchange requires the setting of prices for the products being exchanged through a social mechanism. The third type of market exchange is that which occurs through the operation of a price-fixing market. Thus, although market exchange occurs under a variety of circumstances, it is only in the last instance, that the market serves as a means of articulating social and economic processes in a society and achieving a high degree of economic integration. Regularized exchange through price-fixing markets is a finely tuned mechanism for co-ordinating and integrating the activity of large numbers of individuals acting independently. But to be effec-

Reflection of a consumer culture: giant supermarket trolley for a St Patrick's Day parade on Hollywood Boulevard, USA.
Source: L.A. Daily-Yee-Liaison/Gamma.

tive, this system requires that individuals, in terms of the goods and services at their command to dispose or acquire, respond appropriately to price signals, otherwise there would be no economic integration.

Integration through price-fixing markets is characteristic of the capitalist mode of production. This mode of production requires that all factors of production be treated as if they were commodities like oranges and apples and be brought to the market to be bought or sold at a price. The price for land is, of course, its rent; for labour, it is the wage or salary; for capital, it is the interest paid on it; and for entrepreneurship, it is the profit it can command. For each of these factors of production, it is its exchange value rather than its use value that is of interest to the market. Exchange value, expressed in prices, is, however, an abstract quantity determined through the functioning of a market system based on money as a measure of value. However, for a society to be able to convert all its factors of production into marketable commodities (commoditization) with a price tag attached to them a major social transformation with unprecedented traumatic effects is involved (Polanyi, 1944).

With respect to land, for instance, it involved bringing it out of the web of communal or kinship ownership and individualizing it. Sometimes this entailed bringing together the scattered holdings of the group, ascertaining what each individual member had, and re-allocating the group land as new consolidated plots to the individual members on a pro rata basis. Such individualized plots of land then need to have their boundaries clearly demarcated through surveying. The ownership right of the individual also has to be formally conferred through a title deed for which the authorities have to provide a registry where this can be validated, formally registered and preserved as incontrovertible evidence of the power of the individual to alienate the plot of land if needs be. Similarly processes of commoditization transformed the services of self-employed, free men into wage-earning labour and their savings into investable, interest-earning capital. And against all religious injunction against usury, profit came to be accepted as a due reward for enterprise.

This market-exchange system, so central to the capitalist mode of production, operates under conditions of resource scarcity. But, as Harvey (1973:113) observed, the concept of scarcity, like the concept of a resource, is not an easy one to comprehend although people constantly make reference to it when they talk of the allocation of scarce resources. The fact is that this concept only takes on meaning in a particular social and cultural context; for in a sophisticated mode of production such as the capitalist economy, scarcity is socially organized in order to permit the market to function. This takes place through fairly strict control over access to the means of production and control over flow of resources into productive processes. Similarly, the distribution of output has to be controlled in order for scarcity to be maintained. Consequently, scarcity must be seen as something socially organized, produced and controlled in society because without it price-fixing markets could not function.

The institution of individual property rights is the mechanism whereby scarcity is socially organized. Under capitalism, this institution is not only protected by the state, but it is guaranteed by the legal system and the social mores and norms of behaviour. With respect to land, for instance, this institutionalization of private property ensures that owners possess monopoly privileges over pieces of space. Or, as Marx (1967, vol. 3:615) puts it:

> *Landed property is based on the monopoly by certain persons over definite portions of the globe, as exclusive spheres of their private will to the exclusion of others. With this in mind, the problem is to ascertain the economic value, that is, the realization of this monopoly on the basis of capitalist production. With the legal power to use or misuse certain portions of the globe, nothing is decided. The use of this power depends wholly upon economic conditions, which are independent of their will . . . Whatever the specific form of rent may be, all types have this in common: the appropriation of rent is that economic form in which landed property is realized.*

It is thus the existence of this institutionalized relative scarcity that gives rise to the market mechanism of supply and demand and that determines the value or price at which an individual commodity or factor of production is exchanged.

Such a market economy, according to Polanyi (1944:68), assumes that the supply of goods (including services) available at a definite price will equal the demand at that price. It assumes the presence of money, which functions as purchasing power in the hands of its owners. Production will then be controlled by prices, for the profits of those who direct production will depend upon them; the distribution of the goods also will depend upon prices, for prices form incomes, and it is with the help of these incomes that the goods produced are distributed amongst the members of society. Under these assumptions order in the production and distribution of goods is ensured by prices alone and is entrusted to this self-regulating mechanism. Self-regulation implies that all production is for sale on the market and that all incomes derive from such sales. Accordingly, there are markets for all elements of industry, not only for goods (including services) but also for labour, land and money, their prices being called respectively commodity prices, wages, rent and interest.

Certain advantages derive from this operation of a self-regulating, free market economy. First, the price system which it effectively promotes at the interface between supply and demand serves to successfully co-ordinate a vast number of decentralized decisions and thereby to integrate a vast array of activities into a coherent social and spatial system. Second, the competition for access to scarce resources, on which the market depends, stimulates and facilitates technological innovations as a means of creating new property rights, albeit of an intellectual type. Third, the competitive nature of market organization also fosters a tremendous increase in the total product available to society. And finally, by promoting overall growth in the economy, the self-regulating free market system offers the challenging prospect of transforming the material condition of human beings throughout the world.

This general description of the evolution of the capitalist market system, however, says very little about how the system works or how its capacity to produce wealth can be investigated, analysed, understood and made to guide future policy decisions. The question of the relation between the supply and demand for a particular commodity and the equilibrium price at which the market will be cleared is only the introduction to a very complex area of societal life. This, indeed, is a major area of concern for the science of economics and the details need not concern us here. But to the extent that some of the principles, concepts and theories developed in that discipline have helped to illuminate critical aspects of human locational processes and the human use of land, and other resources, it is important to recognize the strong intellectual link.

Market Forces, Distance and the Location of Economic Activity

The shift from the idiographic to nomothetic approach in human geography was closely associated with a growing awareness of empirical regularities in spatial structure. Patterns of land use and settlement in particular attracted the attention of a new generation of quantitative geographers who began to make their mark in economic and urban geography in the late 1950s. As verbal description was replaced by measurement, a new style of explanation emerged in which it was assumed that general processes could be read off the patterns themselves, with the aid of theoretical constructs imported largely from economics.

Attempts to explain the evolution of spatial structures in the universalist style of location analysis or spatial science rely on some version of the competitive market model with its origin in neoclassical economics. The distance variable is added to those of combination of factors (or inputs) and scale of production in a competitive process in which the market forces of supply and demand still allocate resources across geographical space as well as among firms and sectors of the economy. Free markets are supposed to operate in such a way as to generate an optimal outcome, based on general criteria such as the maximization of profit, utility or welfare, or by such geographical objectives as the minimization of aggregate travel or distance covered. This can be illustrated by agricultural land-use theory, its urban variant, industrial location theory and central place theory. All of these have been discussed in detail in chapter 2, Geographical Theories, and will only be treated cursorily here.

The usual point of departure in the development of location theory is the work of Johann Heinrich von Thunen (Hall, 1966), originally published in 1826. Von Thunen used observations on his own estate in Germany to develop a general model of agricultural land-use in an imaginary isolated state in which a central city market on an homogeneous plain would be served by the surrounding territory. Individual farmers would be willing to supply or give over their land for a particular use (demand) on the basis of which use enables them to maximize their profits in the form of rent (price). Different activities or use would have different rent-producing capacities (or bid price), depending on the importance of proximity to the market, represented by different gradients of rent plotted against distance from the city. Thus, market gardening, involving perishable products, would command high rent close to the market, falling away on a steep gradient with increasing distance from the city. Grain or timber production, on the other hand, would generate lower rents close to the market, but their gradient will fall away shallowly and extend well beyond the distance where market gardening could not provide rent. In the well-known graphic analysis reproduced in numerous textbooks, the allocation of land to the activity producing the highest rent would generate a pattern of concentric zones of different land uses around the city.

There has been some success in finding actual patterns of

concentric zones of agricultural land use explicable in terms of the von Thunen model. For example, Blaikie (1971) observed small farmers in northern India adjusting land use to distance from their village, applying greatest effort to land close to it and using outlying fields less intensively. Similarly, Horvath (1969) found concentric zones of land uses around the city of Addis Ababa in Ethiopia. Peet (1969) has made use of von Thunen's scheme at a global scale.

A similar model of urban land use has been elaborated by Alonso (1964). This derives concentric zones from the different rent or bid-price gradients of, for example, retailing, industrial and residential activities in the order of their capacity to pay high rents for the advantage of using land close to the city centre. The underlying process is again the maximization of profit under competitive market conditions. Modifications of the original urban land-use model include an explanation by Bunge (1971) of successive zones of slums, middle-class housing and affluent suburbs, typical of a North American city.

Concentric-zone patterns have been validated in numerous empirical studies of western cities. However, they can be complicated by such local conditions as topography and transportation lines, which may encourage more of a wedge structure, and by metropolitan growth with multiple nuclei. Indications of concentric zones have also been found in other parts of the world, for example, in South-East Asian cities (McGee, 1967). There have even been attempts to identify such patterns in the socio-economic differentiation of some eastern European cities as restructured under socialism, though actual patterns often appear to be more of a mosaic or patchwork quilt than broad zones (Smith, 1989).

Turning to industrial location, the basic model for the single unit of production goes back to the work of Alfred Weber (1929), originally published in 1909. This derives the least-cost (and maximum profit) location from the spatially variable costs of acquiring materials from fixed sources and sending the finished product to a market point, with a source of cheap labour and economies of agglomeration as added complications. What came to be known as neoclassical location theory has subsequently been extended to incorporate other considerations and has been applied to the analysis of a range of cases in which patterns of particular industries as well as single plant locations have been interpreted with some conviction (for case studies, see Smith, 1981). It has also been suggested that Weber's theory may have some applicability at a global scale (Dicken, 1977).

The complement to this variable-cost approach is a focus on spatial variations in revenue as the principal determinant of profit maximization, based on analysis of competition among firms for spatial market share. This line was originally developed by economists in the 1930s when they first recognized geographic space as a source of local monopoly, and hence imperfection in the idealized markets of production theory. However, conceptual and practical difficulties have made this approach hard to apply to the interpretation of actual location patterns (for further explanation, see Smith, 1981).

A better known application of market-area analysis is in the central place theory initiated by Walter Christaller (1966), originally published in 1933. From some simple propositions about the threshold of a good (the minimum volume of sales required for an enterprise to be viable) and its range (the maximum distance consumers will travel to purchase it), he derived the well-known hexagonal patterns of a hierarchy of central places (market towns or cities) and complementary regions (hinterlands or market areas), specifying the spatial structure of the provision of goods and services satisfying particular criteria of optimality. August Losch (1954) took this scheme a step further in 1940, setting out in both algebra and geometry the economic landscape which would satisfy the neoclassical conception of general equilibrium under which no participant would have anything to gain from change. This marked the peak of elegance and sophistication achieved by the extension of conventional production economics into geographical space, though Walter Isard (1956) broadened the scope still further with a more intuitive graphic synthesis of central place structure, industrial location theory and von Thunen's agricultural zones.

Attempts to explain real-world spatial structures in terms of central place theory, range from Christaller's own detailed analysis of settlement in southern Germany to the various applications which represented the early flourishing of geography's newly discovered numeracy and model-building capacity in the early years of the quantitative revolution (see Berry, 1967). There were also refinements prompted in part by the examination of service hierarchies within cities (Beavon, 1977). Not surprisingly perhaps, reality was found to fit theory best in conditions most closely resembling the isotropic plain assumed in the theory and uncluttered by physical geography. And it should be recalled that explaining the real world was not necessarily the primary goal of the spatial extensions of economic theory. As Losch (1954) made clear, the focus was on what would be optimal under the prevailing assumptions of economic rationality rather than what could actually be observed.

Behavioural Modifications

The limited capacity of the kind of models outlined above on which regular spatial patterns depend to account for observed reality is explained only in part by the actual topography which inevitably distorts free movement or identical cost gradients in all directions. It also arises from their particular behavioural conception of humankind dominated by the conventional economic rationality of *homo economicus*, which requires single-minded dedication to profit maximization on the part of producers and to utility maximization on the part of consumers. The underlying neoclassical economic theory makes no serious allowance for other behavioural considerations which may impinge on decision making, within what is implicitly assumed to be the independent sphere of economic activity somehow removed from other aspects of life which might be labelled political, social and cultural.

Modifications of the rigid economic determinism imported into location theory from production economics were being proposed from the earliest days of geographical engagement with spatial economic analysis. Geographers, it seems, were less satisfied with the abstractions of economic theory and its unrealistic assumptions, which were less in tune with the messy reality of actual human behaviour. However, what was accomplished represented only a partial humanization of *homo economicus.*

Of special significance in a geographical context was the introduction by Eric Rawstron (1958) of the spatial margin to profitability. This is a line (or lines) containing the area (or areas) within which profitable operation is possible. This area is defined by the equality of total cost and total revenue with respect to a given scale of productive activity. This was one of the most original contributions ever made by a geographer to spatial economic analysis. Its significance was that it drew attention away from the single, unique point of profit maximization which real-world entrepreneurs might not even seek, never mind find, towards the spatial limits to locational choice which must be respected if viability is to be achieved. With the margin(s), enterpreneurs would be free to indulge in sub-optimal behaviour, abandoning strict profit maximization for such legendary preferences as a factory location conveniently situated in relation to a golf course or other amenities.

The introduction of the concept of the spatial margin coincided with the explicit recognition of the importance of sub-optimal behaviour in mainstream spatial economics. Losch (1954:224) conceded that, all things considered, entrepreneurs would establish their enterprises at places they liked best. Melvin Greenhut (1956:175–6) developed a more detailed argument along these lines: the non-pecuniary satisfaction which an entrepreneur may derive from setting up in a particular location may be regarded as psychic income, the overall objective being to maximize total satisfaction derived from both pecuniary and psychic income. In the first graphic elaboration of the concept of the spatial margin, Smith (1966:108) suggested that a point of psychic income might divert an entrepreneur from the maximum profit location just as a low labour cost location might divert Weber's entrepreneur from the least transport cost point, providing that the plant remained within the operative profit margins.

A further step towards a more behavioural approach involving the spatial margin was taken by Allan Pred (1967). He devised a behavioural matrix within which individual entrepreneurs would be assigned a position along two dimensions representing respectively the quantity and quality of information at their disposal and their ability to use it. Well-informed and/or highly able entrepreneurs would be most likely to find a location near the profit-maximizing optimum; those with more limited information and/or ability would tend to be further away or even in unprofitable locations beyond the margin.

The subsequent development of a behavioural approach to location analysis moved away from simple modifications or extensions to established models. Increasing attention was given to the observation of the actual decision-making process in the hope of finding empirical regularities in the way entrepreneurs appraised the environment within which possible new locations are identified, evaluated and eventually occupied or otherwise. And, as geographers came to recognize the increasing complexity of contemporary industrial structure, with its multi-plant, multi-product, and often multinational scale of operation, the focus shifted to the nature of the organization itself. It was then a short but significant step to seeing these organizations as the outcome of the development of capitalism, a particular mode of production with its own dynamics and its own historical and geographical specificity.

The engagement of Marxism with economic geography, and with human geography in general, finally undermined the accepted belief, implicit though it may have been, that the models of the era of locational analysis or spatial science captured some universal phenomena. In particular, their abstraction from actual social relations obscured their specificity under capitalism. However, when it came to the interpretation of actual economic decision-making, *homo marxicus* appeared initially to be no more flexible than *homo economicus* had been. Such substitutes for market forces as the law of value or the logic of capital were no lesser sources of determinism, allowing no more room for the exercise of cultural preferences. It was a resurgence of humanistic concerns, associated in part with the growing appreciation of the role of the human agency along with structural forces, which eventually began to repopulate human geography with more credible characters and explanations.

Cultural Considerations

Culture is a very broad concept subject to changing understanding in general usage and in the discipline of geography. From restricted references to certain kinds of human activity (such as agriculture), its general meaning has been extended to incorporate a whole range of practices. These include not only people's interaction with nature in the process of material production but also their artistic and intellectual refinements associated with a cultured individual character and a collective cultural life (including its popular version). In geography, a traditional preoccupation with material culture and its landscape expression has expanded into a conception of culture as, to paraphrase Johnston, Gregory and Smith (1994:116), an active force in social reproduction, the negotiated process of discourse through which people signify their experiences to themselves and others so that all human activity is culturally encompassed. Recognition that culture is a realm of conflict between different understandings is a feature of postmodernism with its rejection of universalism and its celebration of difference. Culture is thus an arena in which different groups seek to exercise power in the interpretation of human experience (Jackson, 1989).

A cursory examination of a range of economic geography textbooks from the era of spatial science confirms that they give slight attention to culture, even in its narrow traditional meaning. Most recent texts take culture more seriously and it is worth quoting from one which is particularly sensitive to the societal context of economic activity. According to de Souza (1990:48):

> *culture – the customs and civilisation of a particular people or group – is the result of learned behaviour. People learn to eat only certain foods; dress in certain ways; speak in certain languages and dialects; assign various roles and status to women, men, children and to different races; and hold certain concepts about life and death. Culture affects demographic characteristics, influences the structure of production and consumption, promotes or handicaps economic progress, and shapes views about other countries of the world.*

While this definition does not capture all the nuances of post-modern understandings of culture, it is obviously very much broader than the behavioural characteristics associated with sub-optimal decision-making in the spatial science perspective. Furthermore, it focuses on groups of people with shared attributes, differentiating them from other groups as opposed to the emphasis on individuals whose conduct departs from some abstract, universal ideal. It does have explicit economic content, however, not only in references to structures of production and consumption but also in the link with economic progress – a matter to which this discussion will return.

The particular cultural characteristics stressed by de Souza (1990) – nationality, language, race, gender and religion – are in fact similar to those identified in the limited number of earlier texts which covered such matters. For example, the British edition of a book originally published in 1974 (Boyce and Williams, 1979) explains the cultural influences of institutional arrangements under the headings of religion, ethnic group, language and stage of technology, synthesized in a map of world cultural blocks. A more recent text with a chapter on social and cultural content (Healey and Ilbery, 1990) illustrates contemporary recognition of the continuing importance of cultural factors in the spatial structure of societies in different parts of the world. They stress that economic systems are created and continuously modified within particular social and cultural settings.

For example, in agriculture, profit maximization may be subordinated to such considerations as prestige of owning and working the land, personal or collective attachment to particular crops and technical practices, and religious attitudes to certain animals. Particularly striking features are cultural islands of distinctive activity associated with local groups (such as the Mennonites) who have perserved a distinctive identity. In the industrial realm, local attitudes of labour militancy or gender divisions of labour originating in one era have a bearing on subsequent development, as Massey (1984) has explained in an elaboration of the interdependence of social relations and spatial structure. Other well-known examples of the impact of cultural considerations include religious attitudes to land (e.g., on the part of Australian aboriginals) as something other than a commodity, periodic markets as a departure from central place orthodoxy, and spatial variations in the tastes and preferences of what economic theory usually assumes to be identical consumers.

A particularly interesting and revealing context for the introduction of cultural considerations in human geography is that of development and underdevelopment. De Souza (1990:434) suggests that, with the possible exception of the eco-political approach which stresses the use of local resources to satisfy the basic needs of the poor, all major perspectives on development emphasize the economic and play down the cultural dimension. This is reflected in the call by Agnew (1987) for geographers to bring culture back into development studies, stressing that every part of the world has its own particular and peculiar relationship to the evolution of the world economy. For example, the success of Japan and other south-east Asian countries, the slow growth of the South African economy under apartheid and the rejection of western-style development in some parts of the world cannot be explained solely in economic terms.

The quotation from de Souza (1990), above, highlights the role of culture in relation to what he terms economic progress. The conventional view, which still tends to dominate development studies, is that particular (backward, even primitive) cultural attributes impede economic development, while others promote it. This readily becomes an expression of the superiority of western culture, as highly organized in ways which facilitate the efficient pursuit of material objectives by rational means associated with production for profit. Non-western (or eastern) culture might be portrayed, in contrast, as loosely organized, fatalistic, passive and concerned with production merely for use rather than for exchange. The property of enterprise is commonly (if erroneously) viewed as a facet of western culture. Thus, the attitudes and values associated with traditional societies in the underdeveloped world must be replaced by those modern, western ideas and institutions which will generate high material living standards. The stages of growth popularized by Walt Rostow (1960) provide a well known if, now, somewhat discredited expression of this thesis. This kind of thinking generated an influential approach to development planning, incorporating some of the models of spatial science. Growth impulses were expected to follow the spatial diffusion of modernization (or westernization) from core to periphery (within nations as well as among nations), articulated by the development of an urban hierarchy connected up by effective means of transport and communication.

The problem with this approach is not only that it misunderstands a process of economic development whereby the supposed growth impulses entailed capital investment, the profit from which might return to the core rather than being reinvested in the periphery. The problem is also that it assumes the superiority and indeed the inevitability of the particular mode of production of capitalism (Rostow's book was subtitled *A Non-Communist Manifesto*), along with the social relations (class structure), institutions (like private property), cultural practices (in particular the commodification and pecuniary evaluations), and the conception of the good life), and the prioritization of individual material consumption) that go with it. This is not to say that there was, or is, an alternative path to development with greater evidence or prospects of success, whatever successful development may be. The point is that one kind of society tended to be naturalized as a universal panacea when it was in fact a geographically and historically specific form of human practice which embodied its own by no means unproblematic conception of economic (and indeed human) progress. In short, it represented a particular culture. As de Souza (1990:434) points out, it was the European cultural system of exchange and value dating back to medieval times that paved the way for the modern economy. So, as capitalism steadily extended its spatial scope, a form of cultural diffusion (some would say imperialism) came to encompass much of the globe and most of its population. With the demise of what passed for socialism in eastern Europe and the former Soviet Union, it is now extending its scope still further, as is evident in the current Chinese embrace of the allocative role of market forces. Thus, the universalization of a particular culture comes ever closer to being a fact.

However, it is important to recognize that contemporary spatial economic structures are significantly different from those which preoccupied geographers in the earlier period of location analysis. In those days, the modern capitalist economy was viewed largely as an *industrial* system, with patterns of industrial location and regional development which were assumed to be well-behaved and predictable (Martin, 1994:22). With few exceptions, the focus was on production rather than on consumption, with the service sector treated as a separate sphere of activity. It is not only that what might still be thought of as the industrial economy which has been changing, with a shift from Fordist production lines to more flexible forms of organization, with regional and local impact focusing attention on the process of restructuring (see, for example, Scott and Storper, 1986; Scott, 1988; Storper and Walker, 1989). It is also that the distinction between industrial (in the traditional sense of manufacturing) and service activity has become blurred and increasingly inconsistent with a world in which the term product is as readily applied to a financial service or leisure experience as to the material output of a factory.

Various changes in both the organization of production and the practice of consumption have helped to dissolve not only the traditional distinctions between industrial and service activities but also those between economic, political, social and cultural aspects of life. This goes further than underlining the interdependence of the economic and the political, manifest, for example, in the influence which major corporations can bring to bear on governments. As Thrift (1994) explains, the activities of economic agents, such as transnational financial elites, have to be understood as socially structured or embedded in institutions and networks within which face-to-face interaction remains important even in these days of instant electronic communication. Thus, international financial centres, such as the City of London, involve people bound together in social networks and a culture involving specific class, gender and ethnic relations as well as discursive practices assigning meaning to money. The social and cultural determinants of economic processes, never far below the surface, have become increasingly apparent as well as important.

Culture has become increasingly significant in another sense. As Martin (1994:24) puts it, the mass-consumption culture of the post-war period has exploded into a new culture of consumption, involving new industries based on the commodification of the visual, aesthetic and symbolic. Thus, culture as sets of meanings of people and places has become part of what is produced and consumed in forms such as residential environment and entertainment, at least in richer parts of the world. Heightened consumption of material products and life-styles has become more of a source of personal identity, caricatured by the axiom 'I shop, therefore I am'. The dissemination of the culture of mass consumption via the transnational communication media generates affiliation, via aspiration, of populations of large parts of the world excluded by poverty from active participation. All this suggests a rather different world from that of the geometry of spatial structure which preoccupied traditional location analysis.

The Market Economy as Culture

Earlier in this chapter, it was suggested that theories and models of spatial structure developed in the spatial science tradition, while demonstrating some capacity to elucidate the real world, are deficient to the extent that they have difficulty incorporating the actual variability of human decision-making as well as the earth's physical character. The discussion has helped to reveal a deeper but less frequently recognized problem with universal models which claim to account for locational choice and aggregate patterns in terms of market forces. This is the fact that markets themselves, as human creations, are historic and, indeed, cultural phenomena and as such subject to change.

A brief elaboration of certain features of the culture or way of life traditionally associated with a market economy under capitalism should help underline its specificity. Central to the structure formalized in neoclassical economics is a particular conception of human identity strongly associated with liberalism, that of autonomous individuals connected together by market exchange rather than by ties of kinship or community. There is nothing like society (as former British prime minister, Margaret Thatcher, famously asserted), except in the very limited sense that individual preferences may be aggregated into a social welfare function which purports to represent collective preferences with respect to bundles of goods and their distribution. Neither negotiation nor conflict is recognized. The thought processes imputed to individuals somehow enable them to undertake constant appraisals of the relative pecuniary advantage, or more generally the utility, to be derived from alternative strategies of production or consumption, and instantaneously to adjust their behaviour accordingly. The prevailing social relations comprise a simple distinction between those who have somehow come to own means of production in the form of capital, land and other property, and those who live by selling their labour power to others. This is a distinction which is implicitly assumed to be natural, unproblematic and uncontentious. Such a culture is, to say the least, peculiar.

Another important feature of culture is the role of the state. The state is expected to intervene only in the event of free market imperfections or inadvertent failures and in other supposedly exceptional circumstances (such as the generation of external effects) beyond the capacity of market-pricing control. In the relationship between the individual and the state, the emphasis is very much on protecting the individual from the power and potential tyranny of government. The institution of private property plays a crucial role in this respect as explained by Hayek (1944:78):

> *The system of private property is the most important guarantee of freedom, not only for those who own property but scarcely less for those who do not. It is only because the control of the means of production is divided among many people acting independently that nobody has complete power over us, that we as individuals can decide what to do with ourselves.*

As to distribution, an exponent of the libertarian philosophy which is closely associated with free market economics (Nozick, 1974:149–50) claims:

> *There is no* central *distribution, no person or group entitled to control all the resources, jointly deciding how they are to be doled out. What each person gets, he gets from others who give to him in exchange for something, or as a gift. In a free society, diverse persons control different resources, and new holdings arise out of the voluntary exchange and actions of persons.*

In addition to the specific features of the (strong) individual, the (weak) society and the (minimal) state, the culture of the neoclassical market economy also has a distinctive morality. Markets themselves, regulated by the impersonal forces of supply and demand, incorporate the ideal of impartiality sometimes regarded as the hallmark of modern reason. Adam Smith's hidden hand resolves matters without fear or favour: resources and rewards are allocated by the competitive forces of supply and demand rather than on the basis of kinship or compassion. Again, as Gray (1993:19) observes, the autonomy of the individual is a primary value:

> *The free market enables the individual to act upon his [sic] own goals and values, his objective and his plan of life, without subordination to any other individual or subjection to any collective decision procedure. It is from its role as* an enabling device *for the protection and enhancement of individual autonomy that the ethical justification of the market is ultimately derived.*

Clearly, these somewhat idealized accounts of the culture of market economies require modification in the light of actual practice. For example, the hand of major transnational corporations is far from hidden in the manipulation of markets to their own corporate ends. And the freedom of individuals to realize their own conceptions of the good life, central to liberalism, may be severely constrained not only by resources required to buy into life-style fashions but also by the choices actually made available within an increasingly global mass culture in which sovereign power lies ultimately with the producer rather than the consumer. In these circumstances, traditional free market economics, with its defence of individual liberty and its supposed welfare-maximizing properties, might be regarded as part of the belief system upholding and, indeed, constitutive of the dominant culture. Involving market forces in defence of particular outcomes, thus, becomes rather like calling on the will of a god: the ultimate, universal source of arbitration.

If what has been attributed here to the culture of market economies does not seem peculiar, perhaps even natural, to those living under capitalism, it is instructive to compare it with a different kind of culture. Take what might be described as the traditional or premodern African way of life. Shutte (1993) encapsulates this in the Xhosa proverb *umuntu gumuntu ngabantu*, or a person is a person through persons. This common African sentiment captures a sense of mutuality and reciprocity in a form of social life in which people achieve self-realization through personal dependence on one another rather than as autonomous individuals floating free of society. The satisfaction of a person's deep desires – to love and be loved, to understand, to create, to play and so on – depends on personal development in communion with others in a way that is supposed to avoid the imposed conformity of collectivism as well as individualism. To Shutte (1993:90):

> *The more fully I am involved in community with others the more completely I am able to realise my own deep desires to the full. The good of the community (in which I am also involved) will be my highest value, just as traditional African thought would expect. At the same time, the influence of the community on me is what enables me to achieve this form of self-transcendence and self-donation which is the fullest expression of my own self-realisation.*

Such sentiments comprise a by no means unfamiliar relational conception of personal identity and of morality, with obvious echoes of contemporary communitarianism and the current (western) notion of an ethic of care (see Tronto, 1993).

It is therefore possible to envisage a quite different culture or way of life from that of market economies in their traditional neoclassical form or as actually manifest under contemporary capitalism. Something like this certainly existed historically and quite widespread geographically: people living in and through localized communities, bound together by reciprocity and mutual interdependence rather than by impersonal market relations, controlling and using productive assets for the common good, concentrating on use values to satisfy basic needs rather than exchange values for profit, and with the pecuniary calculus (if it existed at all) subordinate to rather than dominating life. And such societies had their own economic and settlement geographies, explicable in the terms of the cultures concerned.

Even if a particular (perhaps postmodern) version of the capitalist market economy may seem to be on the way to universalization, it must be recalled that this system itself is dynamic. There are changes in the organization of production (for example, towards greater flexibility), in the distribution of outcomes (towards greater socio-economic polarization), in the role of the state (away from overt regulation and comprehensive welfare provision), and in sources of personal identity (towards the differentiating affiliations of ethnicity and gender as well as of resurgent nationalisms). Insofar as these developments will be spatially specific, impacting selectively on national, regional and local cultures, there will remain scope for differential effects on spatial structures, modifying the tendencies of the dominant universalizing, if still historically specific, culture of the international capitalist market economy.

Conclusion

Reverence for the market mechanism within the culture or ideology of capitalism has no doubt led to the exaggeration of its significance in understanding the world as well as its value in actual human conduct. The positive role of models based on neoclassical economic theory in the explanation of real-world spatial structures is constrained by the degree of abstraction from actual human behaviour as well as by physical geography. The normative interpretation of market outcomes as both efficient and equitable rests not only on dubious assumptions as to how markets actually operate but also on contentious moral principles associated with libertarianism and utilitarianism. Insofar as all these assumptions and principles may be culturally specific, the relevance of models based on market forces will be restricted despite the expanding spatial scope of capitalism.

But this does not necessarily mean that these models should be discarded. They still have a useful role in planning the location of human activity in circumstances where their assumptions are likely to be fulfilled and where the optimization sought is related to broader societal objectives. And this is not confined to single industrial-plant location problems. For example, patterns of the utilization of health-care facilities can be analysed by distance decay functions and a version of Weber's model of industrial location can be used to find the optimal location for a hospital in relation to population need (Smith, 1977:310–19). If compassion, in terms of care for the weak and disadvantaged, was ever to displace profit-seeking as society's prime motive force, then these kinds of applications would help to reveal how to care at least cost. In the meantime, a positive universal theory of spatial economic structure remains as elusive as ever.

References

Agnew, J.A. (1987) Bringing culture back in: overcoming the economic–cultural split in development studies, *Journal of Geography*, 86:276–81.

Alonso, W. (1964) *Location and Land Use: Toward a General Theory of Land Rent* (Cambridge, MA: Harvard University Press).

Beavon, K.S.O. (1977) *Central Place Theory: A Re-interpretation* (London: Longman).

Berry, B.J.L. (1967) *Geography of Market Centers and Retail Distribution* (Englewood Cliffs, NJ: Prentice-Hall).

Blaikie, P.M. (1971) Spatial organization of agriculture in some northern Indian villages, *Transactions of the Institute of British Geographers*, 52:1–40; 53:15–30.

Boyce, R.R. and Williams, A.F. (1979) *The Bases of Economic Geography* (London: Holt, Rinehart & Winston).

Bunge, W. (1971) *Fitzgerald: The Geography of a Revolution* (Cambridge, MA: Schenkman Publishing Company).

Christaller, W. (1966) *Central Places in Southern Germany*, translated by C.W. Baskin (Englewood Cliffs, NJ: Prentice-Hall).

de Souza, A.R. (1990) *A Geography of World Economy* (Columbus, Ohio; Merrill Publishing Company).

Dicken, P. (1977) A note on location theory and the large business enterprise, *Area*, 9:138–43.

Gray, J. (1993) *The Moral Foundations of Market Institutions* (London: Institute of Economic Affairs).

Greenhut, M.L. (1956) *Plant Location in Theory and Practice* (Chapel Hill: University of North Carolina Press).

Hall, P. (ed.) (1966) *Von Thunen's Isolated State*, translated by C.M. Wartenberg (Oxford: Pergamon).

Harvey, D. (1973) *Social Justice and the City* (London: Edward Arnold/ Blackwell re-issue).

Hayek, F. (1944) *The Road to Serfdom* (London: Routledge & Kegan Paul).

Healey, M.J. and Ilbery, B.W. (1990) *Location and Change: Perspectives on Economic Geography* (Oxford: Oxford University Press).

Horvath, R.J. (1969) Von Thunen's isolated state and the area round Addis Ababa, Ethiopia, *Annals of the Association of American Geographers*, 59:308–23.

Isard, W. (1956) *Location and Space-Economy* (Cambridge, MA: MIT Press).

Jackson, P. (1989) *Maps of Meaning: An Introduction to Cultural Geography* (London: Unwin Hyman).

Johnston, R.J. Gregory, D. and Smith, D.M. (eds) (1994) *The Dictionary of Human Geography*, 3rd edition (Oxford: Blackwell Publishers).

Losch, A. (1954) *The Economics of Location* (New Haven, CN: Yale University Press).

Martin, R. (1994) Economic theory and human geography, pp. 21–53 in D. Gregory, R. Martin and G. Smith (eds) *Human Geography: Society, Space and Social Science* (London: Macmillan).

Marx, K. (1967) *Capital*, 3 vols (New York: International Publishers Edition).

Massey, D. (1984) *Spatial Divisions of Labour: Social Structures and the Geography of Production* (London: Macmillan).

McGee, T.G. (1967) *The Southeast Asian City* (London: Bell).

Nozick, R. (1974) *Anarchy, State, and Utopia* (New York: Basic Books).

Peet, R. (1969) The spatial expansion of commercial agriculture in the nineteenth century: a von Thunen explanation, *Economic Geography*, 45:283–301.

Polanyi, K. (1944) *The Great Transformation: The Political and Economic Origins of our Time* (Boston: Beacon Press).

Polanyi, K. (1968) *Primitive, Archaic and Modern Economies: Essays of Karl Polanyi* (edited by G. Dalton) (Boston: Doubleday).

Pred, A. (1967) *Behaviour and Location: Foundations for a Geographic and Dynamic Location Theory* (Lund: Lund Studies in Geography, Series B, 27 Part 1).

Rawstron, E.M. (1958) Three principles of industrial location, *Transactions of the Institute of British Geographers*, 25:132–42.

Rostow, W.W. (1960) *The Stages of Economic Growth: A Non-Communist Manifesto* (Cambridge, MA: Harvard University Press).

Scott, A.J. (1988) *New Industrial Spaces* (London: Pion).

Scott, A.J. and Storper, M. (eds) (1986) *Production, Work, Territory* (Boston: Allen & Unwin).

Shutte, A. (1993) *Philosophy for Africa* (Cape Town: University of Cape Town Press).

Smith, D.M. (1966) A theoretical framework for geographical studies of industrial location, *Economic Geography*, 42:95–113.

Smith, D.M. (1977) *Human Geography: A Welfare Approach* (London: Edward Arnold).

Smith, D.M. (1981) *Industrial Location: An Economic Geographical Analysis*, 2nd edition (New York: John Wiley).

Smith, D.M. (1989) *Urban Inequality under Socialism: Case Studies from Eastern Europe and the Soviet Union* (Cambridge: Cambridge University Press).

Storper, M. and Walker, R.A. (1989) *The Capitalist Imperative: Territory, Technology and Industrial Growth* (Oxford: Blackwell).

Thrift, N. (1994) On the social and cultural determinants of international financial centres: the case of the City of London, pp. 327–55 in

S. Corbridge, R. Martin and N. Thrift (eds) *Money, Power and Space* (Oxford: Blackwell Publishers).

Tronto, J.C. (1993) *Moral Boundaries: A Political Argument for an Ethic of Care* (London: Routledge).

Weber, A. (1929) *Alfred Weber's Theory of the Location of Industries* (translated by C.J. Friedrich) (Chicago: Chicago University Press: reprinted 1971, New York: Russell and Russell).

8 Mode of Production and Spatial Differentiation

In the early stages of the development of geography the importance of the natural environment and its variation over the earth's surface encouraged the notion of a causal interrelationship found among phenomena occupying a particular region of the world. Although the principle of environmental determinism which this conception of causal interrelationship fostered came to be discredited, the areal differentiation of the world as a result of both environmental and cultural factors still carried some undertone of causal relationship. Indeed, some earlier geography textbooks defined culture as the sum total of a people's adjustment to their environment.

With the rise of capitalism in the late eighteenth century in Europe and its global explosion through colonialism in the nineteenth century, this simple perception of the forces shaping the character of regions in different parts of the world no longer became tenable. The potency of the self-regulating capitalist free market to integrate the economies of territories in diverse parts of the world is perhaps the most distinguishing feature of the twentieth century. This dominant mode of production thus provides a point of departure for a new understanding of the areal or spatial differentiation of the world.

Defining the Mode of Production

The concept of a mode of production was introduced into the analysis of societal processes by Karl Marx:

> *The totality of the relationships of production constitutes the economic structure of society, the real foundation, on which arises a legal and*

This chapter is based on a contribution by Milton Santos.

political superstructure and to which correspond definite forms of social consciousness. The mode of production of material life conditions the general process of social, political and intellectual life. It is not the consciousness of men that determines existence, but their social existence that determines their consciousness . . . Changes in the economic foundation sooner or later lead to the transformation of the whole immense superstructure. In studying such transformations it is always necessary to distinguish between the material transformation of the economic conditions of production and the legal, political, religious, artistic or philosophic – in short, ideological forms in which men become conscious of conflict [in the economic basis] and fight it out. (Karl Marx, 1970)

Although this definition gives considerable weight to the economic element, it is usually stressed that this element is not regarded as the *only* determining one. Whilst the economic situation is the basis, the other elements of the superstructure, notably the political forms of class relations, the juridical, philosophical, religious and intellectual norms, all exercise their influence upon the course of the historical struggles within a particular society and in many cases preponderate in determining their form.

There is general agreement that, in any society, social life, taken as a whole, is characterized by its mode of production especially the element of the ceaseless renewal of productive forms and the relations of production. Every society thus exhibits a particular blend of elements, a particular mix of activities and a particular patterning of social relationships. All of these, when taken together and insofar as they contribute to the production and reproduction of real life, constitute the mode of production. Each mode of production, therefore, represents a stage in historical evolution and is manifested by the appearance of new instruments of labour and new social practices. Thus, Marx identifies some four or five modes of production – notably the primitive communalist, the ancient, the Asiatic, the feudal and the capitalist. How to produce material goods and how to produce space are synonymous in the context of each mode of production and it is this that gives rise to differences in the structure of space. Thus, space can be defined as an ensemble that is indistinguishable from the set of material objects produced through the instrumentality of labour and the system of action or social practices within a given mode of production. In other words, modes of production and geographic space evolve together, transformed by a common logic.

Mode of production must, however, be distinguished from the prevailing social formation. While a mode of production specifies the social relations of a definite system of production and distribution, the social formation refers to a distinct and determinate form of economic class relations, its conditions of existence and the forms in which those conditions are secured. This distinction is worth stressing because in the life of a given society no one historical epoch is the exclusive domain of one mode of production, even though a particular mode may be clearly dominant. Society, and therefore the corresponding social formation, always contains within itself potentially conflicting modes of production. Or, as Lukàcs (1970:45) puts it:

> *A particular mode of production does not develop and play an historic role only when the mode superseded by it has already everywhere completed the social transformations appropriate to it. The modes of production and the corresponding social forms and class stratifications which succeed and supersede one another tend to appear in history much more as intersecting and opposing forces.*

At the dawn of history, modes of production and social formations may be said to be one and the same. The manner in which modes of production acted upon space was practically unmediated. However, with the expansion of the capitalist mode of production from the eighteenth century onwards, the possibility of vast intercontinental and transoceanic exchanges appeared involving plants, animals, peoples and their ways of life and know-how. Modes of production that had been distinct up to then began to converge and all social formations involved in this unifying movement became, from this point of view, part of a common historical experience. The spatial progress of capitalism thus denotes an enlarging and deepening of its logic as it embraced each time a greater and greater number of societies and territories. In this sense, the capitalist mode of production has been singular in human history.

But the way in which the capitalist mode of production has acted upon different territories has been mediated by the social formations constituted under previous modes and sustained under the aegis of their nation states. With increasing globalization, however, the extent to which the latter can continue to sustain non-capitalist-oriented social formations has been called to question. Those who believe that the nation state is dead and that national borders are meaningless would answer that the position of the state is no longer tenable. In point of fact, the mediation of the state, of national civil society and of their territorial configuration are fundamental realities. They all have a critical role to play in explaining the differences in the impact of the capitalist mode of production on various regions and countries. Nonetheless, it is indisputable that in no other period of history has a single mode of production been so widespread and been able to impose its presence in such a profound and efficient manner on societies throughout the world. Capitalism has thus become a global mode of production and a major explanatory factor in the current geographical reality.

Globalization of the Capitalist Mode of Production

Although the globalization of the capitalist mode of production is incomplete and is unfolding in an uneven manner, it nevertheless marks the advent of a new era in human history. No other historical epoch has manifested itself in such a dominant fashion throughout the entire world or succeeded in stopping the progress of other modes of production in order to assert its presence. Some of the new vectors of development show a rapid and extensive dissemination (as is currently the case with respect to information technology) whereas other vectors propagate themselves more slowly and selectively (as is the case with morals and ethics appropriate to the current situation). The result is that the globalization process reinforces distortions and inequalities whilst creating new forms of dependence and shortages. This is perhaps most obvious in the internationalization of financial capitalism.

The whole world has now become structured as a single, united, spatio-temporal whole (Bach, 1980). This is why the theses of autonomous geographical regions traditionally elaborated in the discipline are no longer pertinent. Dicken (1992:95) indicates this with regard to the theory of trade and localization, and Michalet (1993:95) demonstrates it with regard to the traditional model of the international economy. The former process of internationalization has certainly attained a new level of articulation now that capitalism has really reached the world scale (Ianni, 1992:36–9) and ecology has come to be viewed as a global problem. The situation, in fact, recalls the assertion of Paz (1990:20) that the new is not exactly the modern, except that it carries a doubly explosive charge, namely that it is the negation of the past and the affirmation of something different.

A new combination of factors different from those which had governed previous epochs is playing a fundamental role in the new global situation. We are witnessing a period where spectacular progress has occurred thanks to the alliance between science and technology. This alliance has the power to transform for good the material bases of our life on a scale which was unimaginable just half a century ago. It intrudes forcefully into all aspects of life and is a situation that one meets in all parts of the world. And it is this combination of science, technology and information that constitutes the new key variable in respect of not only a new temporal system but also a new organization of space. The new capitalism is not only global; it is also a scientific and technical mode of production.

In consequence, the transformation of society and geographic space that is currently underway in all parts of the world can be examined and understood in terms of three constitutive factors, each of which is both a cause and an effect and all of which together operate jointly on a world scale. These three factors are:

1 technological uniqueness and its universality;
2 global convergence of moments as a result of the universal simultaneity of perception;
3 unity of the driving force of social life with the universality of surplus value.

Technological uniqueness refers to the fact that current technology forms a system on a world scale with each location accommodating interdependent pieces of this veritable 'mecano universel', to borrow Moles's expression (1971:82). All parts of the world take part in this globalization of technology, even if this occurs in different places with different degrees of presence and complexity. In the past, technological systems were local or regional in scope. Indeed, at the beginning of history, one could say that there were as many technological systems as there were places. When they displayed similar characteristics, there was no contemporaneity between them nor functional interdependence. The history of humanity is thus the history of the progressive diminution in the number of existing technological systems. The movement towards integration, which has been accelerated under capitalism, has reached its apogee with the dominance everywhere of a single technological system that undergirds the globalization process.

With respect to the convergence of moments, human history emphasizes the fact of events being based on divergent moments, with history itself being the sum of scattered, disparate and incoherent evolutions. On the other hand, history in recent times is a record of converging moments in which events in one place are simultaneously communicated everywhere, thanks to the near unification of time and space on a global scale. This instantaneity of globalized information brings places closer together, renders possible an immediate awareness of simultaneous events and creates between places and events a convergent relationship on a world scale. Thus, on a daily basis, events everywhere have become interdependent and are part of a single global system of relationships. More than this, the technological capacity that enables the earth to be photographed via satellites also permits us an empirical vision of the totality of objects located on the earth's surface. As the photographs follow one another at regular intervals, a snapshot view of the very evolution of events throughout the world is acquired. Photographic simultaneity is a new and revolutionary development in the acquisition of knowledge of the real world and is bound to have a tremendous impact and lead to major paradigmatic modifications in the framework of the social sciences. Empirical knowledge and the understanding of its interdependent significance is a determining factor in historical realization (Santos, 1984). How such realizations will be affected by the simultaneity of information provision is a factor the implications of which are only gradually being appreciated. Already, this factor is encouraging hegemonic actors in the economic, social and political life of different countries to choose the most appropriate place and time for their actions, thereby condemning the situations of other actors to marginal significance.

The unity of the new driving force of social life results from the establishment of a global market based on global exchange and a law of universal values (dos Santos, 1993:3). Surplus value is thus being created and accumulated on a world scale through the convergent activities of large organizations, both private and public, national and transnational. The creation of this surplus value, through a production process that has truly become global and is integrated by a global financial system, constitutes the driving force of economic and social life throughout the entire world. The principal agencies of this process are the transnational enterprises and the financial institutions. In a competitive situation, the search for the highest possible profit has no limitation save for the simple ability to create and utilize productive and organizational innovations. Consequently, these enterprises seek to outdo one another by engaging in activities on a global scale with a view to creating the largest possible surplus value. The supreme irony, however, is that this urge to create enormous surplus value cannot be truly measured but it has become the principal lever, indeed the single driving force, of the most characteristic development of the globalized economy.

Technological uniqueness, the convergence of moments, the unity of the driving force of social life – these are the present realities the interdependence of which promotes the current process of globalization. But the technological phenomenon, which is at the root of these great transformations, has been insufficiently used as a point of departure for geographical explanation of global spatial differentiation.

The Mechanization of Spatial Structures

Just as Bucheenschutz (1987) regrets that archaeologists rarely deal with technological problems or with the technical processes of material vestiges left behind by past human activities, Sigaud (1981) poses the question as to why geographers systematically avoid the study of technologies which are at the centre of human–environment relations. Although themes on technology have appeared in the works of authors such as Max Sorre (1950), P. Gourou (1973), P. George (1974), Ph. Wagner (1960), A. Fel (1978), J. E. Sanchez (1991), and many other geographers, it is rare for such themes to benefit from use as explanatory variables in the elaboration of a theoretical framework. Nonetheless, the very idea of the human–environment relation is inseparable from the notion of a technological system. Thus, the history of the development of any area on the earth's surface can be roughly divided into three stages, namely the natural, the technical and the informational.

During the stage of the natural environment, human beings chose those resources of nature that were considered basic to their survival. They gave differential value to such resources according to their location and culture. These natural resources constituted the material base of the existence of the group. However, since the end of the eighteenth century with the industrial revolution, there has emerged an increasing mechanization of territory. Space has become more compact and ordered with the presence of machines and technology. And as Siegfried (1955) observed, this was the point at which the technical environment began to emerge. This new environment superimposed itself on the natural environment in many places and has sought to replace the latter. Today we are already facing a situation in which this technical environment is also being replaced in different parts of the world by a new landscape arising from recent advances in informational technology.

In the same way that they contributed to the emergence of new productive processes and the creation of new species of plants and animals, science and technology in league with modern informational processes are at the very base of the production, utilization and functioning of current spatial structures. Thus, a situation has now emerged in many parts of the world whereby the natural environment is being forced, often brutally, to retreat. According to Gellner (1989), 'nature has ceased to be a significant part of our environment'. The idea of an artificial environment which was first advanced by Labriola in 1896 but only published in a later collection of his works (Labriola, 1947) is now taken generally for granted. Technology, in creating an increasingly dense space, has projected itself into the very centre of existence of a good part of humanity (Rotenstreich, 1985:71). It is thus possible to speak of the manner in which the human landscape has been transformed by science and technology. With regard to information technology, it is necessary to note not only how information is present in the objects that make up space but also how it determines the acts carried out by or on these objects. In this sense, information is the fundamental factor of social processes and spatial structures. Territorial units are simply means of facilitating its dissemination. Such spatial structures that are defined in an informational context must be recognized as responding above all to the interests of the hegemonic actors in the global economy, culture and politics and are, therefore, fully incorporated into the new global currents. In this way, the informational environment can be said to be the geographical face of globalization.

At the same time as fixed assets (such as roads, ports, silos and plowed farmland) and constant capital (such as machinery, vehicles, specialized seeds, fertilizers and pesticides) are increasing in importance, the need for their mobility also grows. The amount and importance of the flows of goods, people and money have grown exponentially in recent times, offering a special perspective on changing spatial relationships. Pre-existing equilibria have been ruptured and new ones are being established especially with regard to the quantity and quality of the population and their employment, the capital utilized in the production process, and the forms of organization and social relations. In geographical terms, the consequences of this development are that while the space reserved for direct production processes is becoming increasingly limited, the space for other levels of production such as circulation, distribution and consumption continues to grow. The process of specialization, in creating distinct areas where the production of certain commodities are specially advantageous, increases the need for exchange, which then tends to take place over much larger spaces.

The organizational and technological possibilities of long-distance transfer of products and orders result in these productive specializations operating jointly on a world scale. As more exchange value is produced each time, specialization is soon followed by the need for more circulation. The role of the latter in the transformation and production of space thus becomes fundamental. One of its consequences is precisely the deepening of productive specialization which strives in turn to give rise to greater circulation. This circle of relationships depends on the fluidity of networks and the flexibility of regulations.

The dynamic aspects of globalized spaces presuppose a permanent adaptation of forms and norms. Geographical forms, that is the objects required to optimize production, become operative only on the basis of juridical, financial and technical norms adapted to the requirements of the market. These norms are created at different geographical and political levels. Given global competitiveness, global norms, induced by both supranational organization and the market, can be expected to configure public norms at national and lower levels.

The new sub-spaces are more or less able to make productive output profitable. Each level has its own production logic, inducing specific forms of actions from economic and social agents. Hegemonic actions are established and realized by means of hegemonic structures. In this context, space and actions are conjoined through information to give rise to what is referred to as spatial productivity, a concept that emphasizes the ensemble of activities in a particular place. Without minimizing the importance of cultural conditions, such a space is an artificial creation expressing the level of the informational technological development of a particular society. Spatial structures the characteristics of which are thus determined are, at present, unevenly distributed by continent, country and regions of countries. In some cases as in western Europe, the greater part of the country is already occupied by such an informational landscape. In other places, such as Brazil, such spaces though covering significant areas are far from encompassing the entire territory. In many other developing countries, they are limited to a few places or spots, more often than not, in the capital city.

Implications for Patterns of Regional Configurations

In the present global circumstances, therefore, spatial structures do not configure as regions did in the past, that is as a continuous areal expanse. Rather, they often appear as a set of discrete locations which are linked together, thereby constituting a space of regulatory flows. Consequently, the possibility of segmenting and partitioning spaces leads to the recognition of two new aspects of spatial relations designated as horizontalities and verticalities. The former refers to relations on the basis of the existence of a continuous expanse of areas as in the traditional conception of a region; the other relates to relations linking together discrete locations, separated from each other, but having to function as integral segments of society and economy. Horizontalities, for instance, define spatial relations emanating from agricultural production undertaken in the extensive countryside around a city. The city, on the other hand, which, among other things, serves as the centre for regulating various aspects of production including that of agricultural labour, constitutes part of societal verticalities and is linked with other urban centres by means of circulation, exchange and various regulatory processes. The reality of the network structure so constituted is fundamental for understanding the current dynamics of a territory.

According to Bakis (1993:4), network structure can be appreciated in at least one of three ways: first, as the polarization of centres of attraction and diffusion, as is the case for urban networks as described above; second, as abstract forms necessary to facilitate understanding, as is the case with the imaginary lines of latitudes and longitudes on the map of the world; and third, as the concrete projection of lines of relationships and liaisons such as with stream networks, utility networks or even Hertzian telecommunication networks where there are no visible lines and the physical structure is limited to nodes.

But what is a network? Two types of definition can be proffered. One emphasizes only the material reality whilst the other takes the social element also into account. With regard to the former, Curien (1988:212) defines network as all infrastructure permitting the transportation of matter, energy or information which fall within a given territory where it is characterized by the topology of its points of access or its terminal points, its transmission segments and its nodes of bifurcation or communication. But a network can also be social and political, comprising of individuals and the messages and values that criss-cross from their interactions across space. Indeed, without this second type of network, and in spite of the concreteness with which the former imposes on our consciousness, the concept would be of limited value except as an abstraction. This is probably why Dollfus (1971:59) suggests that the use of the term network be limited to systems created by human beings whilst leaving to natural systems the term circuit.

Networks thus refer to meshed space or to the deliberate construction of such space as a living environment that responds to productive incentives in all its material and immaterial forms. Or, as Durand, Levy and Retaille (1992:21) would put it, the stakes (through networks) are not to occupy areas but to activate nodes (points) and their links (lines), or to create new ones. Networks are transmitters of information in the form of products, merchandise, ideas, money and personal messages. Their basic function is to ensure linkages and interactions among discrete centres. This is their strength which is of considerable importance given the sheer variety of ways in which communications can be conducted and the technological possibilities that are now available.

Networks are both global and local. They are global because they envelope the whole of the settled portions of the world (the oecumene) and constitute the principal instrument for their unification. But they are also local since each place, through its technical and informational structure, takes in a small or a large part of the global networks. In any given location, therefore, network connections benefit labour and productive capital and determine their nature. National networks are thus critical for the international division of labour since they tend to structure the content of cooperation (Silveira, 1994:75–6). Nonetheless, owing to technological advances and the current pattern of economic relations, networks increasingly tend to be global in terms of production, commerce, transportation and information. Kayser and Brun (1993) show how, in many countries today, rural spaces, even in seemingly marginal areas, are completely integrated into the global socio-economic system. But the most complete and efficient form of the global network is that constituted by activities on the financial market, representing as it does the dematerialization of money and its instantaneous and generalized use (Goldfinger, 1986).

In a sense, networks are incomprehensible if we look at them only from the perspective of their local and regional manifestations. Yet, the latter are indispensable for any understanding of how networks function spatially on a world scale. Thus, as Braudel (1979:57) observed, for any movement of interest, we can discover movements at the global level all of which are contemporaneous, coexisting, mixed together, adding their movements or passage to the oscillations of the whole system. Space is the arena of these movements with their different contents, intensities and orientations. The entirety of space is formed by all of these movements and by all existing objects including people, enterprises, organizations and actions. The use of space is, however, selective. What one often and inaccurately calls the space of movement (Castells, 1989:348) is in fact nothing other than a sub-system of total space, a sub-system formed by objects and actions deliberately endowed with a higher level of technicality, intentionality and rationality. Both objects and actions in this case have denser information content than other sub-systems of the same space.

Only hegemonic actors such as transnational corporations use all networks and territories. This is why national territories become for them no more than a nodal point in the international economy. In consequence, the development literature is already replete with such references as space without borders or capitalism without borders (Ellul, 1967; Masuda, 1982). These are indicative of current tendencies for transnational enterprises to appear to bypass states in the articulation of production processes (Petrella, 1989). It would, of course, be erroneous to think that governments no longer play a vital role in these processes. The emergence of transnational organizations could, in another sense, be said to do no more than reinforce the role of the state in development, making its regulatory role more indispensable than before (Boismenu, 1993:13; Giddens, 1984; Groupe de Lisbonne, 1995; Silver, 1992). It is in this same postmodernist vein that the negation of the idea of regions finds an echo. Although it is a fact that the character of regional entities has changed in fundamental ways, no one can declare that the phenomena has ceased to exist.

Regions still exist but they are based on organizational arrangements deriving their impulses from distant origins. They are no longer the expression of the absolute territoriality of specific groups whose mode of occupance defines the identity, exclusivity and borders of the regions. The difference between areas is no longer due to this direct relationship between groups of people and their environment. The speed of global transformations since the end of the Second World War has made it such that the regional configuration of the past has collapsed. Indeed, some people talk of the death of the regions because of the difficulty in grasping the nature of the new regional constructions. The expansion of hegemonic capitalism throughout the world may give the appearance of having eliminated regional differentiations and even prevent the continuation of thinking in terms of regions. But there can be no doubt that regions and their networks of places remain the foundation of global relationships. They are functional and serve as spheres of mediating global forces. With the rapidly developing international division of labour and the exponential growth in the volume and rate of exchange, we are witnessing increasingly frequent changes in the form and content of regions. Because regions have always been conceived of as a subspace of stable construction, we have not adjusted to the idea of rapidly changing regional entities. We would need to appreciate that what makes a region is its functional coherence rather than the stability of its form. No part of the world today can escape the joint process of globalization and fragmentation, that is, of integration and regionalization. Regions continue to exist but at a level of complexity never before witnessed. The universal interdependence of place is the new reality of territory.

The globalization process thus presents different amounts of technological, informational and communication content to different parts of the world. Places are defined on the basis of density of these three attributes. These attributes in turn tend to interpenetrate one another giving rise to fused characteristics. Technological density relates to the varying degree of sophistication of technological facilities available to promote human productive activities. At one extreme, one can conceptualize a nature reserve, a wild ecology never touched by human beings; at the other extreme would be a humanized sub-space with sophisticated gadgets. The latter fairly describes the situation in business centres of virtually all large cities in developed countries where interactions with the rest of the whole world now tend to be almost instantaneous. Informational density derives in part from the technological one. Technical objects with high informational potential may, nonetheless, remain inactive while waiting for someone to activate them. Information is only possible with action. Informational density thus indicates the degree of external relations of places and the realization of the propensity to engage in such relationships. Communication density results from what Berger (1964:173) has called the human nature of the time of action. Particular events may be seen as involving intersubjective relations or as entailing trans-individual exercise. Such shared occurrences of daily life often represent the conflictual time of coexistence. They occur in geographic space and are indicative of the obligatory interdependence necessitated by face-to-face interactions. Thus, unlike both technological and informational densities, communication density relates to a social environment and has a more geographical flavour to it.

Spatial Differentiation and the Typology of Sub-Spaces

Clearly then, globalization emphasizes a contemporary world economy characterized by a single but highly dynamic division of labour and operating over a complex mosaic of cultural systems and environments. None of these cultural systems may be understood simply in their own, purely local terms or considered to be autonomous and completely self-contained. The geography of any particular part of the world must take full account of its relationships with the operation and organization of the world economy.

Nonetheless, the temptation to see the spatial differentiation of the world as drawn around such a single centre is increasingly shown to be flawed. In the period up to the early 1980s it was possible to conceive of a global geography shaped around a core and periphery of nations and countries. The core comprises most countries of western Europe and North America where traditionally a complex variety of economic activities such as mass-market industries, commerce of both local and international range controlled by an indigenous bourgeoisie and a relatively progressive agricultural middle-class have had the effect of mobilizing accumulated capital and promoting a high level of rural and urban productivity. Under competitive capitalism, however, this situation soon leads to the emergence of a few monopolistic enterprises whose activities force other firms to seek for their profit elsewhere. The result is a strong urge towards colonial adventures with a view for capital to seek new markets in less developed areas of the world, otherwise conceived of as the periphery. This process of capital transfer from the core regions, however, distorts the form of development that takes place at the periphery. Through its external control mechanism, it constrains enterprises at the periphery to engage in specialized but complementary functions, notably of primary produce exportation. The continued monopolization of trade, credit provision and employment opportunities in the periphery by institutions of the core region under colonial and neo-colonial conditions prevents equalization of wealth between the core and the periphery. This relationship is further strengthened by state allocations of resources and systematic ethnic, religious and linguistic discrimination, which combine to produce permanent backwardness in the periphery (Wallerstein, 1979).

Developments since the 1980s have, however, forced a rethinking of the pattern of spatial differentiation as a result of the global dominance of the capitalist mode of production away from the dichotomous or dualistic models which recognize core–periphery or even north–south division of the world. Different explanations have been provided for the more variegated pattern of spatial differentiation that has emerged in the world in the wake of the globalization process. One explanation, for instance, is based on the inherent antagonistic relation between capital and labour in a capitalist economy (Peet, 1986). It notes that this relationship has developed unevenly in time and space. Times of intense conflict give way to times of lesser conflict; regions of high conflict lie next to regions of lower conflict. Particularly in the capitalist centre of the world or the core region of the dualistic model, the late 1960s and early 1970s saw high levels of capital–labour conflict, increasing unionization, higher real wages, and greater constraints on entrepreneurial freedom of action. The situation continued to the early 1980s with the notable exception of Germany and Japan which had remarkably low levels of industrial conflict. The newly industrializing countries (NICs) of southern Europe had lower wage levels but higher levels of conflict. The NICs of Latin America had moderate wages and apparently low conflict levels, kept low by state repression in most cases. Africa had low to moderate wage levels in countries with industrial traditions but very low wage levels elsewhere, while the level of conflict appeared to vary tremendously within the continent. South Asia had the lowest wage rates in the world but with a tendency for quite high levels of industrial conflict. Finally, East and South-East Asian countries were remarkable for low (but increasing) wage rates and industrial peace.

Not unexpectedly, therefore, transnational corporations searching the globe for advantages in a competitive struggle to expand, or looking for low-cost sources of products, thus, found a mosaic of differing relations with labour. East and South-East Asia in particular offered conditions especially suited to labour-intensive manufacturing. Initially in the 1960s, most of these countries were exclusively exporters of primary commodities. Hong Kong, Korea and Singapore together exported the same value of manufactures as a small European country like Austria. By the 1980s, each had manufactured exports equivalent to a medium-sized European country, with the production of sophisticated products (machinery) increasing rapidly and primary exports decreasing in importance. Thus, according to Peet (1986:93), geographical variations in capital–labour relations and the exigency of inter-capitalist competition have structured a decline in industrial rates of growth in western Europe and North America of sufficient magnitude to be termed industrial devolution and an increase in industrial growth rates in peripheral regions where a precarious industrial revolution is occurring. Time, of course, has shown that this revolution is not as precarious as was thought at the time. But, more than this, the rise of East and South-East Asia as regions having characteristics of the traditional core regions underscores the weakness of the duality concept in examining the pattern of global spatial differentiation.

Another explanation for the more complex pattern of spatial differentiation following from the globalization process is provided by Storper and Walker (1989). They see the development as in the very nature of capitalist development. According to them, the systematic expansion of capitalist territory underlies startling shifts in global financial, mercantile and industrial centres as innovative industrial ensembles leapfrog established growth centres to emerge as prominent new centres of capitalist development. This process of growth and spatial differentiation is generated from an accumulation of above normal profits within new industrial territories and new industrial complexes. To quote them (1989:19):

> *industries are able to create a productive capacity that did not exist before, often without very much regard to the previous conditions of the place in which they are situated. To a large degree, they provide their own impulses toward development, endogenously, in place, that is fundamental to the nature of industrialization as a world-historical – and world-geographical – process.*

Storper and Walker, indeed, posit an alternative to the existing model of industrial location and regional development to cope with the expansion, instability and differentiation that characterize the inconstant geography of capitalist industrialization. This model argues that, contrary to neoclassical theory, the capitalist economy is fundamentally a disequilibrium system, driven to grow and to change through its own internal rules of surplus generation, by investment to expand capital, by fierce competition, and by technological change to extract more surplus value from human labour. This almost inexorable drive to compete and accumulate is what constantly disrupts existing conditions of production and makes possible the rise of new centres of growth. This model of industrial and territorial development thus rests on three basic and related forces: capital investment involved in the restless cycle of investment–production–realization, strong competition driving capitalists to consistently revolutionize production, and the dynamism of technological change.

These three forces define the ways in which industries produce regions as sites of economic activity. Innovative firms and industrial sectors capitalize upon the opening of locational windows of opportunities to generate their own conditions of growth. The above normal profits that often result from the activities of such enterprises help to attract factors of production to such locations or cause factor supplies to come into being where they did not exist before. Thus, capitalism is shown to be capable of escaping from the past to create new localizations of industry. The geographic cluster of new activities and the development of vertically-disintegrated production complexes of linked firms form the basis for the new territorial industrialization. These agglomerated territorial complexes create their own locational specifications. Networks of intermediate buyers and sellers, users, and suppliers develop. Labour supplies are created and new rhythms and imperatives of work become habituated as place-specific employment relations. Technological knowledge is also localized and embedded in organizational structures that are specific to the complex generating a technological milieu that forms the basis for regional-specific technological trajectories.

Storper and Walker attempted to incorporate in their territorial model the distributional relations between classes and the relations between investment and consumption. They argue that the objective territorial differentiation is reflective of more than the shifting balance of class forces. The development of territorial capitalisms based upon innovative territorial production complexes affects the very constitution of classes themselves. It impacts directly on the territorial shape of the labour markets and political organization whilst inducing a restructuring of class relations and the salience of class struggle. Consequently, as the development of new territorial capitalisms based on novel technologies, production organization, work practices or employment relations puts pressure upon older territories, a reverse diffusion of the social and political conditions of growth is liable to occur as what began as a specific politics of place comes to be a politics of wide areas of a nation or of the capitalist world in general.

Conclusion

The concept of a mode of production has been important in helping to situate patterns of global development that are divergent in spite of the ostensible operation of an internationally dominant capitalist economy. It has been shown that mode of production, social formation and space are interdependent categories. All the processes which together make up the mode of production – namely production, circulation, distribution and consumption – are historically and spatially determined in a movement of the whole through the social formation.

The social formation represents a technical-productive structure expressed geographically by a specific distribution of activities of production. It contains a complex of different technical and organizational forms of the productive process which correspond to various existing relations of productions. As a new and dominant mode of production, capitalism seeks to express itself in different parts of the world through a struggle and interaction between it and the old pre-capitalist modes that define the historical, cultural and social heritage of particular peoples. That it has not completely succeeded in replacing these pre-capitalist modes in many places is one of the most important factors accounting for the range of spatial differentiation noticeable in the world today. However, the fact that the pattern is more a mosaic than a series of systematic gradations puts a different complexion on the global trajectory of capitalist development.

The emerging new territorial industrial complexes of East and South-East Asia are already affecting the configuration of development in that part of the world. Their growth and success underscores the fact that the world is going through a new and qualitatively different phase of capitalist development, associated with, among other things, the advent of new information and communication technologies, more differentiated and individualized patterns of consumption, the renegotiation of boundaries between transnational enterprises and nation states, and the redefinition of relations between public and private spheres of economic activity. All of these highlight a world that is increasingly becoming a system of linked nodal structures with variegated interstices rather than an orderly world of transitional regional characteristics. It is a world in which geography is having to rediscover its age-old interest in areal differences, less conditioned by the given environmental resources and more by the level of articulation within the social formation of technological, information and communication processes that have become the life-wires of a globalizing world.

References

Bach, R. (1980) On the holism of a world-system perspective, pp. 289–310 in T.K. Hopkins and I. Wallerstein (eds) *Processes of the World-System* (Beverly Hills: Sage).

Bakis, H. (1993) *Les reseaux et leurs enjeux sociaux* (Paris: Presses Universitaires de France).

Berger, G. (1964) *Phenomenologie du temps et prospective* (Paris: Presses Universitaires de France).

Boismenu, G. (1993) Polycentrisme, dissymetrique et strategie defensive dans la transformation du rapport salarial, *Cahier du Gemdev*, 20:23–39.

Braudel, F. (1979) *Le Temps du Mode, tome III, Civilisation Materielle, Economie et Capitalisme, XV–XVIII siecle* (Paris: Armand Colin).

Buchsenschutz, O. (1987) Archeologie, typologie, technologie, *Techniques et Cultures*, 9:17–25.

Castells, M. (1989) *The Informational City: Information Technology, Economic Restructuring and the Urban-regional Process* (Oxford: Blackwell).

Curien, N. (1988) D'une problematique generale des reseaux a l'analyse economique du transport des informatione, pp. 221–28 in G. Dupuy (ed.) *Reseaux territoriaux* (Caen: Paradigme).

Dicken, P. (1992) *Global Shift: The Internationalization of Economic Activity* (London: Paul Chapman).

Dollfus, O. (1971) *L'analyse geographique* (Paris: Presses Universitaires de France).

Dos Santos, T. (1993) Quelques idees sur le systeme monde, *Cahier du Gemdev*, 20:55–66.

Durand, M.F., Levy, J. and Retaille, D. (1992) *Le Monde, espaces et systemes* (Paris: Presses de la Fondation Nationale des Sciences Politiques et Dalloz).

Ellul, J. (1967) *La metamorphose du bourgeois* (Paris: Calmann–Levy).

Fel, A. (1978) La geographie et les techniques, pp. 1082–110 in *Histoire des Techniques* (Paris: Encyclopedie de la Pleiade).

Gellner, E. (1989) "A psicanalise enquanto instituiçao social," São Paulo: Folba de São Paulo (23 September 1989)

George, P. (1974) *L'ere des techniques: constructions ou destructions* (Paris: Presses Universitaires de France).

Giddens, A. (1984) Space, time and politics, an interview with Anthony Giddens by Derek Gregory, *Environment and Planning, D: Society and Space*, 2:123–32.

Goldfinger, C. (1986) *Geofinances* (Paris: Seuil).

Gourou, P. (1973) *Pour une Geographie Humaine* (Paris: Flammarion).

Groupe de Lisbonne (1995) *Limites a la Competitivite, Pour un nouveau contrat mondial* (Paris: La Decouverte).

Ianni, O. (1992) *A Sociedade Global* (Rio de Janeiro: Civilizacão Brasileria).

Kayser, B. and Brun, A. (1993) La place de l'espace rural sans une politique d'amenagement du territoire (mimeo 6 pp.).

Labriola, A. (1947) *La concezione materialiste della storia* (Bari).

Lukàcs, G. (1970) *Lenin* (London).

Marx, K. (1970) *A Contribution to the Critique of Political Economy* (New York: International Publishers Edition).

Masuda, Y. (1982) *A Sociedade da informacão* (Rio de Janeiro: Rio).

Michalet, C.A. (1993) Globalisation, attractivite et politique industrielle, *Cahire du Gemdev*, 20:129–49.

Moles, A.M. (1971) Teoria de la complejidad y civilizacion industrial, pp. 77–94 in *Los Objetos* (Buenos Aires: Tiempo Contemporaneo).

Paz, O. (1990) *Los hijos del limo. Del romaticismo a la vanguardia*, 3rd edition (Barcelona: Seix Barral)

Peet, R. (1986) Industrial devolution and the crisis of international capitalism, *Antipode: A Radical Journal of Geography*, 18, 1:78–95.

Petrella, R. (1989) La mondialisation de la technologie et de l'economie, une(hypo) these prospective, *Revue Futuribles*, 135:3–25.

Rotenstreich, N. (1985)*Reflection and Action* (Dordretch: Martinus Nijhoff Publishers).

Sanchez, J.E. (1991) *Espacio, economia y sociadad* (Madrid: Sigio XXI Editores).

Santos, M. (1984) The rediscovery and the remodelling of the planet in the technico-scientific period and the new roles of sciences, *International Social Science Journal*, 36, 4.

Siegfried, A. (1955) *Aspects du Xxe siecle* (Paris: Hachette).

Sigaud, F. (1981) Pourquoi les geographes's interessent-ils a peu pres a tout sauf aux techniques? *L'Espace Geographique*, 4:291–3.

Silveira, M.L. (1994) Os novos conteudos da regionalizacao: lugares modernizados e lugares letargicos no planaito nortpatagonico argentino, *Finisterra*, XXIX, 58:65–83.

Silver, H. (1992) A new urban and regional hierarchy? *International Journal of Urban and Regional Research*, 15, 4:651–3.

Sorre, M. (1950) *Les Fondements de la Geographie Humaine* (Paris: A. Colin).

Storper, M. and Walker, R. (1989) *Capitalist Imperative: Territory, Technology and Industrial Growth* (Oxford: Blackwell).

Wagner, Ph.L. (1960) *The Human Use of the Earth* (Glencoe, IL: The Free Press).

Wallerstein, I. (1979) *The Capitalist World-Economy* (Cambridge: Cambridge University Press).

9 Transportation, Trade, Tourism and the World Economic System

The increasing globalization of the phenomenon of exchange has resulted in the tremendous mobility of peoples and goods. Based on major technological advances, the process has brought about a profound transformation in trade relations, considerably reinforced the control of global markets by developed countries, and widened the economic gap between them and the developing world. This emergent global economic system is undergirded by new structures of trade. These are themselves the product of new processes and conditions of transportation which have altered significantly the relationship between space and time. Thus, along with the increased movement of goods and services there has emerged a remarkable growth in the mobility of leisure. Human beings now move around the world in such large numbers in search of recreation that tourism has become one of the more important sectors of national economies.

Accessibility is thus central to both trade and tourism. New technologies of transportation and communication, starting from the era of the industrial revolution and continuing up until today, have made distances in real life essentially a relative concept. From the era of the steam engine to that of petroleum and nuclear energy, from communication by messenger or semaphore to the immense traffic of messages generated by advanced telecommunication and information technology, transactional distances between and within societies have narrowed in the sense of both travelling times and of mental maps. The remarkable development of tertiary and quarternary economic activities in developed countries since the 1980s has contributed a lot to accelerating the mobility of goods and people. It has brought about an undreamt of control over global

This chapter is based on a contribution by Gabriel Wackermann.

economic activities by powerful entrepreneurial groups in developed countries. This unprecedented growth in the movement of goods and information across the world has, however, created a remarkable feedback phenomenon in the emergence of the international competition of newly industrializing countries, notably those of the Pacific Rim.

This chapter reviews the close historical links between transportation, trade and the development of the world economy. It traces the passage from nationally-based economies to the international structuration of the world and ultimately to globalization. It then analyses the genesis and evolution of the increasing mobility of people in industrialized countries which have been facilitated by the rise of tertiary economic activities among which transportation and communication have acquired privileged positions. Together, these two tertiary activities have propelled tourism into one of the strategic sectors of the free-market economy on the international scale. Today, tourism has become an economic sector in its own rights, impacting as much on production as on services.

Transformation in Global Mobility

The notion of time span or duration has become a more decisive factor in trade relations than simply that of distance. This is why efforts to increase productivity and promote technological innovations are essentially aimed at reducing the costs of transportation time. This is why the introduction of highly specialized and advanced

Trading in silk and ginger in the city of Quengianfu, China. Miniature of about 1410 from the illuminated parchment manuscript of Marco Polo, Book of Marvels (1298). *Source:* Bibliothèque Nationale, Paris.

technology has accompanied the globalization of trade and communication to such an extent that the gap between those areas considered to be economically profitable and those that are not has widened to a very disturbing degree. Areas at the core and those that are peripheral or marginal are thus closely associated with this notion of timely or speedy accessibility.

The present situation contrasts sharply with the internationalization that accompanied the industrial revolution. Global trade at the time depended on railroads, steamboats and the construction of transoceanic waterways such as the Panama and Suez Canal. Later developments during the period included the construction of extensive networks of motorways to facilitate long-distance trucking, the use of diesel for all modes of transportation, the emergence of the great transatlantic steamship lines, the development of aviation on a world scale and the opening up of the under-developed world through the construction of airports, and the establishment of commodity markets for agricultural produce and manufactured wares.

By contrast, the globalization era is distinguished by the predominance of tertiary activities in productive sectors of the global economy. These activities rely heavily on advanced technology giving rise to significant innovations in expanding sectors of the economy. The considerable increase in the speed of terrestrial, maritime and, above all, airborne transportation is certainly the most significant phenomenon of this period (Bauchet, 1988). It has brought about the integration of national economies into new supranational systems and weakened those territorial entities whose productive organizations have not yet been able to participate in the increasingly well co-ordinated and orchestrated global economy.

In the area of rail transportation, high speed trains are responding to expectations of mobility over long distances and to concerns over the need to reduce congestions in air corridors or on the highways. The capacity of multimodal forms of transportation, combining rail and road networks, has been greatly enhanced by the containerization and palletization of freight and the use of trans-shipment platforms which allow cargoes to be transferred from one mode to the other without having to split them. All this has been possible because of sophisticated management practices based on advanced telecommunications and information technology and have drastically changed the transportation system of the industrial era. Supersonic jets and cruise vessels have replaced the transatlantic steamship lines, at least those, such as the prestigious S.S. France that have not themselves adapted to the cruise vogue. They attest to the vital importance in the post-industrial era for transportation to be able to respond effectively to the demands of speed, advertised timetables, prices, safety, comfort and profitability. River transportation has also come back into favour with the development of a large-scale network of navigable stretches along which goods can be transported to different continents.

All these developments, however, have been possible because of major changes in the nature of exchange. The introduction on a world scale of advanced technology in the production of highly diversified goods and services has accelerated the emergence of a new international division of labour to meet the demand of mass markets. This new restructuring of activities has reinforced ties of solidarity between regions producing raw materials and those areas that consume services and capital goods (Dollfus, 1994). These ties have further accentuated the urgency for seeking ways and means of reducing the cost of transportation. More significantly, this new solidarity belies Weber's theory of industrial location since, instead of locating at the site of raw materials, the emphasis increasingly is to develop large port complexes as Export Processing Zones (EPZ) within which processing industries can be located for transforming raw material imports from developing countries. In addition, the considerable reduction in the volume of heavy cargo that is moved across long distances in favour of semi-finished goods has helped to promote technological innovations in favour of freight with high value-added. This is expected to support higher transportation costs per ton-mile particularly when shipment time can be made shorter and more reliable. The result has been increased competitiveness among modes of transportation including barge transportation on rivers.

Competition has become even more vigorous given the rising cost of energy following the succession of oil shocks since the 1970s. The increase in wage-bill arising from labour agitations in the transportation sector has also been a factor in the pattern of response to competition in the industry. It became imperative to give primacy of attention to how to increase transport capacity and/or competitive speed. Thus were developed the Boeing 747 cargo planes capable of transporting 90 tons of freight over long distances, the 2500 (TEU) container ships able to cruise at speeds of up to 40 knots per hour, automatized supertankers, airbuses with a seating capacity for 300 passengers, lorries able to haul between 32 and 38 tons of freight throughout Europe or 100 tons in North America, high speed trains going at 300 km/hours, and so on. In order better to appreciate what all these mean in terms of the relationship between the freight transported and the distances covered and to allow for serious comparisons between different modes of transportation, concepts such as those of ton/kilometre and traveller/kilometre are often used in the industry. This also enables the distance/time relationship to be expressed in cost terms and induces effort to improve further on performance. This is particularly important in the current climate of deregulation when the cost structure is expected to internalize the external costs relating to infrastructural and environmental considerations. Over time, therefore, the constraint of distance has been dealt with through increasing recourse to technological innovations. These take into account the economies of scale required by the enhanced mobility deriving from the globalization process.

The Change in Scale

In the period since the industrial revolution, internationalization, followed by globalization, have profoundly modified the scale upon which exchange is based. The multiple and continuous interpenetration of influence on a world scale are a major driving force of the new economy and the evolving global society (Harrington and Warf, 1995). The automobile industry is an example, *par excellence*, of this interpenetration and globalization of national markets. New manufacturers from the Far East – first Japan and, more recently, South Korea – burst onto the international scene in the early 1970s. They made a strong impact on trade between the OECD countries as well as on transactions between the latter and the rest of the world. Despite the global crisis, world-wide production of automobiles has increased at an average annual rate of around 2.5 per cent over the past fifteen years. Close to 36 million vehicles have rolled off factory assembly lines in 1991 compared to little more than 28 million in 1981. If western Europe remains the largest automobile-producing region, Japan and South Korea show the highest rates of growth. The proportion of Japanese automobiles on the European market has gone from 5 per cent at the beginning of the 1980s to around 13 per cent in the mid-1990s. This penetration is considerably higher in the United States where the proportion of new Japanese cars on the market is over 36 per cent, including those manufactured in US plants (Beaufort, 1995).

The increasing restructuring of this global market rests on complex international networks of subcontracting and finance (Dezert and Wackermann, 1991). It also depends on numerous interfirm agreements promoting joint effort in research, transfer of technology and the production of components and/or whole vehicles. This mechanism of global interlinking of tasks by multinational corporations which share out the market between them is reinforced by the establishment of joint venture enterprises, forming partnership which may or may not involve owning shares in the company and alliances which may significantly transform the very structure of the market. The logic of market globalization which encourages such intensification of interfirm relationships on a world scale contributes inexorably to the obscuring of the identity and national origins of multinational conglomerates.

This same logic undermines the existence of markets oriented towards smaller-scale, local, national, sub-regional and regional production. The organization and management of mass transportation, both passenger and freight, which is possible on an international scale because of advanced technology, encourages large scale, multilingual media action, opens up prospects of expanding industrialization and reduces scope for independent retailers and service providers. Production at the level of nation states which not long ago used to be the determinant of operations at the international scale has become subordinated to the degree to which it is integrated into supranational mechanisms of control over global space. More than a simple ranking of national power and influence, the increased mobility of people and freight has encouraged the emergence of new spatial structures with multinational vocations. Thus, one can distinguish between the countries of the Triad which are the driving force of the world economy, the intermediate countries that are emerging out of underdevelopment into the spiral of sustainable development, and the peripheral or marginalized countries where the local, national and regional scale of production is still dominant and constitutes a handicap to economic growth.

Trade activities within and among the Triad illustrate the complexity of the present pattern of exchange (Nehme, 1993).

Comprising North America, Europe and Japan, the Triad gives the appearance of three distinct and clearly demarcated realms. In reality, the two leading world economic powers, namely the United States and Japan, are already in the process of forming an American–Japanese complex. American and Japanese companies have forged alliances and initiated joint capital ventures throughout the world, notably along the Pacific Rim. In order to escape high taxation in their respective countries of origin, American and Japanese firms have profited greatly from the enticements of the fiscal paradise provided by countries like Singapore where they have created subsidiaries, organized joint-venture enterprises, purchased shares in each other's companies and collaborated in R&D and in the transfer of technology. The American–Canadian capitalist melting pot, which also involves Japanese and South Korean enterprises, has, along with Mexico, recently constituted itself into a new multinational block through the formation of the North American Free Trade Area (NAFTA). Within Europe, the world's third largest economic power, Germany, has also forged close ties with Japanese conglomerates, more so than any other of its European partners. Since the 1980s Düsseldorf has become the European capital for Japanese firms. If the German automobile industry has not experienced as many difficulties as that of France, it is indisputably because of this German–Japanese connection. This has enabled Germany to maintain its grip on the German car market and induces other European multinationals to seek to expand to the other countries of the Triad.

The newly industrializing countries (NIC) of the Pacific Rim which are following in the wake of Japan have, in recent times, been referred to as the new dragons. They present another example of this complex, overlapping of scale in trade relations. The flow of capital from the Triad has enabled these countries to take off economically. Their increasing global competitiveness is already adversely affecting the economic fortunes of the older industrialized countries. Markets are being divided and redivided as a new balance of power is taking shape. The emerging new configuration of economic forces is being driven by a new logic of global profitability. This has undermined the previous logic based on territorial sphere of influence and technological innovativeness. This new logic, initiated by the great entrepreneurial groups of the Triad, has now spread, from the point of view of investments, to the more dynamic of the intermediate countries of the NICs and of central and eastern Europe. More importantly, it now dominates the process of international exchanges to the extent that those countries which are unable to submit to its demands find themselves inexorably relegated to marginality.

The logic significantly influences the eventual success of markets. It constitutes a stimulus for the deployment of raw material and for freight transportation. It is also the reality behind the notion of the information superhighway and the critical factor influencing the increasing importance of tourism in the global economy. All of this contrasts sharply with the logic of space which requires that consideration be given to the adjustments necessary between the supply of resources to companies and environmental circumstances.

Transportation – The Basic Conditions of Current Exchange

The collapse of socialism highlighted its inability to adapt to the socio-economic and cultural demands of the current system of international exchange. As a result, it has reinvigorated the capitalist market economy and has made it all-powerful to the point that it has, to some extent, become unbridled and uncontrollable. This remarkable expansion of the free-market economy has accentuated technological and organizational innovations and optimized the means of production. Producers of goods and services are now required to be capable of measuring and quantifying their gains in productivity with maximum precision. They are obligated to oversee the maintenance of quality and security during the entire duration of transportation of both the equipment and goods and services produced, right up to the moment of delivery to the consumer. Consequently, transportation is forced to become a critical factor in the processing, packaging, delivery and after-sale services of industrial products. Obliged to seek to be flexible and competitive in its costing, producing firms are thus constrained to subcontract their transportation requirements to other enterprises.

In this situation, transportation involves constantly seeking to reduce the effect of space through shortening the time required to cover a given distance. Industries heavily dependent on transportation thus seek to minimize such dependence by actively engaging in the diversification of their products, intensifying trade flows and undertaking partial takeovers of companies that help to enhance interdependence and the complex of overlapping of firms.

Such a situation entails careful consideration of the terms of exchange at all levels of space. It involves a rigorous management of the logistical support of production organizations provided by transportation and communication.

Logistics thus impacts significantly on the whole exchange economy. Starting with activities at the place of origination and production of goods and services, it remains a critical factor in the process of ensuring that these services get to the places of consumption. Its aim is to ensure that a desired product is available at the appropriate place, at the best value for money and within an optimal period of time. To achieve this aim, four seemingly contradictory conditions have to be met:

1. the product must take account of the characteristics of the consumer as well as of competition in the product market, during different seasons and at different destinations;
2. the routing or itinerary of the product must, irrespective of the wishes of shippers or consumers, take into account the imperatives of the overall price, of problems relating to the crossing of borders, of the nature and degree of the manner in which transportation is organized, of processing, and of the insurance of the goods and people in transit;
3. the delivery time must give due recognition to the fact that the best route is not necessarily the shortest either in distance or in time. It must give due consideration to other factors such as the quality of handling, attention to product characteristics, security and the reduction of waiting time; and
4. the cost which, apart from its dependence on the three preceding conditions, must take into account international competition, the value of money, and the number of persons or the value of the product that is being transported.

In general, however, the cost of transportation is a function of speed. In land transportation, for instance, trucking is ubiquitous despite its slowness relative to rail transport. Its flexibility of access to places makes it specially favoured for the movement both of freight and passengers. However, in the area of international transportation especially at the global level, maritime transport is dominant both upstream and downstream. Indeed, three-quarters of all tonnage traded in the world is transported by sea, especially as 80 per cent of the major world metropolitan centres are located on the sea coast and possess sizeable ports.

In the early 1980s, issues of logistics were concerned largely with movements of goods and services by individual producers. With globalization, logistical considerations now cover movements resulting from the totality of productive enterprises. Various logistical agencies are currently engaged in how to improve operational techniques to meet the aggregate of these demands. And it can be expected that logistical strategies and techniques will have to be further developed to greatly enhance and facilitate the flow of goods and people world-wide. Indeed, as Pache (1994) observed:

> *the success that the market for logistical operations is currently enjoying is due very much to the global approach to trade relations (within and between productive organizations) which it has fostered . . . Operations and isolated tasks situate themselves in such a way that they are best tackled in a systemic manner which, a priori, proves superior to the sum total of local solutions, even if the latter, taken singly, may appear more satisfying.*

Internationalization, followed by globalization, has made logistical considerations central to trade relations. Because of trade expansion beyond national borders, operators within the logistical system have had to forge reciprocal ties with a view towards maximizing market share. Furthermore, because much the same considerations are involved in the movement of freight as of passengers, it has been necessary to establish a complex but integrated network of enterprises of various nationalities comprising, in the case of merchandise, of shippers, freight or forwarding agents and carriers, and in the case of travellers and tourists, travel agencies, tour operators and carriers. Logistics thus implies the association of complex tasks in a world in which mobility poses increasingly daunting challenges and raises problems that go beyond the strict notion of transportation. These challenges, however, represent only a small share of logistical costs. On the other hand, logistical issues associated with the strategies of new firms entering into the system always compel a continuous recomposition of the transportation landscape. In short, logistics has become a powerful instrument of competition and lies at the very heart of international and transnational economic preoccupations. In concert with the issue of logistics, the concept of scale economies has also assumed some considerable importance in transportation. Aided by the new climate of deregulation, this concept has relegated to a secondary position the question of returns to scale that used to be important in the era of mono-production. Transportation is now directly affected by scale economies. Maritime transport has become less costly because a shipper can combine various goods and commodities or an enterprise can replace a traditional obsession with growth in the volume of shipments with co-ordination and a strong interconnection of activities. Trucking also has become a multi-product activity. In this case, routing economies has overtaken the economies of scale. The former relates to the efficient combination of different centres of exchanges, whereas the latter can occur without any concern with co-ordination and through development on a particular link in the transportation chain.

Routing economies also allow for the integration of different modes of transportation so as to reduce the duration when transport capacity is not being utilized. They tend to encourage enterprises and well-informed transport auxiliaries to orientate themselves towards profitable rhythms and frequencies of movements. This development naturally imposes often difficult revisions of standard ways of doing things, with a view towards the refinement of paths already taken and the search for new objectives of growth.

Transportation thus contributes significantly to meeting the challenges of the new imperatives of a global market (Merlin, 1991). Those involved can no longer ignore the fact that international exchanges are now to a large degree managed by some 20 or so transnational groups which control the movements of major raw materials. These groups have, for the most part, gone beyond their initial activities of trading in basic commodities. Owing to the trend towards forming conglomerates, they have become transnational industrial enterprises, integrating on a world scale entire sections of their trade in raw materials and on a basis uncompromisingly aimed at controlling the market. In addition, they have transformed themselves into transnationals in the service sector, integrating into their commodity trade and transportation, services such as banking, research and management of vast logistical networks.

New developments of this magnitude cannot maintain intact transportation structures inherited from another epoch. They are products of a new cultural and financial system which is driving global competition and accelerating the rate of transformations currently underway. The present situation in the field of international trade is well characterized by the Nobel laureate, Maurice Allais, who in the early 1990s observed:

> *Each day, some $420 billion . . . changes hands throughout the world. The portion of that exchange which actually corresponds to trade and which represents real needs accounts only for $12.4 billion. The rest is short-term speculation whose effects are destabilizing and harmful.*

Network Axes and Logistical Chains

Logistical imperatives now necessitate a reconstruction of the spatial arrangements for international exchange relations. New nodal points have emerged and continue to evolve within the framework of this changing environment. They constitute not only points of transfer or breaking bulk but also transportation centres equipped with hi-tech infrastructure for managing the flow of goods or people in an optimal period of time and under conditions of optimal security and comfort. Freight movement, for instance, has overcome the need to break bulk by a simple change in traction towards containerized or palletized transportation. The resulting introduction of new nodal structures, notably the platforms, often take on imposing dimensions depending on the position in the spatial hierarchy of movement. Vast reception facilities for bulking and breaking bulk as well as elaborate packaging services channel and determine the rhythm of movements of goods from city centres to their periphery. They also assist in no small measure in alleviating traffic congestion and bottle-necks and promoting the most efficient inter-modal forms of transportation.

These new nodal structures sometimes attain quite impressive proportions. A good example comprises the group found around the huge Rungis wholesale market near Paris, with its SOGARIS and NOVATRANS platforms for road and rail transportation and its proximity to Orly Airport. There is also on the north-eastern end of the Paris region the complex near Roissy-Charles de Gaulle Airport. This has continued to expand. In addition to air-freight platforms, the complex now boasts a GARONOR centre for bulking and breaking of bulk as well as high-speed rail (TGV) connections across the English Channel and to south-eastern France. The growth and prosperity of this important nodal point has been further enhanced by the surrounding technopole cities and various research centres just as the recreational city at Marne-la-Vallee in the Paris suburb has been stimulated by Disneyland.

Similar nodal structures have become part of the equipment of large, modern airports which have, thus, become major hubs of activity on an international scale. They are now points of contact between two or three modes of transportation (airports, ports and railroad stations), drawing to themselves regional flows of goods and passengers and acting as world-wide relay stations. New York, thus, constitutes a point of convergence for the movements of goods and people inside the United States and serves as a centre of departures and arrivals as well as an international gateway. These hubs thus serve as veritable air-cities and windows on their entire country.

Platforms are linked with one another by strategic ties known as logistical axes. Because of their importance, these axes attract to themselves an increasing number of actors in the transportation field and in the related financial sector. All of them benefit from the complementarities emerging from the improvement of transportation infrastructures and the growing volume of movements of goods and people. These logistical axes are thus evolving as dynamic poles of attraction for both established firms and new, state-of-the-art enterprises and, with time, are likely to become a veritable grid of overlapping carrier axes. When they do come to join together, these axes give the appearance of a chain or an international joint networks of exchange. They involve sets of complementary firms which are benefiting from the existence of publicly and privately provided infrastructural facilities constructed essentially to sustain a large and continuous flow of traffic (Johnson and Johnston, 1994). These chains enter into competition with one another. In other words, they promote competition between groups of firms which are operating on an international scale and in the most diverse of sectors. They are also frequently associated with the organization of networks which are intended to give a certain rhythm to the market, depending on the politics and strategies of supply and the availability of purchasing power. The chains depend upon public and private facilities, particularly on infrastructure, which enhance transportation within the most developed areas. In such areas, nodal points such as ports, airports, sorting stations, mono- or multi-model platforms as well as road, rail and water and wide-gauge canal routes serve to boost further the volume and intensity of traffic both quantitatively and qualitatively.

The growth of these chains and the scale of the grids that result from them are both the stimulus and the reflection of global economic confrontation. The global command centres of these networks are naturally to be found in the metropolitan centres of countries of the Triad. Here, markets are highly active, the axes are well structured and transportation policies, both national and local, are well articulated. Promising commercial niches are also vigorously competed for and failure to secure such niches often leads to the search for business incentives further afield.

The emergence of axes and chains in the transportation systems of the world has introduced the need for a new interpretation of the pattern and relationships of mobility in space. It has revealed the existence of forces and tensions that are imprinted in the spatial fabric of the global economy through the operating mechanism of global mobility. It has also indicated how much each country is now dependent on external forces from which it is no longer possible to escape, at least not indefinitely. The end result of the interactions, both complementary and contradictory – among chains on a logistical axis – is a network of movements some of which are the spontaneous response to market demands within a given system and others which are the product of deliberate decisions by powerful pressure groups that have assumed economic control over large regions of the world. Organizing and operating such networks attest to the fact that the whole world has become the field of play and the scale of reference for groups of large firms and international organizations of all kinds. This development of network structures has been made possible by the conjunction of a business culture in its most advanced form and information technology dependent on telematics. The latter performs at ever higher levels of efficiency from hertzian networks, through cabling and fibre optics to satellites. Its development thus reinforces interdependence whilst marginalizing initiatives that remain too individualistic or locally based.

Kansky (1963) defines a network as an assembly of geographical locations interconnected in a system through numerous links. The network of movements of goods and people that represent mobility are based on these links, on infrastructural facilities and modern equipment grafted onto them, and on services called into existence by the needs of these amenities. The resulting efficiency in the operations of the network is, however, a function of how far the relations between the activities are compatible, harmonized and integrated into a technically coherent whole with well-harnessed synergy.

Networks form themselves into branches or circuits. They can be linear, orthogonal, radial or sinuous. They can be simple or complex; connected with each other, parallel or superimposed; tight or in loose order; used densely or very little. It is possible to use cartographic techniques to visualize the nature, scale and degree of profitability or the capacity for pollution inherent in a given network (UNEP, 1994). Such techniques also facilitate comparisons, the depiction of evolutionary trends, and the extent of openness to the world of various territories.

Networks are shaped by major centres of activities. They emerge, evolve, develop and transform themselves in the framework of the reality presented by these centres which prevent them from being static. Mostly of local dimension for over one millennium, networks have in the last two centuries taken on a national character and have become internationalized on a massive scale (Durand, Levy and Retaille, 1994). Territorial limits have evolved as a result of the catalytic power of the network relations at their centre and the vigour with which this centre asserts itself on the socio-cultural and economic level in the country involved. The catalytic power of the Anglo-Saxon economic and cultural force in North America has, for example, extended far beyond the borders of the United States and Canada to give rise to what is now described as 'the American way of life', the American model of production and consumption, as well as the American system of financial structures. To push back these limits right now is a very difficult task. This is because the manifold and powerfully overlapping networks already developed by the United States across the globe tend to reinforce one another. In this connection, they depend on two other sets of economic and cultural networks developed by the second- and third-ranking world powers, namely Japan and Germany. These other networks were fostered by the United States which had then worked closely with them to build and direct a complex network system that has fused the whole world into one single, global space.

Nonetheless, obstacles to the progressive expansion of the territorial limits of networks are numerous. Japan, for instance, reveals the extent to which it is still a difficult country to penetrate even when its technologies are basically the same as those of North America. Cultural dynamics, which are intimately linked to the expanding economic system, have been used here to erect a formidable barrier to penetration. In Europe, the impact of the satellite television system has been constrained from having an extensive span by cost, commercial legislation and differences in culture and life-styles among countries.

Despite these difficulties, the globalization process has continued unabated. For example, the arc represented by European countries on the coast of the Atlantic has progressively regained its central place in international trade after a period of relative marginalization. It has also become the focal centre of international networks. The setting in motion of a new operational system along the Atlantic coast and its integration into the international system is a function of the expanding world of international networks. This underpins the emergent structures of the future and defines the relations between firms and the socio-economic environment within which they have to operate (Wackermann, 1995b). The progressive systematization of logistical chains leads to a veritable division of labour between groups involved in the production of goods and those engaged in the delivery of services. It also promotes the growth in the number of schemes involving the participation of large, highly capitalized enterprises whose interest is a function of their link into material supply- and service-provisioning networks.

Networks are, however, not a neutral mechanism of international economic relations. They do structure systems and sub-systems which emanate from them and the future of which they consequently determine. Since the collapse of centrally planned economies, for instance, free-market economies have become larger through the inclusion of numerous, new sub-systems, representing the economies of these former communist states. Within these sub-systems, competition has also been unleashed. The growth of economic activities within these countries gives the impression of a landscape in a state of flux, involving the constant collision and blending of whole sets of multiple movements of newly-produced commodities and people. The emerging world of new network systems is already giving the appearance of being more real and concrete than the previous divisions of the area into regions and states or supranational blocs.

The Special Role of Maritime Transport

The rhythm of international commercial exchanges is set primarily by maritime transport. This mode of transport is generally situated upstream and/or downstream of the transportation chain and is inserted in between sequences of land transport activities. Around 75 per cent of the volume of international transport is carried in and out on sea lanes and waterways. This corresponds to a little more than two-thirds of the value of global merchandise transactions. Dry merchandise transported in bulk, notably coal, iron, cereals, bauxite, aluminium and natural phosphates, makes up 22.8 per cent of maritime freight, with petroleum accounting for another 39.4 per cent, and other products 37.8 per cent (OECD, 1995). These figures show the importance of tanker traffic in international commercial relations, with all the geopolitical and environmental consequences that this entails. The current global economic crisis has, however, reduced the rate of increase in tonnage transported. In recent years, this has hardly been greater than 3 per cent per annum. This relatively modest rate of increase says little about the absolute growth in tonnage, even though this is less than in the recent past. The world's fleet currently posts an annual movement of freight of the order of 450 billion gross registered tonnage. On the other hand, while disarmament has receded and the tonnage of ships in storage remains low – down to only one per cent of the world's fleet – deliveries of new naval vessels and sales for demolition have increased. Despite the growth in activity of naval demolition, tonnage continues to grow due to increase in new construction. The generalized over-capacity throughout the world has nevertheless weakened future prospects for the development of the naval industry.

It must, however, not be assumed that stagnation across the board is imminent. For instance, with a view towards a traffic volume of a billion tons by the year 2000, China has been accelerating its maritime capacity. The improvement in transportation currently underway in China is such that the rate of freight space utilization there is among the highest in the world. In the coming decades, China expects to disburse a further $111 billion in this sector. This orientation is due in large part to the fact that maritime transportation is of primordial importance to China. Its coastal provinces supply more than 70 per cent of industrial production and close to 90 per cent of exports. They also receive nearly 90 per cent of the country's foreign investment. In 1994, China ranked tenth in the world in terms of harbour facilities, with 90 per cent of its foreign trade being transported by water and its ports handling some 678 million tons (OECD, 1995).

Whereas, in 1955, Asian shipowners possessed only 5 per cent of the world's fleet, today they account for close to a third. In effect, of the ten most important ship owners in the world – all categories of ships included – there are eight Asians and only two Europeans. Their ranking in descending order of importance is as follows:

1	NYK (Nippon Yusen Kaisha)	Japanese
2	Mitsui OSK	Japanese
3	China Ocean Shipping Group	Chinese
4	Kawasaki Kisen	Japanese
5	China PR (Provincial Shipping Consolidated)	Chinese
6	Navix	Japanese
7	A.P. Moller (Maersk)	Danish
8	World Wide Shipping	HQ in Hong Kong
9	Bergesen	Norwegian
10	India (State Shipping Consolidated)	Indian

If the west is lagging behind in this sector, Japan is not in a leading position either. The Japanese share of the market has multiplied only by a factor of three whilst the non-Japanese Asian fleet has increased its share 18 times. This breakthrough by the non-Japanese group occurred in two stages. The first was between 1955 and 1965 and involved a systematic chipping away at the west's share of the market; the second was during the 1970s through recourse to the use of flags of convenience in order to deal with the problem arising from the collapse of freight rates and shipping tariffs in the wake of the oil shocks. The west, on the other hand, had been rather reluctant to resort to the use of flags of convenience, preferring to preserve employment for its citizens through the maintenance of national merchant marines. After a decade, however, western countries were able to preserve the use of their national flag carriers whilst meeting the operating costs of flying flags of convenience. The French Kerguelen line is an example of this development. The strategy enabled the west to stem its loss although not to reverse the situation.

Thus, by the beginning of the twenty-first century, East Asia, which is already prepared to become a key actor in world markets, could contribute to a stiffening of global competition and ensure that maritime traffic in the OECD zone grows less quickly than elsewhere in the world. Policies in regard to this growing competition are viewed differently in different countries. The OECD's (1995) annual report on maritime transport for 1993, for instance, observed that the interaction between trade policy and policy in regard to competition will become increasingly important.

In this regard, the OECD Maritime Transport Committee has been launched on a programme to do something about the disorderly situation in international shipping. Given the pre-eminence of shipping as a mode of transportation, this disorder is having a negative effect on the totality of international exchange. Though it emphasizes the need to reinforce the liberalization of international maritime transport, the committee insists on the necessity of interventions to counter the growing disorder that is particularly evident in the domain of maritime practices. Such interventions include the promotion of greater respect for safety standards and the introduction of measures to prevent pollution. The process is expected to entail greater scope of dialogue including countries of central and eastern Europe, the new independent states of the former Soviet Union, and the dynamic Asian economies. This is with a view to agreeing to a commonly defined set of rules applicable to all actors and thereby facilitate the evolution of a new international maritime order.

This new approach calls into question a previous basis of decision-making in respect of modal distribution. The bursting onto the international scene of a combined, pluri-modal system of transportation has already brought about a revolution that is now recognized as a necessity by numerous actors in the field, even if it has taken time to take shape and be concretized. According to the type of merchandise, modal disparities, in effect, have always indicated the need for some collaboration. European maritime transport offers a manifest demonstration of the imperative for such collaboration. Following on the boom years of the 1950s, it became complacent and delayed revamping its structures. It did little to adapt these to the demands of mass transportation especially for wet and heavy goods which was to propel in the 1970s the sharp growth of the South-East Asian and Japanese fleet as well as those of other developing countries. These fleets benefited from the 40/40/20 system endorsed by UNCTAD. This stipulated that cargo for maritime transport be divided up in the ratio of 40 per cent for the fleet of the exporting country, 40 per cent for that of the importing country, and 20 per cent for the outsiders, that is, fleets of non-member countries of the maritime conferences. In the whirlwind of this decision, European merchant marines were driven to concentrate on traffic between the European islands and the continent. Meanwhile, containerization and competition on a world scale were reinforcing rivalries between North America, Japan and Taiwan and forcing them to achieve greater and greater efficiency in their operations. Europe was, thus, placed in a position where it had to struggle hard to catch up, not only to be able effectively to confront this competition but also efficiently to connect its numerous large, hinterland ports to the rest of the world.

A distinction, however, is emerging between organization and direct production of transportation services. The former provides a framework for constituting multi-modal chains through encouraging traditional modal operators to develop their own activities as multi-modal forwarding agents. This has led maritime shipowners to assume responsibility for pre- and post-rail transportation and to seek to do the same for inland waterways, especially as the railways can be assisted by inland waterways to become a more effective alternative mode of transport. The saturation of traffic on the highways and the impossibility of the railroads picking up the available additional freight because of prevailing congestion, bottle-necks and environmental effects, provide transcontinental inland waterways a unique opportunity for filling an important niche in the network system (Vigarie, 1993). Coastal navigation is also destined to increase in importance. The European Union, in particular, favours this system of transportation. To increase their volume of traffic, European ports are striving to take market share away from land transportation and are participating in the growth of coastal navigation, which has already increased to almost one-third of intra-European freight. It is true that for a lasting development of coastal navigation, it will be necessary to increase the cost of trucking phenomenally. This is a condition of success which is not usually well appreciated by those concerned. In the long run, in any case, the need for greater accuracy in costing every mode of transport and, in particular, integrating the cost of the externalities of environmental impact, cannot be resisted. Even maritime transportation will not be able to avoid this since it will be in tune with the dictates of the market. Such a development could come to necessitate a reappraisal of the relative importance of different modes of transportation and significantly transform existing patterns of movement.

The World System of Markets

Globalization has profoundly modified and altered both the structures and the functioning of exchanges on national, regional and international levels. Since the end of the Second World War, a progressive and permanent intermeshing of scales has become noticeable in world trade. Some trading activities now cover the whole world. The movements of goods and peoples have come to be articulated and orchestrated by big trading powers who have themselves become creators of scale. In this connection, it is necessary to distinguish between attempts to integrate into a system to achieve a defined spatial scale of influence and attempts to control a system within the globalized whole. The evolution of these two processes has taken place in stages. These have tended to upset the traditional order of importance on the world scene and encourage the emergence of a new one. In the course of the first stage, Japan emerged as one of the leading economic powers. This projected the country to membership of the oligopoly of world decision-makers, a position reinforced by its role as the first of the Asian dragons. The second stage witnessed the emergence of the four little dragons – the newly industrializing countries (NICs) of South Korea, Taiwan, Hong Kong and Singapore, which were all resolutely oriented towards the export of finished goods. The third stage involved both those countries traditionally engaged in the export of raw materials, as well as those others referred to as baby tigers. This heterogeneous group includes in the first set westernized countries such as Australia and New Zealand as well as developing countries such as Indonesia and Malaysia, which export oil and rubber respectively. The second set are those countries such as Thailand and Vietnam, which are attempting to imitate the strategy of the NICs and which are therefore referred to as baby tigers (though they could, of course, become big ones). The fourth stage involves two types of countries, namely those without major resources. The countries may be considered as following the wrong path to development: for example, the countries of South America whose only development strategy is import substitution industrialization, and those which are experiencing great difficulties in developing, such as most African countries and many island states.

Two types of territories, however, are actively involved in the trend towards global integration of the world economy led by the Triad countries. The first comprises countries of the northern Pacific where integration is already very advanced; the second involves countries in the south-west Pacific which is a peripheral area seeking to be integrated. By contrast, countries in the south-east Pacific, with the exception of Hawaii and the Marquesas Islands in French Polynesia, are well off the beaten track as far as navigation goes and are therefore not involved in the present movement. The same goes for Latin America which seems at the moment to be at a dead end as far as world economic integration is concerned.

The Pacific Rim seems to have become the centre of the new economic system that is now global in scope, almost in the manner of the Mediterranean of Antiquity or the Atlantic from the sixteenth to the twentieth century (Dollfus, 1995). A financial centre such as Bahrain forms one of the relays between the great American and European stock markets on one side and the Tokyo stock market on the other. The latter has become one of the three principal financial markets in the world along with New York and London. The Pacific Rim has thus managed to pull itself up to the level of one of the world's major poles of exchange through its mastery of the use of critical raw materials, an exceptional development of technological infrastructure and a high aptitude for financial trading. The major economic powers of the Pacific Rim notably the United States, Japan, Hong Kong and Singapore, participate closely in setting the pace for international exchange. Three-fifths of investment in Hong Kong and Singapore are of American origin. In Singapore, the number of US–Japanese joint ventures is growing. Thanks to its efficient satellite system in the area of telematics, Singapore transmits, retransmits and receives more messages from multinational firms with decentralized operations in Germany than those firms exchange inside Germany itself.

The question may, thus, be asked: is the Pacific Rim destined to become a new centre of the world? The answer to this will be positive, at least in so far as the northern and western portions of the area are concerned. But there are some major reservations. Perhaps the most important is the fact that Japan's geopolitical weight remains relatively weak and the country has not managed to gain real control over the global flow of information. The secondary centres in the Rim, namely the NICs and the west coast of the United States, are not of sufficient strength to pose an efficient counterweight to Japan. Japanese firms themselves continue actively to engage in building up capacity in these areas so that they can become real economic centres in the Pacific. As far as the other countries of the Pacific are concerned, they remain economically dominated and weakly integrated into the global economy. Their level of development does not provide them with a real opening for playing effectively on the world market. In addition, the future of the Pacific Rim remains somewhat uncertain, given the high instability that characterizes much of the region and the political and economic tensions that are still a feature of global international relations.

Be this as it may, the Pacific Rim accounts for the highest rates of economic growth in the world today. It totals almost a third of global maritime trade and two-thirds of all tonnage transported. Seven of the ten leading container ports in the world are located on the Pacific coast and include such ports as Hong Kong, Pusan and Los Angeles. Eight of the twenty largest ports in the world are also in the Pacific. Kobe is the largest port in the area and, with its 180-odd million tons, is annually placed second largest in the world. Tokyo, with its 80-odd million tons is in seventeenth position. With respect to the overall maritime traffic, Japan is globally in third position. Singapore, with some 140 million tons is placed sixth, though ranks fourth in total traffic. Its particular strength, however, is with regard to container vessels for which it ranks first in the world and operates as the great service station for Asia. In 1970, maritime traffic in the Pacific was about 800 million ton/km, a figure much lower than that of Latin America at the time. By 1990, however, the figure had jumped to around 20 billion ton/km, a figure reached by Europe only in 1985. With respect to global air traffic, the share of the Pacific Rim jumped from 7.8 per cent in 1970 to more than 15 per cent in 1990.

The spectacular economic growth of the Pacific Rim must, however, be situated in an international context. The multitude of links upon which the positioning of its enterprises are based presupposes global thinking. This now involves a complex network of financial interests and other agreements relating to extra-territorial joint ventures. For most of the corporations concerned, it also necessitates designing strategies for global integration rather than simply assertion on a national or sectoral level.

Minerals and raw materials, energy sources, agricultural produce and foodstuffs – all give rise to an impressive volume of trade which is often subject to long-term planning. These goods are often covered by multiyear trade agreements whose goal is to reduce the destabilizing effects of stock market speculations. A determining factor for industrial production and mass consumption, these goods are monopolized by a small group of large, oligopolistic firms. They receive due attention from public authorities who, to a large extent, help them to create protected markets. In particular, trade in agricultural products and foodstuffs has witnessed a substantial transformation since the process of globalization began. Industrialized countries with large populations, high living standards and low agricultural production are today the leading importers of foodstuffs.

On the whole, global exchange is today more tied to large areas where trading actors are based than to bilateral agreements between countries. In this regard, the three countries of the Triad are mutually affected by the bulk of international exchange. Amongst them, transnational corporations play a determining role in the face of the economic weight of the United States and now the North America Free Trade Area (NAFTA). The European Union accounts for some 40 per cent of world trade whereas the share of the United States has declined to 14 per cent, at a time when Japan and South-East Asia are taking ever greater shares of the market and causing real concern in American financial circles. As with production, trade unites many closely related services through their being linked by telematics. This includes, in the most important instance, advertising. It also involves suppliers of services, sales and after-sales services, promoters of quality and efficiency, financial circuits and long-distance transactions. Globalization accelerates cultural interaction and favours the new vogue of large trading systems which have the Americans and the Japanese in the lead. Traditional commercial methods find it difficult to resist the grip of the big trading powers. Only local markets with somewhat particularistic identities and limited to relatively modest levels of exchange in relation to the total volume of global transactions have been able to escape their dominance. These local markets are thus regarded as the last place of refuge in the face of the competitive whirlwind fuelled by businesses through their large, international chain of retail outlets for commodities for which consumer taste has been heavily influenced by massive advertising campaigns. Small-scale commerce is thus dependent willy-nilly on the movement of goods and people on a world scale. Representing, as it does, the very last link in a vast international-trading chain, it is particularly vulnerable because it is really not integrated into the network.

The Challenge of International Tourism

Tourism, together with its associated recreational activities, constitutes a rapidly growing component of the tertiary sector of the global economy. With its increasing income of hard currency, it is considered a component of the foreign trade activities of countries. When it brings in the hard currency, it is treated as an export earner; when money leaves the country, it tallies in the import column. The contribution of tourism to national economies is closely monitored in order to determine its share of the gross national product. In some countries, it has come so to dominate the economy that other sectors are both weakened and made dependent on the hazards of an activity that can, at any moment, fall victim to an economic recession. In short, the recreational economy, of which tourism is the most visible component, is one of the last arrivals among the productive sectors of economic activities with tremendous international reverberations. Its importance has been greatly enhanced through the complementary and contemporaneous development of whole new areas of electronics engineering.

The mass production of tourism dates from the aftermath of the Second World War and especially after the 1950s when the mobility of goods and people increased tremendously (Wackermann, 1991). Triggered by the phenomenon of paid vacations and other social legislations, and accelerated among the well-to-do by the rapid rise in income during the three decades of prosperity following the Second World War, the tourism sector quickly became the promoter of numerous ancillary services. It evolved into a complex economic system dependent on computerized reservations, holiday itineraries and so on, and with a capacity to deal with seasonal and customer diversification. The demands of the sector soon led it into land and property speculations, concern with policies and strategies regarding national economies, and finely-tuned practices for locating sites and places propitious for the global development of tourism and leisure. Within a short period of time, other services and the production of tourist-related goods have come to constitute high-performing economic activities, integrated along with the tourist sector proper, into the world economy and the international division of labour.

The industrial level of production of tourist activities on a world scale led to a better appreciation of the fact that tourism constitutes a more economically efficient and higher yielding investment than many, single isolated enterprises. Paying no attention to national and sectoral borders, the massive 'production' of tourists on the global markets has attracted considerable attention on the part of financiers. The trend has been so sustained that not even the world economic crisis has slowed down the continued expansion of tourism whose growth has been steady over the last 15 years.

The transformation that has followed the growth of tourism and the high performance of its transportation and communication infrastructures has resulted in many countries becoming less dependent on the development of their own natural resources and on the constraints of distance and time. The rhythm of technological progress and of the circulation of capital has made possible changes in the specialization and training of operators. The natural environments, attractive though they may be, have had to be subjected to the dictates of the market and all sorts of technological developments. Consequently, the tourist industry, which is already highly sensitive to its own fluctuations, has to take account of changing fashions, changes in exchange rates and competition from newly developed areas.

These vagaries of the sector have helped to reinforce the dominance of large corporations and other large investors. These corporations have inexorably swallowed up the large number of marginal operators in the sector. But they also ensured that the sector more than doubled its volume in international trade. At the present time, tourism accounts for 20 per cent of all trade exchanges (OECD, 1989). The breakdown of how this figure is made up shows that raw material consumption accounts for only 5 per cent of all trade exchanges while services make up 20 per cent. Capital charges take up 40 per cent, the largest part, due to the overwhelming importance of short-term capital. Thus, tourism as a promoter of local activities and investment participates in the growing interlocking of economies into the global market. Its export potential implies a significant capacity to provide services which, in turn, necessitates the presence of an important supporting financial environment.

For tourism to be profitable, there is need for a framework of collective organizational structures (Van Lier and Taylor, 1993). At the global level this framework itself is supported by linkages, networks, chains and powerful pressure groups that guide the development of the tourist market. The recreational economy thus benefits immensely from heavy promotion through the media. The search for quality service is intensified in all sectors of tourism and leisure. Customers' requirements are constantly being raised especially as the proportion of novices among the tourists declines with each passing year. Competition thus becomes a critical factor not only in the realm of economic profitability and management but also in the area of quality of service. Confronted by powerful international pressure groups who know the market, small tour operators began to disappear in droves from the early 1980s onwards.

The rapid rate of internationalization in the sector has been all the more hastened by the phenomena of marginalized regions that have not as yet been integrated into the world's major logistical axes. These regions are drawn into the global market by ties of interdependence based on technology transfers, exchange of knowledge and know-how, and systems of origination, production, transportation and supplementary sales. In addition, these ties come to depend on the expanding logistics of transportation and exchange involving many conglomerates, both industrial and commercial, as well as numerous small and medium-sized enterprises. The system is also massively supported by a network of subcontractors both in the countries of origin of tourists and at their destinations. A trend towards what has been referred to as generalized deterritorialization is emerging which is making the provision of services everywhere almost homogeneous. It is based increasingly on the fuller utilization of transportation capacity, the more rapid processing of information and better co-ordination of services and activities. In this regard, the airline hubs have become very central in reorganizing global space and co-ordinating the dominant poles that drive the tourist industry.

Tourism has thus become fully integrated into the mechanisms that direct the global economy. As such, profitability in the sector has to take account of the fact that states are disengaging themselves as much as possible from productive activities and the global economic crisis is becoming pervasive. It is thus necessary not to over estimate the attractiveness of particular places and of types of leisure or tourism. Nature and the rhythm of climate and seasons must be closely examined. Winter sports resorts, for instance, have had to bear the brunt of the capricious nature of snowfall, making it more likely that future choice of such resorts will emphasize places where ample amount of snow can always be relied on. Similarly, the troubles of the hotel industry, which has been hard hit by the economic crisis, will also induce greater prudence on the part of investors. Expansion in boom periods has not been based on the production of goods but rather on the large-scale, stock market speculation on enterprises in the service sectors which have the potential for high profit margins. This creates an artificial and deceptive impression of well-being that usually weakens sectors that are subject to such speculations, especially when their markets are experiencing an unhealthy degree of competition.

The tourist industry, including high-class tourism, is currently going through a crisis period. The combined effects of obsolescence and inefficient management, added to the difficulties of the stock market, have shaken even prestigious establishments. Investors have had to diversify and to look for more profitable niches such as hotels and golf courses. The French Riviera, for example, was once the world's most prestigious resort. Today, it is a pale image of its former self, with de luxe tourism giving way to pleasure boating. At Cannes, which faces Port Canto (a popular base for yacht owners from the Middle East), the Palm Beach Casino, which was rated number one in France in 1985, is now ranked only 23rd. This has resulted in a sharp fall in the casino's receipts, which is now barely two-thirds of its erstwhile figure. Its cumulative losses since the 1980s now total 170 million francs.

The structuring of the international tourist system thus requires comprehensive knowledge of the sources of tourists and the factors, in the larger global market, that determine their likely availability in any year. In the absence of adequate information, the leaders of the industry who set the limits of its growth and expansion, have the responsibility of compensating for the most notorious failures of expectations, especially in developing countries. Their insensitivity on this score has given rise to much disorder in the tourist industry and opened it to charges of neo-colonialism or even a veritable act of aggression. This situation is especially serious given the powerful means of communication, which large corporations have at their disposal, and the effect of negative images on their investments. Given that communication underlies the operations of the global economy, of which the tourist industry is an increasingly important sector, the overall impact of telecommunications has been decisive in the conquest of market share. Designed to reach everyone, telecommunications and telematics now rely upon satellite systems to efficiently induce global participation in the bouquet of programmes offered to potential tourists.

The tourist industry shows a high propensity for spatial expansion. This is the main reason for the increasing collaboration between enterprises which generally come together to counter the excessively threatening postures of other groups of firms. For example, British Airways, following an uninterrupted series of alliances and acquisitions, took the offensive to become a formidable competitor in July 1992 when it acquired a share of the capital of US Air, the fifth largest American airline. Airlines such as United and Virgin Atlantic reacted in order to stop this aggression of British Airways in the US market. The United Airlines, the number two American carrier, for instance, decided to answer this provocation by proposing to Air France that it be allowed to build up its operations at Paris's Charles de Gaulle-Roissy Airport, which would bring potential passengers with onward destinations in Europe to the French carrier. The economic windfall to France of United Airlines' transatlantic flights was calculated at 300 million francs a year with an additional 2500 jobs in all sectors concerned, notably hotels, transportation, services, and tourism.

Nonetheless, because of the inadequacy of supply and demand, the difficulties in totally mastering the economic and sociological parameters of the sector, and a market that has not been easy to dominate effectively, the tourist industry has sought to protect itself through courses of action organized around versatile teams and structures, in which people work together. This has given rise to strategies that bring together enterprises involved in touristic and leisure-oriented activities. This strategy requires that organizations and institutions engaged in the promotion of tourism must be prepared to provide a totality of services and facilities to ensure that the industry develops in a coherent and harmonious manner. The short-term perspective of this strategy has certainly proved to be of major importance but it is the medium and, above all, long-term horizons that are likely to be the more decisive.

Transmitting and Receiving Tourist Areas

The internationalization of tourism is based on the interplay of actions and reactions from areas that send and areas that receive leisure-seekers and holiday makers. Industrial societies, flush with capital and the technology that facilitates the general promotion of holiday making among all sections of the population, were responsible for unleashing these short-term, mass movements of individuals (OECD, 1994). The great fairs and congresses that are associated with commercial exhibitions in the area of leisure have encouraged the growth of creativity and the propensity to experiment with new experiences in recreation on an international scale. Thus, increasingly, the originality of tourism lies in travelling considerable distances in the pursuit of leisure and seeking new experiences outside the home, involving overnight stays.

When mass tourism first emerged, the existing poles of attraction were based on media-induced powerful images of places long frequented only by the privileged and wealthy. Such places included the French Riviera and the luxury resorts of California. Thus, when paid vacations were first legislated in France and significant reductions in rail fares were instituted, a small number of workers in the 1936–9 period descended on the Riviera in the hope of rubbing shoulders with – and thumbing their noses at – members of the moneyed class whom one expected to find there. But the infrastructures in such places were not the same for everyone. To be able to cater for large numbers of tourists, such centres first had to put in place new facilities, against what they had when only receiving small numbers of people with relatively high levels of purchasing power. The growth in demand speaks eloquently of the remarkable changes that took place within a relatively short time. Whereas in 1950, the number of international tourists was hardly more than 25 million, by 1960, the number had jumped to 71 million, by 1972 to close to 200 million and by the 1990s it had surpassed 370 million (OECD, 1989). International tourism is expected to continue its exponential rate of growth despite the global socio-economic crisis, with the number of people wanting to take a holiday outside their own country eventually surpassing 800 million.

In the period 1950–80, many countries put in place the necessary infrastructure to attract tourists from the countries of the Triad, adding their names to the list of areas of tourist attraction. In addition to areas with a Mediterranean climate, European tourist centres included areas with thermal spas, beautiful cities of historical interest, picturesque countryside and mountainous regions with winter-sports resort facilities. Internationalization of the industry entailed the extension of the supply sources into developing countries. Their exotic nature was meant to appeal to tourists who were in search of something different. However, unprepared both culturally and economically for a massive influx of tourists, the economies of these countries have become increasingly fragile and the threshold of their spatial freight capacity has quickly been reached. This hard tourism – as opposed to the soft tourism that displays more respect for local identities – has provoked various reactions. It has shown that globalization cannot go on without taking into consideration the socio-cultural roots of different societies called upon to receive large numbers of tourists.

Internationalized leisure has thus contributed to a heightened awareness of different milieux and their ecology and has sought to integrate them into the overall global environmental concept. Thus, the environment, far from being just a physical issue, now involves above all concern for individuals and entire societies through an appreciation of the need to preserve the essence of their culture within the ecological base that serves as its framework. Notwithstanding, the majority of tourists continue to come from the countries of the Triad. With some 350 million overnight stays away from home, France is the premier tourist destination of the world, followed by the United Kingdom which boasts of some 200 million tourists annually. In sum, Europe both sends and receives the largest number of visitors each year, indicating the continued importance that tourists attach to the curiosities of the Old World.

Conclusion

This discussion of transportation, trade and tourism within the context of a global economy emphasizes the extent to which international economic networks have become decisive in the complex interplay of existing mechanisms. It has also ensured a better appreciation of the developmental lag between countries, above all between those of the Triad on the one hand and developing countries on the other. This is often the source of serious imbalances and has given rise to the contention that transnational corporations cannot be an effective instrument of development in these countries. Indeed, if anything, it is feared that their role may be further to widen the geopolitical fault lines and intensify the collapse of social life in most parts of the Third World.

Clearly, trade, in all its forms, including tourism, and the transportation system and other infrastructures which support it take on a special significance in the restructuring of global geography. If humanity currently has at its disposal the technology and know-how necessary for its progress, it has not as yet succeeded in effectively regulating the flow of goods and people. It has also not mastered the mechanisms for effectively managing the operations of the global economy in a manner to promote either sustainable development or beneficial global change. Increasingly, the world is having to learn that the only form of development which can minimize the socio-spatial harshness of today's unbridled competition is development that is sustainable. In this regard, the present system of international exchanges needs to recognize more seriously the weaknesses in global restructuring, which have taken place since the end of the Second World War. Redressing their negative impact both on the environment and on the level of global conflict will be a veritable test of the extent to which an international political will is developing to direct and manage the inexorable forces of globalization.

References

Bauchet, P. (1988) *Le transport international dan l'économie mondiale*, (Paris: Economica).

Beaufort, H. de. (1995) *Guide mondiale d'économie et de géopolitique* (Paris: Editions Le Cherche Midi).

Dezert, B. and Wackermann, G. (1991) *La nouvelle organisation internationale des échanges* (Paris: SEDES).

Dollfus, O. (1994) *L'espace monde* (Paris: Economica).

Dollfus, O. (1995) *La nouvelle carte du monde* (Paris: Presses Universitaires de France).

Durand, K., Levy, J. and Retaille, D. (1994) *Le monde, espaces et systèmes* (Paris: Dalloz).

Harrington, J.W. and Warf, B. (1995) *Industrial Location: Principles and Practice* (Berlin: Reimer).

Johnson, M. and Johnston, G. (1994) *European Infrastructure 1994* (London: Sterling Publications Limited).

Kansky, K.J. (1963) *The Structure of Transportation Networks* (Chicago: University of Chicago, Department of Geography, Research Paper No. 84).

Merlin, P. (1991) *Géographie, économie et planification des transports* (Paris: Presses Universitaires de France).

Nehme, C. (1993) *Le GATT et les grands accords commerciaux mondiaux* (Paris: Les Editions d'Organisation).

OECD (1989) *National and International Tourism Statistics, 1974–1985* (Paris: OECD).

OECD (1994) *Tourism Policy and International Tourism in OECD Member Countries, 1993* (Paris: OECD).

OECD (1995) *Maritime Transport 1993* (Paris: OECD).

Pache, G. (1994) *La logistique: enjeux strategies* (Paris: Vuibert).

United Nations Environment Programme, Industry and Environment Programme Activity Centre (UNEP/IE/PAC) (1994) *Transport and Environment*, no. 1–2 (Paris).

Van Lier, H. and Taylor, P. (1993) *New Challenges in Recreation and Tourism Planning* (London: Elsevier).

Vigarie, A. (1993) *Transports et Marchandises sur les grands axes europeens–recherche de routes alternatives terre-mer* (Brussels: Rapport a la Direction Generale de la Commission des Communautes Europeennes, 2 vols).

Wackermann, G. (1991) *Tourisme et transport* (Paris: SEDES).

Wackermann, G. (1994) *Loisir et tourisme: une internationalisation de l'espace* (Paris: SEDES).

Wackermann, G. (1995a) *De l'espace national à la mondialisation de l'espace* (Paris: Ellipses).

Wackermann, G. (1995b) *Le transport de marchandises dans l'Europe de demain* (Paris: Le Cherche-Midi).

10 Spatial Financial Flows and Future Global Geography

The importance of money in the modern world can hardly be overstated. Human geographers, however, only really became interested in the geography of money and finance in the early 1980s. There were four main reasons for this belated discovery. The first was that, until the publication of David Harvey's *The Limits to Capital* in 1982, with its central chapter on money, credit and finance, the discipline had shown little interest in money, as distinct from capital, as a factor of spatial transformation. The second reason was that geographers could not help but notice the dynamic expansion of financial instruments, financial institutions, financial services and financial workforces that took place world-wide over the course of the 1980s. All these had to be factored into prevailing geographical explanations of growth and decline at particular spatial scales. The third reason was that the 1980s was also a period in which the negative effects of financial booms were dramatically underlined. It was a period overshadowed by periodic crises of insolvency and debt in both developed and developing countries. The fourth and final reason was that the geography of money and financial institutions had become an ever more obvious part of the fabric of economies, a grid of power that overlay the globe.

Within geography, research work on money and spatial financial flows has proceeded on three major fronts. The first is the geography of monetary exchange, credit and debt. Geographers interested in finance have concentrated on the relationship between monetary forms and territorial structures with particular emphasis on the relationship between nation-states, and financial systems. This is clearly of the utmost importance since it encompasses issues linking the operation of new types of money, the changing borders between

This chapter is based on a contribution by Andrew Leyshon and Nigel Thrift.

different kinds of financial activities and the regulation of all kinds of financial institutions.

A second area of work has been concerned with cities. Money and credit have to be produced and distributed. This production and distribution takes place in a few world cities such as London, New York and Tokyo, major urban centres where financial institutions and intermediaries gather together. Geographical work has tended to concentrate on the financial economies of these cities but recent research has broadened its remit, both by focusing on financial circulation and flows within smaller urban centres and by stressing the social and cultural, as well as the economic, determinants of these flows.

The third area of work is the most recent of all. It is concerned with processes of financial exclusion and the growth of alternative financial institutions and economies, which both mount a challenge to conventional financial systems and are more likely to echo social and cultural concerns. Institutions like regional banks, community banks, women's banks, rotating credit associations, local economic-trading systems, combined with the growth of more general trends like ethical investment, constitute an alternative financial system which geographers have only just begun to research.

This chapter considers these various areas of interest in spatial financial flows and their implications for future global geography. The first part addresses the geography of monetary exchange, credit and debt; the second explores the spaces of cities and finance; whilst the third considers the credit-determining role of the state. The fourth, then, looks at international financial flows and the global role of cities as social networks and the fifth section considers the growth of the geography of financial exclusion. What hopefully becomes clear from all this is that whereas the landscape of finance used to be considered the exclusive realm of economists, and to a lesser degree, sociologists and anthropologists, it is now also the preserve of geographers.

Geography of Monetary Exchange, Credit and Debt

The history of money and credit has also been its geography, and that geography has been constitutive of what money and credit now are. However, the importance of geography in the evolution of money has not always been recognized. In particular, there has been a

Postcard from French colonial Africa, showing peasants counting cowrie shells, used as money.
Source: Roger-Viollet.

general tendency among monetary historians to be insensitive to the interplay between money, space and place, to ignore the fact that monetary forms, practices and institutions are contingent to both space and time, and to fail to acknowledge that money has often evolved in order to solve the general problems of time–space co-ordination and enable social relationships to be extended across both space and time. To properly understand money, it is necessary to consider its historical geography. Each monetary form has its own geography. The transformation from one monetary form to another has important geographical implications. It is, therefore, possible to identify different geographies of money, associated with different types of money.

But before embarking on an outline of the different types of money and their associated geographies, it is also necessary to answer the rather important question: what exactly is money? In theory and practice money can be, and indeed has been, a wide range of physical objects from shells and teeth to metal and paper (Angell, 1930; Davies, 1994; Einzig, 1966; Galbraith, 1975). However, the material form that money takes is not so important as its ability to perform two key roles in the process of economic exchange. These are, first, to act as a medium of exchange and, second, as a store of value. In performing these two roles, money necessarily takes on two additional roles. These are as a unit of account and as a means of payment. The utility of money is, thus, that it acts both as a lubricant of exchange and as an independent expression of value. This duality of functions, while advantageous in many ways, has also served to introduce an important dynamics to monetary forms and practices. Indeed, it has been noted that the ability of different types of money to perform the functions of a medium of exchange and a store of value tends to be inversely proportional. Money forms which perform admirably in the capacity of the former tend to perform less well in the capacity of the latter. Thus, Dodd (1994:xviii) observed that:

> *legal tender notes tend to lose value over time as a result of inflation, and so are best used chiefly for exchange and payment purposes. Assets which store value stably over time or even appreciate in value, on the other hand, are linked to securities or other investments which make them difficult to convert into a form suitable for payment or exchange, perhaps losing value on conversion or carrying a time constraint delaying conversion.*

It is the differential performance of each and every type of monetary form in this regard which has introduced a dynamic element into the evolution of money. Monetary systems are thus characterized by a range of alternative and complementary forms of money. Among these can be identified: primitive money, commodity money, credit money, state credit money and virtual money. It is necessary to provide some brief description of each of these forms of money as a basis for emphasizing the continuing human innovativeness and development within the field.

The origin of *primitive money* is shrouded in time. Before the human invention of the use of money, exchanges revolved around barter. The advantages of money over barter are, of course, legion but there is a general consensus amongst historians and anthropologists that money did not arise in the first instance in order to circumvent the cumbersome awkwardness of barter. Rather, the origins of money are cultural, in the sense that the use of money arose in processes of exchange that were firmly non-economic in their orientation. The economic use of money occurred almost as an accidental oversight, a social discovery that followed on from well-established cultural practices. According to Davies (1994:23–4):

> *The most common non-economic forces which gave rise to primitive money may be grouped together thus: bride-money and blood-money,*

> *ornamental and ceremonial, religious and political. Objects originally accepted for one purpose were often found to be useful for other non-economic purposes, just as they later, because of their growing acceptability, began to be used for general trading also.*

In other words, money evolved from relatively narrow, culturally specific uses, later to take on a much broader range of social and economic functions commonly associated with it today.

Mapping the geography of primitive money would reveal a patchwork of discrete monetary systems scattered widely over space and time. Each system reflected specific social and cultural conditions. Their incidence, when mapped over time, reveals that the systems that survived longest, in many cases well into the modern period, were part of societies and cultures which maintained some degree of isolation from economies using modern money forms. Thus, the survival of primitive money is inversely related to the degree of contact between the societies in which it circulates and western culture. It is therefore no coincidence that it was in Oceania and parts of Africa that primitive money systems appear to have survived longest or long enough at least to have been documented by western anthropologists. Of course, it is also recognized that such documentations invariably sounded the death-knell for many primitive money systems, signalling as they did the end of isolation and the increasing degree of contact between such societies and a more powerful cultural form, which used modern money (Gewertz and Errington, 1995).

The range of objects and materials that have been used as primitive money is extremely large. The movement towards a more generalized money form, which gained a wider currency beyond the very specific social and cultural conditions that gave it birth, is generally argued to be linked to the growth in use of metallic-based primitive moneys. These were often made of a precious metal such as silver, which had a culturally determined economic value. This made them particularly suited to serving as a medium of exchange. Primitive moneys existed initially merely as lumps of silver, uneven in size and form. They readily served as the raw material for the manufacture of a whole series of metal artefacts. Thus, these metallic moneys had a potential use value as well as an exchange value. The first tentative steps towards the creation of modern money came with efforts to standardize the appearance of these primitive moneys

(Dalton, 1965). This process of standardization led to the eventual creation of coinage.

The development of coins led to the emergence of a set of monetary practices that revolved around the notion of ***commodity money***. In a system of commodity money, money functions as a medium of exchange and a store of wealth. The value of coins emanates from their embodiment of the value of the precious metals that they are made from. The early history of commodity money is thus in large part one of incursions and invasions since the development of coinage made it far easier to mobilize military action at a distance. As Davies (1994:108) describes it, Coins followed – indeed accompanied – the sword, so that payment for troops and for their large armies of camp-followers was generally the cause of minting. The Greek and Roman armies were paid in coin and well paid too in order to maintain their loyalty. It has been estimated that it would require 1.5 million silver denarii per annum to support a single Roman legion. Thus, the majority of silver flowing into Rome from mines scattered throughout the empire was transformed into coins to support the Roman army.

At the same time as the armies went on their military campaigns, they became vehicles of monetary expansion and incursion. The coins they took with them in time became used as money in the territories they conquered. The effect of these actions was to bring about a degree of financial homogenization over space. Thus, one way in which a monetary system gets established in a given territory is through the use of direct force. More powerful states (or colonizing powers) impose their money on weaker states, thereby easing economic integration and eliminating the uncertainties associated with monetary exchange. As Davies (1994:76) noted just such an episode occurred in Greece in the middle of the first millennium BC:

> *In 456 BC, Athens forced Aegina to take Athenian owls and to cease minting their own turtle coinage. In 449 BC, Athens in furtherance of greater uniformity issued an edict ordering all foreign coins to be handed in to the Athenian mint and compelling all her allies to use the Attic standard of weight, measures and money.*

This is an example of financial homogenization occurring across space already dominated by commodity money. But of more importance to the argument being presented here was the extension of commodity money networks into spaces previously dominated by exchange based upon primitive moneys.

As the volume of global trade gradually increased, it came to involve movements often across large geographical spaces. The inherent inconvenience of conducting exchanges at long distances using heavy coinage became evident. Indeed, until about the fifteenth century, the history of money includes heroic efforts to transport commodity money over long distances. However, during the fifteenth century, a new monetary era began due in large part to a series of social innovations which caused a radical shift in the time–space co-ordinates of the financial system. Again, as Davies (1994:174) puts it:

> *The modern monetary age . . . began with the geographic discoveries, with the full fruition of the Renaissance, with Columbus and El Dorado, with Leonardo da Vinci, Luther and Caxton; in short, with improvements in communications, minting and printing. A vast increase in money, minted and printed, occurred in parallel with an unprecedented expansion in physical and mental resources. The invention of new machines for minting and printing were in fact closely linked in a manner highly significant for the future of finance. At first, the increase in coinage was to exceed, and then just to keep pace with the increase in paper money; but eventually and inexorably paper was to displace silver and gold, and thereby to release money from its metallic chains and anchors.*

Beginning in the eleventh century then, and increasing in importance in the twelfth and thirteenth centuries, money began to take on a new form as money of account. This new monetary practice derived its name from its function. It is a measure of value used almost exclusively for accounting purposes.

The origin of this new form of money lay in the problems that monarchical rulers were facing in raising funds in a medieval monetary system based on commodity money. Under such a system, the supply of money was ultimately constrained by the European supply of silver and gold. Money of account emerged as a way of boosting the supply of money beyond the limits of minting. The emergence of money of account is thus linked to the growth of credit money. Its roots can be traced back to the focus on fiscal and monetary policies within the medieval monarchical state. However, the major force driving the development of credit money was the increasing complexity of commercial trade. The advent of *credit money* served to reshape the geography of the financial system. The clearest sign of the birth of this new monetary era was the increasing use of the bill of exchange. When the first bills of exchange appeared, presenting the first means of distanciating credit, distance was, not surprisingly, the crucial factor in calculating the maturity of a bill. Thus, the usuance or usance of a bill, that is, the period between its creation and its maturity, was simply an acknowledgement of distance. Kindleberger (1984:39) commented on this fact as follows:

> *Mails . . . took time. Bills were payable at sight, at usuance, or sometimes half-usuance or double-usuance. Usuance was the standard credit period for a given trade. From Geneva at the beginning of the sixteenth century it ran five days for Pisa, six for Milan, fifteen for Ancona, twenty for Barcelona, thirty for Valencia and Montpelier, two months for Bruges and three for London. From London, usuance was one month to Antwerp, two to Hamburg and three to the northern Italian cities. It was seldom changed: the one month between London and Antwerp lasted from the fourteenth century to 1789.*

As money and capital markets become more extensive, so geography was again crucial. There are numerous examples of the constitutive role of space in the development of money. The need to finance highly profitable but also risky long-distance voyages led to the invention of perpetually transferable shares. More important even than the invention of such new financial instruments was the way in which the constraints of space were overcome so as to make these instruments tradable over greater and greater distances. This involved marrying improvements in transport and communications to specific market nodes, usually large urban centres, so as to produce an increasingly compressed financial space (Castells, 1989; Harvey, 1989).

Money, Time, Space and the Urbanization Process

Credit money also came to represent social power and the ability to use money to make more money, most typically through a form of circulation or financial flows. Money is used to buy commodities which, when combined within a particular labour process, produce a fresh commodity to be sold at a profit. Unlike 'usuance', the special interest in this case is the time it takes the labour process to enable capital to complete its circulation from money back to money plus profit. Each labour process has its turnover time and increasing fragmentation in the division of labour poses serious problems of co-ordination under conditions where profit is the sole objective. These problems are also overcome through the *credit money* system which serves to co-ordinate divergent turnover times. The acceleration of turnover time entails increased productivity of labour and therefore yields competitive advantage. It often becomes an objective of technological change, encouraging the production in the aggregate of too much capital relative to the opportunities available to employ that capital. What happens to this surplus capital in the context of spatial financial flows led Harvey (1985) to propose the existence of three circuits of flows: primary, secondary and tertiary. The primary circuit occurs during the initial period of creating the surplus capital on the basis of the existing relations between capital and labour.

The tendency to produce surplus capital or over-accumulate in the primary circuit, at least on a periodic basis, creates the condition for switching capital flows into a *secondary circuit* to convert them into fixed assets, whether in the form of fixed capital that further enhances the production process (notably machinery) or fixed capital that functions as a physical framework for production or consumption. Fixed assets comprise the built environment: factories, warehouses, public offices, shops, highways, canals, docks and harbours, sewers, residential houses, schools and hospitals, and urban infrastructure generally. Fixed capital in the built environment is immobile in space in the sense that the value incorporated into it cannot be moved without being destroyed. In other words, investment in the built environment represents capital flows through a secondary circuit and entails the creation of a whole physical landscape for purposes of production, circulation, exchange and consumption.

A general condition for the flow of capital into this secondary circuit is the existence of a functioning capital market and, perhaps, of a state willing to finance and guarantee long-term, large-scale projects with respect to the creation of the built environment. At times of over-accumulation, a switch of flows from the primary to the secondary circuit can be accomplished only if the various manifestations of over accumulation can be transformed into money capital that can move freely and unhindered into these forms of investment. Since the production of money and credit is a relatively autonomous process, the financial and state institutions controlling the process must be conceived of as a kind of collective nerve centre governing and mediating the relations between the primary and secondary circuits of capital. The nature and form of these financial and state institutions and the policies they adopt are critical in checking or enhancing flows of capital into the secondary circuit of capital or specifically transportation, housing, public facilities and so on. An alteration in these mediating structures can, therefore, affect both the volume and the direction of capital flows by constricting movements down some channels and opening up new conduits elsewhere.

The *tertiary circuit of capital* comprises, in turn, investment in science and technology. This has the objective of harnessing science and technology for purposes of enhancing productivity and thereby contributing to the processes that continuously revolutionize the productive forces in society. It also embraces a wide range of social expenditure that relates primarily to the processes of reproduction of labour power and its qualitative improvement through investments in education and health.

Pressures within this system of financial flows lead to the systematic pursuit of the annihilation of space by time. Space, however, can be overcome only through the production of space through improved systems of transportation and communication. The cost, speed and capacity of the transport system relate directly to accumulation, because of the impacts these have on the turnover time, of capital. Natural landscapes are replaced by built landscapes shaped through competition to the requirements of accelerating accumulation. Landownership is rendered subservient to money power as a higher-order form of property and land; indeed, it becomes a form of fictitious capital. Thus, control over the production of space becomes integrated into the credit system.

The acceleration of turnover time, in turn, entails the writing off of the value of fixed capital at an accelerating rate (no matter what its physical lifetime) and even to replacing it before its economic lifetime is out. Machinery, buildings, and even whole urban infrastructures and life-styles are made prematurely obsolescent. 'Creative destruction' is necessary if the system of continuous accumulation and financial flows is to survive. But the capacity to set such processes in motion depends upon conditions within the credit system – the supply and demand for money capital, the rate of money growth, and so on. Cyclical rhythms of investment and disinvestment in machinery and in built environments thus connect to interest rate movements, inflation and growth of money supply, and to phases of unemployment and expansion. Time horizons thus come to be defined more and more tightly via the credit system.

The social implications of spatial financial flows are thus of infinite interest. Within the context of the built environment, fixed spaces and times can be overcome only through creative self-destruction. Thus, behind the material solidity of buildings and other infrastructural facilities lurks an insecurity deriving from their position within the circulatory system of capital. That insecurity relates to the issue of how much time such fixed capital can be allowed to occupy its relative space within the capitalist free-market economy. Processes as diverse as suburbanization, deindustrialization and restructuring, gentrification and urban renewal, through to the total reorganization of the spatial structure of the urban hierarchy are part and parcel of a general process of continuous reshaping of geographical landscapes to match the imperative of accelerating turnover time and processes of capital accumulation through the credit money system. Thus, the urbanization process is shaped by spatial financial flows and the circulation of money capital in time and space.

The Nature of State Credit Money

Initially, the growth in the credit money system led to a decline in the influence of the state over money. Nonetheless, the state still remained important, particularly in the creation and validation of commodity money. Despite the growth of credit money and of money of account, commodity money was still widely perceived as the ultimate source of value within the monetary system. However, until the late seventeenth century, the responsibility for regulating the financial system fell on growing bands of private merchants and bankers who effectively oversaw and supervised the circulation of money and credit, often through informal but closely-knit networks exchanging business and information.

Quinn (1995) noted, in fact, that for much of the seventeenth century, it was a network of goldsmith-bankers in the City of London who used their reserves of gold to develop and supervise markets in short-term debt. The possession of a receipt or note from a goldsmith was evidence of ability to pay and of money in the bank. In this way, these early-banking institutions facilitated the growth of credit-money instruments such as bank notes and cheques, helping them become established as readily acceptable means of payment.

But these privatized financial instruments and the institutions that issued them were never entirely free of the influence of the state. For example, many of London's goldsmith-bankers came to grief in the 1670s when, in the face of their refusal to increase their loans to the Crown in order to fund a further round of naval expansion, Charles II prohibited the payment of royal debt. From the beginning of the eighteenth century onwards, the state's influence over the finance system began to reassert itself. There were three reasons for this. The first was the creation of the national debt (chiefly, it should be noted, through the need to finance wars). The financing of debts by the state had often been haphazard so that the total picture prior to 1700 was best described as chaotic. However, after 1700, markets for debt broadened and deepened and, as a result, government debt was consolidated and extended. States became borrowers on a large scale. For example, as late as 1824, the paid-up capital of all domestic companies trading on the London Stock Exchange was £34 million. This compared with a public debt of over £800 million (Neal, 1991). The second reason for increasing state involvement was the reduction in the powers of most monarchical regimes. Once absolutist states were overthrown, the risk of arbitrary seizure of assets was much reduced and lending to states became a more sober and reliable business. Finally, there was the creation of national banks which in time took on a range of regulatory functions. Public banks had been founded in Europe from an early date. The first state deposit bank had been established in Geneva in 1407. Similar banks followed in Spain and Sicily. Increasing sophistication came with the founding of the Bank of Amsterdam in 1608. The first true state central bank, the Swedish Riksbank, was established in 1656 and was taken over by the Swedish state in 1668. However, it is still generally thought that it was the founding of the Bank of England in 1694 that signalled the most important innovation in state finance (Fay, 1988). The Bank was founded to market the national debt but ended up managing it and regulating the British financial system to boot. Yet, the history of the Bank of England is a history of only grudging acceptance of its role as the focus of the British financial system, a role forced on it by various financial crises. Most important of all, under the Bank Act of 1844, the Bank became a lender of last resort. By 1890, it was acting as a full lender of last resort, arranging to guarantee the liabilities of Barings, in a way that it would not have done in previous years (Roberts and Kynaston, 1995).

The idea of a central state bank acting as a lender of last resort then spread to the rest of the world. In doing so, it produced a new kind of money, what Keynes called *state credit money*, in which the state becomes the guarantor of public debts, based on its ability to issue money. What distinguishes bank credit money from state credit money is the fact that the former defines a private debt whereas the latter does not. Thus, the determining factor is not the private or public character of the institution which is getting spontaneously indebted but the nature of its debt (Cencini, 1988:46). State credit money reached its apotheosis in the years after the Second World War. At that time, an international system of state money seemed to be coming into existence as a result of the Bretton Woods agreement of 1944 and the subsequent post-war settlement. The function of lender of last resort between nations was a role discharged by the World Bank and the International Monetary Fund (IMF), as well as by the central bank swaps which grew up outside the IMF.

Nonetheless, the ascendancy of state credit money was always contested. Even at its height, the state was never unambiguously in charge of money. The internationalization of money, the growth of the power of the banks, and the increase in private commercial lending, as opposed to state lending, provided countervailing forces. Even in 1906, one London commentator, with a degree of prescience if not accuracy, could write that Lombard Street has been more under the control of the Japanese banks than of the Bank of England (cited in King, 1992:283). Central State banks were able to fend off countervailing forces through more active banking strategies and the Bretton Woods agreement seemed to signal the final success of state credit money. However, by the late 1960s and early 1970s, the strength of countervailing forces had been significantly boosted. Much of the international capital market had moved outside of state control through the growth of the Eurocurrency and other capital markets. Other woes piled up thick and fast. The opportunities to create and distribute fictitious capital became much greater because of the invention of new financial instruments (some of which were precisely designed to avoid state regulation). State legislation of successive bouts of deregulating encouraged financial-service companies to move across established regulatory boundaries. Systems of monetary transmission and clearing went electronic, becoming harder to track. The result was that it was no longer possible, as it had still been in the early 1970s, to use the weight of government in the Gross National Product to control or even direct the private sector.

As the clarity of a global geography of state credit money begins to blur somewhat, it is possible to talk about a new kind of money coming into existence which may be defined as *virtual money* or *book entry money.* Money becomes an activated double book entry, a spontaneous acknowledgement of debt that is no longer a commodity. This new system of fleeting instants is based on quasi-private institutions and on the full range of instruments of fictitious capital (Hart, 1986) where money is accepted in the belief that whoever offered it will make it good in the future.

With the rise of virtual money, the national financial space, the territory which has so long been unproblematically accepted as the ground upon or between which economic processes are played out, has been undermined or decentred (Corbridge, 1994) because of a process of deterritorialization. This development has important geographical consequences. National financial spaces have served to channel economic processes in distinctive ways so that the patchwork that is the nation-state system is also, to a certain extent at least, a geography of capitalisms (Agnew, 1994; Christopherson, 1993; Cox, 1986; Lash and Urry, 1994). However, processes of financial homogenization at play within the late twentieth century are conspiring to make national financial spaces increasingly similar, as may be seen in the equalization of short-term interest rates and the convergence of national systems of financial regulation (Moran, 1991). In short, the distinctiveness of national financial space is being eroded, reflecting the empowering of financial capital over space and the disempowering of other economic actors, particularly those rooted in space, such as regulators and planners. As Robinson (1991:13) has aptly argued:

> *With the triumph of flow, it becomes increasingly difficult for urban governments and planners to afford cohesion . . . New kinds of network – physical and virtual – subvert traditional territorial formations, deconstructing and recomposing them in more complex*

ways. In the process, established forms of urban community, culture and sensibility are disrupted. The key issue, then, is whether it is possible to manage these new dynamics of deterritorialization.

The taken-for-granted world of financial space is being undermined in other ways too. For example, the growth of electronic or plastic money is not tied to the nation-state in the same way that fiduciary money is. Indeed, it can be argued that the international credit cards issued by companies such as Visa, Mastercard, American Express and so on represent a new form of post-national money which grants the holders of such money easy movement across different financial spaces, transcending the need to translate different forms of fiduciary money from one national form to the other.

However, as financial institutions and financial capital have increasingly strayed beyond the borders of national financial space, so a new post-national financial geography has begun to emerge. This new geography is detached from a necessary congruence by a set of tightly drawn national borders, as characterizes the traditional state form. The post-national state of the financial system transgresses national borders. At the heart of this geography sits a handful of increasingly powerful and influential financial centres that are made up of a dense network of financial institutions and markets, which increasingly radiate power on a global scale (Thrift, 1994; Thrift and Leyshon, 1994). It is to the geography of these financial centres that we now turn.

International Finance and Cities of Social Networks

The depiction of international finance that is most often given credence is one of a striving, instrumental social world which mirrors a model of money as a quality-less, rationalizing medium of pure exchange and accumulation. This is a formative motif in the work of writers like Harvey (1989), Giddens (1991) and Habermas (1989), following on, especially, from the work of Simmel (1990). As Zeliser (1994:11) puts it, sociologists still accept with a remarkable lack of skepticism the notion that once money invades the realm of personal relations it inevitably bends those relations in the direction of instrumental rationality.

Yet, as in the world of ordinary life, perhaps even more so in the world of international finance, money does not have these effects for four reasons. First, money is socially and culturally defined and used. It is not a fully transferable commodity which can be causally reproduced across diverse social situations. Rather, it has multiple-use meanings depending on the different monetary forms and the different contexts in which these are deployed. Values and social relations reciprocally transform money by investing it with meaning and social pattern. Second, money depends on the constitution of reciprocity and trust. In part, trust in money depends upon money as a formal social institution or money in general. But, in part, it also depends upon trust in the outcome of actual transactions which depend upon a mixture of particular people, monetary forms and regulations, which will often involve social ties. Third, money relies upon (and indeed is) information. As Dodd (1994:26) observes:

> *Information plays a substantive role in the constitution of monetary networks which cannot be reduced to the function of a lubricant within a processing monetary order. The role of information in the transaction of money is not confined merely to observation about an external and independent economic environment. Information is on the contrary the defining feature of money networks, for they are networks of, not just containers of, information.*

Thus, money cannot be deployed without information. Obtaining this information requires social networks which rely, to a greater or lesser extent, on trust. Fourth, money relies on power. But this power cannot be reduced to just the possession of money. To deploy money to best advantage requires communicative power. This demands credible social infrastructures which take time and effort to construct. Money is clearly a part of the construction of these infrastructures but it is only ever a part.

Thus, the business of international finance is perhaps best characterized as being made up of a series of interconnected social networks dealing in particular monetary forms. These rely for their maintenance on the deployment, quite apart from money, of three different but interconnected resources which contribute towards the construction of trust. First, there are the formal knowledge systems based on professional training, especially today in tertiary institutions or in-house training. Second, there is knowledge gleaned from the media (Parsons, 1989; Boden and Molotch, 1995). In international finance, nearly all the relevant media are relatively specific, often circulating to only a few hundred subscribers. Third, there is social knowledge, knowledge about who knows who, who knows what, who is important and so on. Most of this knowledge can still only be obtained through face-to-face interaction. As Kay (1995:7) notes:

> *effective communication depends on non-verbal, possibly even non-visual, cues. The barely perceptible hesitation introduced by a satellite link invited misunderstanding when we are used to interpreting hesitation in other ways. Shared experience and values are a central element in developing trust. The linking of social and commercial relationships increases the penalties for opportunistic behaviour. That is why the lunch room is a central facility of the City of London and extensive entertainment an integral part of Japanese business culture.*

It is this last resource – social knowledge – that we want briefly to note here by reference to the case of the City of London.

In the past, the social nature of international finance has chiefly been made apparent through tightly defined social networks based on a potent mixture of gender, ethnicity and class. Thus, in the City of London, the chief networks were based on masculinity defined in terms of a particular British upper and upper middle-class background or, alternatively, various émigré origins, especially but not only a German refugee background. These networks were closely intertwined by dint of kinship and a series of common social arenas, including chophouses, freemasonry, London clubs and so on. In New York, similar kinds of networks flourished, but were based, especially, on various forms of Jewish ethnicity and social arenas.

Since the breakdown of the Bretton Woods system, these close-knit networks have been subject to various forms of attrition. The marked growth of numbers employed in international finance has produced a dilution of their influence and has also broadened the social base upon which international finance is built. Most especially, it has involved more women and more employees, generally, from diverse ethnic and class backgrounds. In other words, international finance has become more cosmopolitan and more multicultural.

However, the need to obtain social knowledge in order to build up trust, a need rooted in the constitution of social networks, does not seem to have dissolved in this more heterogeneous situation. Indeed, in some senses, the need has increased since trust has to be worked at more now than in the past. Thus, social networking has become an even more important activity because of the increased amounts of knowledge of all kinds that now circulate, the need to interpret this knowledge, the need to keep certain kinds of knowledge exclusive and, therefore, the subject of competitive advantage, and the need to snap up unexpected business opportunities. The result has been what Thrift (1994) referred to as 'an interactional frenzy' in both London and New York. But this frenzy is more and more based on the heterogeneity of the social networks which allow increasingly different experiences and values to be shaped in formalized ways. Such formalization is necessary in order to stabilize and give some basic coherence to the experiences and values garnered in the course of financial interactions. Thus, a more structured space of mediated and face-to-face interaction or interpretation has been constructed in recent years. This draws on the old social practices of the city but has extended and changed them.

Specifically, five formalizations of these social practices can be identified. The first is the formalization of the social encounter. Not only are more and more people trained in self-presentation (such as by the use of video) but the whole encounter is now orchestrated by the use of devices such as business cards which have become endemic. The second formalization has been the extension of old social arenas. Like the business card, the business lunch is now *de rigueur* and has become one of the main ways in which knowledge is exchanged and social contact kept up. The third formalization is the construction of new social arenas. Of these, the most important has undoubtedly been the almost exponential growth of conferences and conventions. A fourth formalization has been in the technological back-up of face-to-face contacts. The telephone, fax, e-mail, pager, and other forms of telecommunication are all now used as ways of supplementing face-to-face interactions, having become cheap and easy to use in this fashion. The fifth formalization is the attempt to build up long-term networks without the benefit of gender, ethnic and class homogeneity as in the past. This has been achieved in two ways: one by the use of the relationship manager, and two, by the judicious use of two- or three-year secondments to other firms with overseas offices.

It is through international social networks, such as those documented for the City of London above, that the international financial system is ordered. The emergence of such social networks, which are at once local and global, are just one indication of an emergent post-national financial system to which reference has been made earlier. This post-national financial system is a more dynamic and fluid entity than the earlier nationally-based financial system. Its borders are constantly in motion, sometimes expanding, bringing new territories into the global process of financial exchange, sometime contracting, when the financial system withdraws from central places and spaces, usually in the wake of an economic or financial crisis of one kind or the other. At present, the borders of this financial system are contracting, withdrawing from spaces that were once incorporated totally into national financial spaces. In the following section processes of financial exclusion, associated with the withdrawal of the financial system from particular territories and places, are considered.

Geographies of Financial Exclusion

In the 1990s there emerged a general sense of unease about finance that, for the most part, was absent during the 1980s. An important reason for this unease is the spectre of debt. This had darkened an increasing proportion of the world's surface during the 1980s including developed countries in the 1990s. In the wake of financial crisis exclusion has followed bringing with it attendant social and economic problems. Although originally a crisis of the 1980s, considerable effort continues to be expended upon documenting the tragic legacies of debt crisis in less-developed countries, a crisis which continues to be endured by large parts of the developing world (Corbridge, 1992, 1993; George, 1992; Golub, 1992; Watts, 1994). The scale and duration of this crisis prompted Castells (1993:37) to argue that a significant part of the world population is shifting from a structural position of exploitation to a structural position of irrelevance as far as the global capitalist system is concerned. This is what Castell terms the Fourth World, a disadvantaged space made up of marginalized economies in the retarded rural areas of three continents and in the sprawling shanty towns of African, Asian and Latin American cities.

But it is not just the Fourth World that is disadvantaged in this way. New spaces of indebtedness continue to appear in other parts of the world. The former communist countries of east-central Europe have been transformed into highly indebted capitalist economies (Gibb and Michalak, 1993; Altvater, 1993). Meanwhile, the early part of the 1990s also saw the outbreak of a developed-countries debt crisis in core capitalist economies (Leyshon and Thrift, 1993; Macdonald, 1992). In sharp contrast to the optimism surrounding the financial explosion of the 1980s (Bond, 1990; Magdoff and Sweezy, 1987), the 1990s have been characterized by a rash of financial crises founded in high levels of corporate debt (Hallsworth, 1991; Fagan, 1990), property debt (Ball, 1994; Coakley, 1994; Daniels and Bone, 1993; Hallsworth and Bone, 1993; Hamnett, 1994; Oizumi, 1994; Pryke, 1994; Warf, 1994) and high levels of personal indebtedness (Ford, 1988, 1991).

The consequences of this latest debt crisis has been a round of financial restructuring that has created new geographies of financial exclusion within economies such as Britain and the United States (Leyshon and Thrift, 1994, 1995). At present, these problems are most acute in the United States where, in an effort to retreat to a more affluent client base, large tracts of US cities have been faced with the closure of banking infrastructures with catastrophic economic consequences for the population abandoned in this way. The overwhelming majority of these communities is African-American and Hispanic, which are already struggling with the problems caused by institutional racism within the financial services industry (Bates, 1991; Bond, 1991; Brimmer, 1992; Canner, Passmore and Smith, 1994; Caskey, 1994; Cloud and Galster, 1993; Dymski and Veitch, 1992, 1995; Galster, 1990; Grown and Bates, 1992; Jenster and Overstreet, 1990; Leven and Sykuta, 1994; Robinson, 1991; Rostein and Duncan, 1991; Squires, 1992; Squires and O'Connor, 1993). Processes such as these are leading to a deepening of uneven development within urban arenas, the economic and racial dimensions of which have been captured by descriptions such as spatial apartheid (Davies, 1992). Whereas in the past, the ghetto of the United States inner city was a space in which all of the institutions of the dominant society were reproduced in a parallel set of institutions, in the contemporary US ghetto, such institutions have largely disappeared (Lash and Urry, 1993:156), including financial institutions.

More hopefully, there are signs that spatially particular, rather than parallel, financial institutions are beginning to emerge in such spaces of exclusion (Barnekov and Jabbar-Bey, 1994; Taub; 1988). Although these micro-financial institutions, such as community banks and credit unions, have brought some relief to areas of financial abandonment in many inner-city areas of the United States, they are clearly not panaceas. This point is recognized by many supporters of the new trend who also make the case for a much stronger regulation of the financial-services industry to prevent areas being abandoned in the first place (Dymski, Epstein and Pollin, 1993).

There is growing evidence to suggest that spaces of financial exclusion are now beginning to open up in Europe as well. Efforts to document their emergence have begun and include, for instance, the studies of Leyshon and Thrift (1995, 1996). There are also proposals to build alternative financial institutions which can provide the beginnings of a bulwark against the worst excesses of financial capitalism.

Conclusion

This chapter has attempted to emphasize that the institutions of money and finance are at the centre of modern economic geography. Thus, it has been demonstrated that the history of money and finance is as much their geography, that money and financial institutions currently operate from well-defined urban locations, and that the result of their current operations is to exclude many people and parts of the world from the emerging global system of interactions.

Furthermore, the geography of money and finance has moved well beyond the level of conceptualization of its early days. The subject has increasingly come to mirror the diversity of its object. This makes it difficult to hazard a guess concerning future directions of research. Nonetheless, it is possible to anticipate that research in the field is likely to follow three main directions. First, increasing emphasis is likely to be placed on the rise of institutional investors as a power in the financial world. These investors have a specific urban geography and in turn they are creating a new geography of investment and exclusion. Second, there is likely to be an increasing emphasis on the financial consumer, and especially on the degree to which these consumers know about and are, therefore, able to exert influence and agency in modern financial systems. In this, they will be aided by new technological developments which hold the possibility of giving consumers more control over their financial lives.

Finally, there will be more emphasis on the culture of money. Increasingly, there is growing appreciation that money is not an abstract instrument, simply available for manipulation through monetary policy. It is made up of many modes of use as well as means and forms of representation. Explaining its culture will mean that the geography of money will increasingly intersect with the current cultural turn in human geography. Just as money gets everywhere so will the geography of money.

References

Agnew, J.A. (1994) The territorial trap: the geographical assumptions of international relations theory, *Review of International Political Economy*, 1:53–80.

Altvater, E. (1993) *The Future of the Market: an Essay on the Regulation of Market and Nature after the Collapse of Actually Existing Socialism* (London: Verso).

Angell, N. (1930) *The Story of Money* (London: Cassell).

Ball, M. (1994) The 1980s property boom, *Environment and Planning: A*, 26:665–95.

Barnekov, T. and Jabbar-Bey, R. (1994) Credit and development financing for low-income and minority communities, *Regions: The Newsletter of the Regional Studies Association*, 188 (December):4–8.

Bates, T. (1991) Commercial bank financing of white- and black-owned small business start-ups, *Quarterly Review of Economics and Business*, 31:64–80.

Boden, D. and Molotch, H. (1995) The compulsion of proximity, in R. Friedland and D. Boden (eds) *Now/Here: Space, Time and Modernity* (Berkeley: University of California Press).

Bond, P. (1990) The new US class struggle: financial industry power vs grass-roots populism, *Capital and Class*, 40:151–81.

Bond, P. (1991) Alternative politics in the inner city: the financial explosion and the campaign for community control of capital in Baltimore, pp.141–68 in M. Keith and A. Rogers (eds) *Hollow Promises: Rhetoric and Reality in the Inner City* (London: Mansell).

Brimmer, A.F. (1992) The dilemma of black banking: lending risks vs. community service, *Review of Black Political Economy*, 20:3–29.

Canner, G.B., Passmore, W. and Smith, D.S. (1994) Residential lending to low-income and minority families: evidence from the HMDA data, *Federal Reserve Bulletin*, February:79–108.

Caskey, J.P. (1994) Bank representation in low-income and minority urban communities, *Urban Affairs Quarterly*, 29:617–38.

Castell, M. (1989) *The Informational City: Information, Technology, Economic Restructuring and The Urban-regional Process* (Oxford: Blackwell).

Castells, M. (1993) The informational economy and the new international division of labour, pp. 15–43 in M. Carnoy, M. Castells, S.S. Cohen and F.H. Cardoso (eds) *The New Global Economy in the Information Age: Reflections on Our Changing World* (University Park, PA: Pennsylvania State University Press).

Cencini, A. (1988) *Money, Income and Time: A Quantum-Theoretical Approach* (London: Frances Pinter).

Christopherson, S. (1993) Market rules and territorial outcomes: the case of the United States, *International Journal of Urban and Regional Research*, 17:274–88.

Cloud, C. and Galster, G. (1993) What do we know about racial discrimination in mortgage markets?, *Review of Black Political Economy*, 22:101–20.

Coakley, J. (1994) The integration of property and financial markets, *Environment and Planning: A*, 26:697–713.

Corbridge, S. (1992) Discipline and punish: the new right and the policing of the international debt crisis, *Geoforum*, 23:285–301.

Corbridge, S. (1993) *Debt and Development* (Oxford: Blackwell Publishers).

Corbridge, S. (1994) Maximizing entropy? New geopolitical orders and the internationalization of business, pp. 281–300 in G. Demko and W. Wood (eds) *Reordering the World: Geopolitical Perspectives in the Twenty-first Century* (Boulder, CO: Westview).

Cox, A. (ed.) (1986) *State, Finance and Industry: A Comparative Analysis of Trends in Six Advanced Industrial Economies* (Brighton: Harvester Wheatsheaf).

Dalton, G. (1965) Primitive Money, *American Anthropologist*, 67, 1:44–65.

Daniels, P.W. and Bobe, J.M. (1993) Extending the boundary of the City of London? The development of Canary Wharf, *Environment and Planning: A*, 25:539–52.

Davies, G. (1994) *A History of Money: From Ancient Times to the Present Day* (Cardiff: University of Wales Press).

Davies M. (1992) Who killed Los Angeles? A political autopsy, *New Left Review*, 197:3–28.

Desai, M. (1988) Foreword to A. Cencini (1988).

Dodd, N. (1994) *The Sociology of Money: Economics, Reason and Contemporary Society* (Cambridge: Polity Press).

Dymski, G.A., Epstein, G. and Pollin, R. (eds) (1993) *Transforming the US Financial System: Equity and Efficiency for the 21st Century* (Armonk, NY: M.E. Sharpe).

Dymski, G.A. and Veitch, J.M. (1992) Race and the financial dynamics of urban growth: Los Angeles as Fay Wray, pp. 131–58 in G. Riposa and C. Dersch (eds) *City of Angels* (Los Angeles, CA: Kendal Hunt Press).

Dymski, G.A. and Veitch, J.M. (1995) Taking it to the bank: race, credit and income in Los Angeles, pp. 150–79 in R.D. Bullard, J.E. Grigsby III and C. Lee (eds) *Residential Apartheid: the American Legacy* (Los Angeles, CA: University of California at Los Angeles, Center for Afro-American Studies).

Einzig, P. (1966) *Primitive Money*, 2nd edition (Oxford: Pergamon).

Fagan, R.H. (1990) Elders IXL Ltd: finance capital and the geography of corporate restructuring, *Environment and Planning: A*, 22: 647–66.

Fay, S. (1988) *Portrait of an Old Lady* (Harmondsworth, Middlesex: Penguin Books).

Ford, J. (1988) *The Indebted Society* (London: Routledge).

Ford, J. (1991) *Consuming Credit: Debt and Poverty in the United Kingdom* (London: Child Poverty Action Group).

Galbraith, J.K. (1975) *Money: Whence It Came, Where I Went* (London: Penguin).

Galster, G. (1990) Racial stereotyping by real estate agents: mechanisms and motives, *Review of Black Political Economy*, 19:39–63.

George, S. (1992) *The Debt Boomerang: How Third World Debt Hurts Us All* (London: Pluto Press).

Gewertz, D.B. and Errington, F.K. (1995) *Duelling Currencies in East New Britain: The Occidentalism, Images of the West* (Oxford: Clarendon Press).

Gibb, R. and Michalak, W.Z. (1993) Foreign debt in the new east-central Europe: a threat to European integration? *Environment and Planning: C – Government and Policy*, 11:69–85.

Giddens, A. (1991) *Modernity and Self-identity* (Cambridge: Polity Press).

Golub, S. (1992) The political economy of the Latin American debt crisis, *Latin American Research Review*, 26:175–215.

Grown, C. and Bates, T. (1992) Commercial bank lending practices and the development of black-owned construction companies, *Journal of Urban Affairs*, 14:25–41.

Habermas, J. (1989) *The Structural Transformation of the Public Sphere* (Cambridge: Polity Press).

Hallsworth, A.H. (1991) The Campeu takeovers - the arbitrage economy in action, *Environment and Planning: A*, 23:1217–23.

Hallsworth, A.G. and Bone, J.M. (1993) How the interest rate cat ate the Dockland's canary, *Area*, 25:64–9.

Hamnett, C. (1994) Restructuring housing finance and the housing market, pp. 281–308 in S. Corbridge, N. Thrift and R. Martin (eds) *Money, Power and Space* (Oxford: Blackwell).

Hart, K. (1986) Heads or tails? Two sides of the coin, *Man*, 21:641–2.

Harvey, D. (1985) *The Urban Experience* (Baltimore: The Johns Hopkins University Press).

Harvey, D. (1989) *The Condition of Postmodernity: An Enquiry into the*

Origins of Cultural Changes (Oxford: Blackwell).
Jenster, V. and Overstreet, G.A. (1990) Planning for a nonprofit service, a study of the United States credit union, *Long Range Planning*, 23:103–11.
Kay, J. (1995) *The Foundations of National Competitive Advantage* (London: ESRC Fifth Annual Lecture).
Kindleberger, C.P. (1984) *Financial History of Western Europe* (Oxford: Oxford University Press).
King, W. (1992) *History of the London Discount Market* (London: Frank Cass).
Lash, S. and Urry, J. (1994) *Economies of Signs and Space* (London: Sage).
Leven, C.L. and Sykuta, M.E. (1994) The importance of race in home mortgage loan approvals, *Urban Affairs Quarterly*, 29:479–89.
Leyshon, A.and Thrift, N. (1996) *Money/Space: Geographies of Monetary Transformation* (London: Routledge).
Leyshon, A. and Thrift, N. (1993) The restructuring of the financial services industry in the 1990s: a reversal of fortune, *Journal of Rural Studies*, 9:223–41.
Leyshon, A. and Thrift, N. (1995) Geographies of financial exclusion: financial abandonment and the United States, *Transactions, Institute of British Geographers*, 20:312–341.
Leyshon, A. and Thrift, N. (1994) Access to financial services and financial infrastructure withdrawal: problems and policies, *Area*, 26:268–75.
Macdonald, H.I. (1992) Special interest politics and the crisis of financial institutions in the United States, *Environment and Planning: C – Government and Policy*, 10:123–46.
Magdoff, P. and Sweezy, P. (1987) *Stagnation and the Financial Explosion* (New York: Monthly Review Press).
Moran, M. (1991) *The Politics of the Financial Services Revolution* (London: Macmillan).
Neal, I. (1991) *The Rise of Financial Capitalism: International Capital Markets in the Age of Reason* (Cambridge: Cambridge University Press).
Oizumi, E. (1994) Property finance in Japan: expansion and collapse of the bubble economy, *Environment and Planning: A*, 26:199–213.
Parsons, W. (1989) *The Power of the Financial Press* (London: Edward Elgar).
Pryke, M. (1994) Finance, property and layers of newspapers ironies, *Environment and Planning: A*, 26:167–70.
Quinn, S. (1995) Balances and goldsmith-bankers: the co-ordination and control of interdebt clearing in seventeenth-century London, in D. Mitchell (ed.) *Goldsmiths, Silversmiths and Bankers: Innovation and the Transfer of Skills, 1550–1750* (London: Alan Sutton Publishing Limited, Centre for Metropolitan History, Working Paper Series No.2).
Roberts, R. and Kynaston, D. (eds) (1995) *The Bank of England: Money,*

Power and Influence, 1694–1994 (Oxford: Oxford University Press).

Robinson, C.J. (1991) Racial disparity in the Atlanta housing market, *Review of Black Political Economy*, 19:87–109.

Rostein, A. and Duncan, C.A.M. (1991) For a second economy, pp. 415–34 in D. Drache and M. Gertler (eds), *The New Era of Global Competition* (Toronto: McGill Queens University Press).

Simmel, G. (1990) *The Philosophy of Money*, 2nd edition (London: Routledge).

Squires, G. (1992) Community reinvestment: an emerging social movement, pp. 1–37 in G.D. Squires (ed.) *From Redlining to Reinvestment: Community Responses to Urban Disinvestment* (Philadelphia, PA: Temple University Press).

Squires, G. and O'Connor, S. (1993) Do lenders who redline make more money than lenders who don't?, *Review of Black Political Economy*, 21:83–107.

Taub, R. (1988) *Community Capitalism* (Boston, MA: Harvard Business School).

Thrift, N.J. (1994) On the social and cultural determinants of international financial centres, pp. 327–55 in S. Corbridge, R.L. Martin and N.J. Thrift (eds) *Money, Space and Power* (Oxford: Blackwell Publishers).

Thrift, N.J. and Leyshon, A. (1994) A phantom state? The detraditionalization of money, the international financial system and international financial centres, *Political Geography*, 13:299–327.

Warf, B. (1994) Vicious circle: financial markets and commercial real estate in the United States, pp. 309–26 in S. Corbridge, N. Thrift and R. Martin (eds) *Money, Power and Space* (Oxford: Blackwell Publishers).

Watts, M.J. (1994) Oil as money: the devil's excrement and the spectacle of black gold, pp. 406–45 in S. Corbridge, N. Thrift and R. Martin (eds) *Money, Power and Space* (Oxford: Blackwell Publishers).

Zeliser, V. (1994) *The Social Meaning of Money* (New York: Basic Books).

Part IV

Spatial Organization and the Globalization Process

11 Exploration, Mapping and the Modernization of State Power

The absolutist states that developed in western Europe in the late Middle Ages, and which consolidated their power and territory during the Renaissance, constitute the political and institutional basis of the different types of democratic and representative states that appeared in the world since the last quarter of the eighteenth century. These states have been distinguished by the long history of their exercise of exclusive territorial power by a monarch who also had a monopoly of power of legitimate coercion, of tax collection and of public administration (Alliès, 1980). For both the absolutist states and the contemporary democratic and representative states, political sovereignty was defined in terms of a people occupying a specific geographical area (Anderson, 1986). However, while in respect of the absolutist state, the sovereign (monarch) was quite different from his territory and subjects, for the representative state, the tendency has been towards identifying the sovereignty with the collectivity of citizens (Escolar, 1995). Indeed, there has been a transition from the former to the latter type of state in the countries of western Europe which came to be extended even to their colonies.

This process of modernizing state power began in the fifteenth century. It was centred on the development of methods and techniques to improve the functioning of the bureaucracy in its control over territory and population. It also involved attempts to legitimize state power and produce knowledge about societal and environmental conditions within and beyond the geographical jurisdiction of the state. Consequently, the process entailed five main aspects:

1 the institutional consolidation of the centralized political and administrative apparatus of state;

This chapter is based on a contribution by Marcelo Escolar.

2 the infrastructural and bureaucratic organization of the territory;
3 the projection and evaluation of alternative strategies for state expansion;
4 the undertaking of alternative offensive and defensive territorial tactics; and
5 the internal and external legitimation of the sovereign's political rights.

These five aspects of the modernization process were structured during the Renaissance period, generalized in the seventeenth and eighteenth centuries, and became part of the legacy of later democratic states. They were transferred to colonial empires and adopted in the post-colonial states of the nineteenth and twentieth centuries.

The historical and territorial continuity of most absolutist states helped to simplify the understanding of the process of modern state formation. This entailed the integration of national territories both with respect to the unification and the homogenization of the state space, and the institutionalization of the legacy of political, cultural, technical and administrative structures. It also facilitated the identification of common features of state development such as the discovering, inventorizing, surveying and mapping of both the boundaries and internal resources of the state's territorial space and those of the world beyond its borders. State building gave impetus to the collection of information and the production of the type of knowledge needed to plan, control, evaluate and manage the territorial space of government and administration. At the same time, this knowledge ensured the emergence of new ways of political domination and the organization of more efficient ways of state penetration of civil society. The close relationship between technical development and the political use of its results should, however, not be construed as indicating that modernization of state power was simply a matter of the mechanical accumulation of information and knowledge. Rather, it should be seen as a means available at most times but whose use depended on the character and perspicacity of the sovereign (Solé, 1976:206).

Cartography thus came to play a key role in the capacity of the state to manage its own territory and to secure a better understanding of the resource implications of global geographical differences. The practical task of producing maps, however, went hand in hand with

Louis XVI giving instructions to La Pérouse before his voyage round the world. Engraving by Pigeot, 1783.
Source: Roger-Viollet.

the intellectual activities of interpreting their results and drawing necessary conclusions for policy and action. The former entailed the collection, inventory and classification of information of different parts of the world being opened up by explorers, whilst the latter relates to the systematic analysis, literary exposition and representation of such areas to the public at large and to state decision makers. Cartography, however, was never a mirror image of the reality represented. It provided no more than a visual scheme of reality and was considerably influenced by the technical means of the period available for making such representations (Jacob, 1992). For this reason, it is possible to attempt a periodization of the changing visions of state power and the world since the fifteenth century. Three major periods are identified. In discussing each period an attempt will be made to provide a broad view of the world at the time, the main features in the process of state formation, the available technical means for collecting spatial information about the state's territory and those of their colonial possessions, and the use made of such information in strengthening state capacity for spatial organization and management.

The first two periods extended from the fifteenth to the eighteenth century. The earlier, usually identified as the Renaissance period, witnessed a slow break with the world view deriving from antiquity.

It experienced the social and cultural impact of the various voyages of discovery and the process of state centralization. The later period, known as the baroque and neoclassical period, was when European absolutist states were at their apogee. It was a period distinguished by the emphasis on metropolitan survey and inspection, the consolidation of initial colonial territories and the development of a vision of a world open to territorial expansion. The third period covers most of the nineteenth century and was marked by the rise of democratic-representative states. It was a period that witnessed different processes of political and administrative transformation of state power following simultaneously on the direct results of the French and American revolutions and the accelerated development of capitalism in western Europe and the United States. It is further distinguished by state promotion of the collection of descriptive statistics and monographic expositions associated with the recognition of geographic differences within the territory of the state itself and of its overseas possessions.

Although this attempt at periodization is to emphasize differences in the processes of exploration, mapping and state power relationships, events do not necessarily fit strictly into each of the three periods. Moreover, although the history of cartographic representation and the expansion of knowledge about the world are related to the process of modern and contemporary state formation, no linear relation needs be assumed. They indicate the political and administrative factors that have helped to shape the pattern of spatial organization and territorial restructuring both at the level of individual states and of the world in general.

The Double Discovery and the Changed World View

At the beginning of the fifteenth century, the general view of the world was as captured in the cartographical visualization of Christian western Europe, best depicted in the *imagomundi*. This map, apart form insisting on a flat earth, did not believe that there were unknown areas of the world still waiting to be discovered. Like other areas of knowledge at the time, this view of the world was meant to be an accepted *auctoritas*. However, the artistic flowering of the Renaissance and its strong naturalistic or realistic emphasis began to undermine this process of knowledge acquisition. This urge to be as realistic as possible also extended to the realm of cartographic representation and came to underscore attempts at engaging in voyages of discovery.

Initial efforts to seek realistic knowledge of the world began with the rediscovery of ancient texts and maps. Those of Ptolemy, the Alexandrian astronomer and geographer of the second century BC, greatly facilitated the development of empirical knowledge about the world at this time. These texts and maps were translated into Latin by Jacobus Angelus in 1409 and retranslated into different common languages after that. They became the basis of the western world's geographical knowledge until its definite denial by the Church in 1570 after the publication of Ortelius's *Theatrum Orbis terrarum* and Mercator's consideration of the Ptolemaic *La Geographia* as a document of significant historical value in his Atlas of 1578. These texts have the intellectual responsibility of opening the way for challenging the Christian *auctoritas* and control over knowledge. They were also responsible for stimulating the urge to explore and empirically investigate the true nature of the world during the fifteenth and the first half of the sixteenth centuries.

Together with information from the Ptolemaic *mapamundis* and regional maps that included the technical novelty of presenting global locations in terms of latitudes and longitudes, the portolanos and nautical charts, referred to as cartograms, came to be used for navigation following coastal courses and geographical points whose orientations were easy to determine. Cartograms were developed in Italy and Catalonia during the fourteenth and fifteenth centuries for commercial and colonization purposes in the Mediterranean. With time, new data were added and the quality of the cartograms considerably improved. This was largely due to the activities of nautical schools established in Portugal and Spain, during the fifteenth and sixteenth century, in the course of the explorations and overseas expansion of these two nations across the Atlantic, the Indian and later the Pacific Oceans (Broc, 1980).

The processes of state formation and territorial organization of these two Iberian powers were greatly influenced by the cartographical representational trends during their consolidation as modern states. The early establishment in Portugal of a centralized monarchy facilitated the rise in that country of a stable government with capacity for a limited and efficient control of its territory during the fourteenth and fifteenth centuries (Wallerstein, 1974). This development, together with the unification of the two Spanish crowns under the Castilian hegemony and their reconquering of the Moorish-occupied kingdom of Granada, promoted a systematic policy of expansion in both countries. This was also associated with the overseas exploration on the bases of increasing technical information and knowledge of navigation and cartography (Boorstin, 1983).

In 1420, the Portuguese monarch, Henry the Navigator, founded at Sagres the *Casa da Guine*, a school for pilots, cartographers, mathematicians and instrument technicians to instruct navigators and explorers about their role in the royal projects. After Henry's death in 1460 the *Casa da Guine* acquired the status of a ministry, as the *Casa da India* and the *Junta dos Matemáticos* (Broc, 1980). Similarly, in Spain, after the first Spanish colonies had been established in the New World following on the explorations of Columbus, there was founded the *Casa de Contratación*, an interdisciplinary entity with economic and scientific responsibilities as well as the *Consejo de Indias*, an institution devoted to political and religious control.

The European courts were frequented by geographers, cartographers and cosmographers, who offered their services and their empirical knowledge to the highest bidder. Their activities helped to disseminate information about the main discoveries of the world. However, the detailed cartography they produced, which permitted easy recognition of terrain and locations within it, was usually kept hidden and, in many cases, was destroyed by the Iberian monarchs (Texeira da Mota, 1976). This treatment of vital information as a state secret impeded the rapid and automatic correction and expansion of knowledge about the known world and ensured that such expansion could occur only through piracy, attacks and conquest in war.

Nonetheless, in this period, the main problems of the absolutist states of Europe outside of the Iberian peninsula were not overseas expansion. Rather, they were the building up of their own centralized state structure. This structure gradually emerged over the ruins of religious and dynastic wars, the disappearance of the hierarchy of feudal lords, the modifications of the feudal productive base and the final failure of the empire-building efforts of the Spanish Habsburgs. Initially, this political process of wanting to deepen the power and domination of the absolute monarch over his territory did not entail any major change to the non-cartographic style of administration of the Middle Ages. But, with the opening of the known world resulting from the Iberian expansion, these monarchs and their officials began to be concerned with the representation of their own metropolitan geography and the possibilities of their own territorial expansion.

The idea of a central state institution projected exclusively over a defined territorial domain was not a novel idea at the time of the epistemological rupture with the medieval period. It became legitimated by the turn of events emanating from this double discovery (Livingstone, 1992). In consequence, the institutional and political intentions of the state came to be closely related to its instruments of legitimation, control and expansion. Different kinds of survey techniques began to be developed for accumulating geographical information for the purposes of spatial organization and the consolidation of state power from the Renaissance period onwards. All the emerging absolutist states began to acquire cartographic means to represent the geography of their states for two main reasons: the first was to facilitate the social legitimation of their dominion through improved communication and propagandistic endeavours; the second was to construct effective spatial coverage of jurisdictional institutions for purposes of administration and governance (Buisseret, 1987; Tilly, 1989).

Related to the first purpose is a set of pictorial-cartographical images that decorate palaces and official buildings and the *mapamundi*, globes and maps of state territory that the monarchs keep to express their effective dominion. The identity of the state and its territorial jurisdiction needed representation of a type that bordered delicately between pictography and cartography. Master examples of this type of map include the strategical-political maps of the *Palazzo Vecchio* in Florence, painted by Enzio Danti between 1563 and 1575, and those of the Maps Hall in the Vatican, painted by Antonio Danti between 1580 and 1583 (Marino, 1987).

With regard to the second purpose, there are different types of cartographical representations distinguished by varying levels of technical competency and devoted to the primitive organization of the land-division system, the establishment of administrative jurisdictions and the inventory of resources within the territory (Alliès, 1980:53).

By the middle of the sixteenth century, the political-institutional and technical-scientific conditions had developed to the point where the need for cartographical representations as instruments of bureaucratic and administrative management, as well as effective control of territory, had become imperative. The development of mercantilist capitalism and the need for expansion to obtain new markets for raw materials gave new dynamism to the productive structures of the absolutist states. It encouraged the process of state consolidation through an evolving centralized system of fiscal administration and revenue collection. The state came to evaluate the costs of overseas expansion in terms of the revenue from taxes and custom duties that could support the bureaucratic and other institutional structures necessary to administer such territories. At the same time, it developed active promotional policy for cartographic representation of administrative units and inventories of resources within the metropolitan territory. Starting with land division, water control and channelling assignments in Venice and, to a lesser degree, in Milan, Florence and the Papal States, cartographic practices became official policy in England and France from the sixteenth century onwards (Marino, 1987).

Following on the Saxton's Atlas of 1579 where England was described in aggregate units of hundreds as against the diocesan unit of ecclesiastical division that had been used until then, information about the population and the location of large individual properties was systematically used by Elizabeth I's Minister Burghley and the main functionaries of the state bureaucracy to organize the reckoning and collection of taxes and other incomes (Barber, 1987). The practice became institutionalized under James I who in 1607 ordered the Great Survey with a view to having a detailed land survey of Crown land. Although this project was not accomplished as desired, the political-administrative and the fiscal exigency that provoked it, led to the establishment in 1610 of the State Paper Office whose responsibility was to provide cartographic information about the royal domain.

Across the Channel in France, the problems were similar. Catalina de Medici, the force behind many of the political decisions of her different sons in charge of the affairs of the French throne in the second half of the sixteenth century, entrusted Nicolas de Nicolai, lord of d'Arfeuille, with the task of producing a general cartographical survey of the provinces of the kingdom in 1560. This decision was not unconnected to information concerning the production of Saxton's Atlas in England. The assignment, however, was never completed. The portions completed remained available as manuscript (Buisseret, 1987). Nonetheless, under Henry IV, Sully undertook further cartographic production to plan the kingdom's infrastructural equipment and effectively contribute to the location of its defensive fortresses and military garrisons (Dockès, 1969). Not until under Richelieu was there a real geographical policy leading to the production in 1624 of a map of thirty sheets by the engineer and cartographer, Nicolas Sanson (Buisseret, 1987). Nonetheless, neither in England nor in France was the monarch's decision to obtain detailed locational information about territories willingly agreed to by local and regional feudal lords who refused to provide the necessary data or help with the surveys of their own properties (Barber, 1987; Buisseret, 1987).

Because of the strong centralizing policy adopted after the defeat of the *comuneros* in 1520, the Spanish monarchy was the only one that accomplished the production of a complete map of the peninsula by 1577. This was part of an atlas started by Pedro Esquivel and finished by Diego de Guevara under commission by Felipe II. This map may be considered the most exact (from a geodesic point of view) cartographic work as it included the largest amount of information in terms of the geographical knowledge of the time (Parker, 1987). However, after the defeat of the Spanish Armada in 1588 and the beginning of the decline of the Spanish dynasty, only cursory attention (largely through employing the services of foreign specialists) was paid to the production of the administrative and political-jurisdictional map of the kingdom, although great interest continued to be shown in the inventory and survey of resources in the colonial territories (Parker, 1987:145).

The accumulated information produced by the great voyages of discovery and the surveys of the local and regional areas of European states were refined and disseminated through the work of cartographic exponents, such as Ortelius, Mercator and Gastaldi, as well as through the work of cosmographers, such as Sebastian Münster and his disciples (Broc, 1980). The most important centres for this dissemination were initially in North Italy. Later, these shifted to South Germany and the Rhine States and eventually to Flanders (Jacob, 1992). Increasingly, both the projections and the material content of the maps were improved. The process led eventually to the break from medieval and Renaissance cartography, which can be considered the most important contribution made by the voyages of discovery to the modernization of state power. This rupture initiated attempts to institutionalize strategies of effective territorial administration and of systematic cartographic delimitation and description of the character and resources of the territory.

The Institutional Process of Survey, Inventory and Description

At the beginning of the seventeenth century, the main centre of cartographic production was in the newly independent political entity of the United Provinces (de Vrij, 1967). The new State, comprising of seven secessionist provinces headed by Holland and Zeeland, had by 1569 successfully revolted against Spanish domination. Based on middle-class bourgeois concern over freedom and independence and espousing a clearly liberal economic policy, this new state had evolved a decentralized administrative system with wide political, religious and commercial freedom (Wallerstein, 1980). Over a period of some 70 years with a budding capitalist economy, this new state had built an overseas empire overlapping and going beyond the Portuguese possessions in the South Atlantic and Indian Ocean and had strengthened its privileged position in commerce and the Levant and North Europe.

The economic liberalism, the religious tolerance and the administrative decentralization characteristic of the new state affected its cartographic representation of the world and the view of its own geographical peculiarities. They ensured an attitude to cartographic representation that was devoid of any interest in authoritarian control over the landscape and that contrasted sharply with the close relationship between cartography, administration and territorial management characteristic of the earlier absolutist states (Alpers, 1983). In Holland, land had more importance than men especially a place built and developed by society as a resource. Because of the weakness of feudalism in the area, it was not difficult for rationalism to sustain the idea of progress based on the development of individual capacities and on an austere morality concerning the dignity of labour. Moreover, the importance attached to land and landed property promoted an interest in the systematic survey of the terrain, not as an act of ostentation by a lord of his possession but as a necessary inventory of a scarce resource. This enabled the state to overcome the strictly emblematic use of cartographic representation as propaganda (Alpers, 1983:90).

The collective participation of different social strata of the population in the continuous process of economic growth and intellectual development in the Netherlands ensured that cartographic representation evolved simply as a more effective visual language for transmitting information about the systematically produced knowledge of the state. In this way the experience in Holland was substantially different from that of the rest of Europe at this time. The liberalism characteristic of Dutch overseas enterprises and their impact on the growing capitalist economy in the state encouraged a considerable use of cartographic techniques not just by agents of the state but by almost the whole population. Effective means of detailed topographic and geodesic information gathering were linked to relatively sophisticated techniques of visual representation. Then an interpretation of reality was available to the public at large.

The state directly supported the various cartographic activities not only for purposes of effective territorial management but also to control infrastructural transformation both at home and in the colonies. Indirectly, it also used them for the protection of Dutch entrepreneurs and for the organization of institutions of civil society. This was especially so in the area of social security and collective services organized to minimize the effect of shortages arising from the rapid growth of population (Klein, 1969). From the period of the big atlas of Mercator and Ortelius at the beginning of the century to the production of the atlas of Blaeu in 1636 and 1663, cartography and chorographic pictorial art followed a parallel course. Cartography and painting finally fused in various urban panoramic views, such as Miker's View of Amsterdam or Ruisdale's Panorama of Amsterdam, its port and the Ijsselmeer. The modernization of the Dutch state power was thus reflected in the socialization of the technical and aesthetic resources for representing its territorial space (Bann, 1990).

The situation in France and England at this time was quite different. In both countries, the cartographic project suffered some reversal of fortune. This was because geography was regarded as falling within the scientific intellectual domain while cartography was regarded as part of the technical sphere of state management and government. If mercantilism, as Heckscher (1935) noted, was neither a clearly defined economic doctrine nor an economic development programme, its conceptualization should be sought in the different institutional organizations that, during the seventeenth and eighteenth centuries, determine public policies in the absolutist states of Europe especially in the field of public finance. The activities of these organizations were directed to seek areas where government could exercise a high degree of commercial monopoly. Two aspects of these activities need to be stressed. The first is the pre-eminence given to the task of unifying the domain of the monarch behind this strategy of development; the second is the establishment of a set of administrative, judicial and economic rules which held sway over the whole territory whilst delimiting areas of exclusive jurisdiction subordinated only to the central power. This relationship between wealth creation and territorial commercial monopoly played a key role in mercantilist thought and politics. It also ensured that the relationship between space and economic thought remained a privileged aspect of the administrative and economic actions of the state (Dockes, 1969:17).

Within this context, and with the exception of the Netherlands where liberalism prevailed over mercantilism, cartographic representation of territory gradually became a component of the political arithmetic, which allowed the absolutist monarch to know his domain better and project his mercantilist enterprise with greater effectiveness and administrative efficiency (Dematteis, 1985:61). This process was, of course, not exactly the same everywhere either in its purpose, its scope or its results. Its operation in any particular country was determined by the inherited medieval conditions and the broad architecture of the state's social and administrative institutional capacity.

Britain, for instance, emerged from the late Middle Ages as the state with probably the highest level of centralized political power (Anderson, 1979). Even with this high level of power concentration, Henry VIII (Elton, 1953) still had to contain the powers of the feudal lords and those of the yeomen (Coleman and Starskey, 1986; Harris, 1963). A premature mercantilism was, during the early sixteenth century, gradually transformed, under Elizabeth I and the Stuarts, into a liberalism promoted and protected by the state. By the middle of the seventeenth century, this enabled Britain to challenge the Dutch state and finally to replace it in pre-eminence in the last quarter of that century (Wallerstein, 1980).

The initial centralization of monarchic power in England was responsible for the early development of a rudimentary cartographic representation of the territories under the sovereign, the most important example of which was Gough's map of England in the thirteenth century. It was also, no doubt, a factor in the production of the first exhaustive European cartographic inventory of the royal territorial possessions by Saxton in 1574. Following on these developments England came to be distinguished by the proliferations of maps and descriptive accounts of its localities and regions as well as of other parts of the world, all produced from a patriotic and naturalistic perspective (Cormarck, 1991). These maps were printed and published for the exposition of discoveries, regional chorographic descriptions, treatises on astronomy, cosmography and navigation. Their texts, appearing at about the same time as the emergence of Baconian science, still emphasized the close relation between science, magic and astrology (Livingstone, 1992). In addition there were reports of travels, by such writers as Richard Hakluyt and Walter Raleigh, aimed at the political announcements of British rights over North America and constituting the first set of geographical works using a colonial expansionist discourse for a patriotic purpose.

The rapid growth of a commercial bourgeoisie in England in the seventeenth century, the triumph of the power of Parliament and the consequent redistribution of power among the different strata of English society gave the English monarchy, after the revolution in the middle of the century, a less centralizing character and imposed an effective juridical limit on its exercise of power. In these circumstances, geography developed as a field of knowledge and cartography as a scientific tool for the description of the world independent of the need of the state (Bowen, 1981). The situation was in many respects similar to that of Holland. While practices in both countries kept a parallel existence with respect to the relation of cartography to public administration, the development of English scientific geography came to be linked with the need to make inventories of the growing scatter of colonial territories. Development of geodesic and topographic surveys, together with their cartographic representation were associated with the establishment of a fiscal, economic, judicial and administrative system from the second half of the seventeenth century to the eighteenth century, without requiring official Royal sponsorship and control as happened on the other side of the Channel. (Cormarck, 1991; Fordham, 1929).

In France, the emerging state after Richelieu and Mazzarino had a strong centralizing character, shaped by direct coercion during the first half of the seventeenth century. The innumerable feudal and municipal prerogatives, the fiscal jurisdictions of the *Intendants*, and the inherited districts of the *Officiers* had been changing since the fourteenth century. This has resulted in the overlapping of functions and of competencies. It has also enabled the monarch to appropriate in varying degrees local and regional powers and to control the robed bureaucracy (Alliès, 1980; Anderson, 1979). The situation was one of institutional and administrative disorganization challenging the capacity of the monarch to unify the country around himself and to exercise effective control over the whole territory.

This resultant centralized state, almost devoid of any restraint on the monarch's authority, was inherited by Louis XIV in 1661 who came to deepen most authoritarian aspects considerably. The *parlements* were silenced in 1663, military garrisons were introduced into the *bonnes villes*, tribunals were reduced to obedience and the high *noblesse* and many provincial functionaries were obliged to live at Versailles (Anderson, 1979). On this basis, the French state easily became the model of the European absolutist state of the period.

It was thus in France that mercantilism organically became a state doctrine. As in England, the emergence of non-official geographical knowledge during the seventeenth century was, thus, largely peripheral because it was basically linked to the popularization of relatively obsolete and quite general cartographic-descriptive knowledge. Examples of such geographical texts include the LeClerc atlas (with several editions between 1619 and 1632), the Melchior Tavernier atlas (re-edited between 1634 and 1637), and others by Tassin, Nicolas Nicolai and Guillaume Sanson (Pastoureau, 1980). All of these, together with the *amour au map* and a number of regional descriptions and travel guides enjoyed popular patronage during the period. They contrast sharply with the Geographie du Roi, which was centred on exalting the expansion of the power of the monarch over the whole territory (Dematteis, 1985).

The ambitious reformist posture of French mercantilism had to face up to the contradictions of a strongly centralized state and its as yet not totally consolidated social basis. It could hardly disregard the cartographic potential for facilitating effective territorial management. French science, although already adopting the Baconian paradigm, did not, as in Britain, share completely in the development of growing capitalism and in the formation of an autonomous economic and political sphere in civil society. The science 'pour la Gloire du Roi', unified geographical knowledge and the official cartographic production in a single project. This statization of scientific and technological knowledge with respect to cartographic production, suffered from being too closely linked to the territorial ambition of the monarch and its centralized bureaucratic management (Brown, 1949).

By 1670, England gradually replaced Holland in the hegemony of the world economy. From then on, imperialist competition between London and Paris for the acquisition of new overseas markets and the enlargement of their colonial possessions became the order of the day. This competition continued until it was resolved with the English victory in the Seven Years War (1756–63) that ended with the British occupation of Quebec and the loss of French possessions in India and south-west Asia. These colonial adventures of European powers were directed mainly at discovering and exploiting new territories, and organizing their capacity for producing raw materials for the capitalist growth of the metropolitan country (Wallerstein, 1980).

Consequently, the second half of the eighteenth century witnessed the expansion of knowledge through the explorations of new worlds as happened during the Renaissance. The travels of Cook between 1768 and 1780, of Bougainville in 1766, La Perouse between 1785 and 1788, and a great number of similar smaller enterprises during the next half century expanded knowledge of the known surface of the earth by as much as 25 per cent according to Walter Behrmann (Dumbar, 1985). The financial support for these explorations by the dominant powers of England and France as well as by other European nations such as Russia, Scotland and Spain distinguished this period as the second age of discovery. The distinctive character of these new explorations was their close link with imperialist and scientific purposes.

Consistent with the prevailing mercantilist policies of the time, overseas expansion was seen as one of the major means of supporting the growth of national power. Such a policy could be successful, however, only if the internal market and the nation's economy are effectively managed and developed. The development process, in turn, cannot be achieved without the construction of a necessary infrastructure to guarantee the growth of production and consumption and the fiscal and administrative system to provide and maintain the ports, roads, channels and general communications (Dockès, 1969). Thus, in France, between 1713 and 1716, the administration of *Ponts and Chaussées* was institutionalized. The *Ecole des Ingénieurs des Ponts et Chaussées* was founded on the basis of work done by the military corps of the *Ingénieurs du Roi* and *Marchaux des Logis* during the reign of Henry IV and the vigorous promotion of Sebastien le Preste Vauban (1633–1707) (Gottmann, 1944). This institution was given the responsibility for planning, designing, constructing, and maintaining roads and channels necessary for improving movements within the kingdom. It was also expected to produce most of the detailed topographic maps needed for the effective execution of its construction works. Since the middle of the seventeenth century, many surveys and statistical works have been produced. The first population census was ordered in 1694 in anticipation of the introduction of the poll tax. This probably encouraged the production of a new series of regional monographs with a statistical background (Broc, 1974). Both the improved statistical information and the regional monographs came to constitute strategies of political arithmetic for guiding the different administrative and political decisions of the government. This set of technical and intellectual activities could not have been supported without an aggressive policy of fiscal, customs and judicial management adopted since the time of Henry IV by a succession of minsters such as Sully, Richelieu, Mazzarino, and reinforced by Colbert and Turgot (Alliès, 1980).

In England, a more decentralized and less statist policy was pursued with regard to infrastructural development and the effective management of the realm. Concessions were given to individuals to build turnpike roads and private trusts were allowed to operate toll roads under supervision by the county (Albert, 1983). Works on social statistics, initiated by the mid-seventeenth century with the essays of William Petty on political arithmetic, were published between 1676 and 1787. The foundation of official cartographic production grew out of the activities of William Roy and the Duke of Richmond and led to the establishment of the Ordnance Survey in 1791 (Gardiner, 1977; Skelton, 1962).

Overseas exploration and the internal survey undertaken by the English and the French absolutist states promoted scientific and technical activities. Institutions for the promotion of scientific activities were founded in both countries with the sponsorship of the state. In England, the Royal Society was founded and in France L'Académie des Sciences, Le Jardin du Roi, L'Observatoire Royal and L'Académie des Inscriptions were all established during this period

(Broc, 1974; Livingstone, 1992). Many projects of surveying the national territory were developed within these institutions.

However, these institutions themselves strove to break this dependence and close link of science with statecraft and state policy (Dematteis, 1985:59). Such tension between the practical purpose and the intellectual interest of science was also evident in the development of cartography (Gallois, 1909). Gradually, a strictly neutral and scientific cartographic discipline developed. Its emergence was related to the epistemological argument that knowledge should be the result of scientific work the applications of which would not be automatically directed towards political goals. The methodology of the discipline also improved with the growing maturity of theoretical and instrumental capacity and their applications in geodesic and topographic surveys (Brown, 1949). Thus, L'Académie des Sciences of Paris and the Observatoire completed in 1745 the *Description Géometrique de la France* comprising some 18 maps. They extended their activities to the production in 1755 of the *Carte Générale et Particulière de France.* This was eventually completed in 1789 with the publication of the 180 sheets of the *Carte de Cassini* or *Carte de l'Académie* where more than 3000 triangulation points were used. Similarly, in England, William Roy in 1745, with the support of the Royal Society, completed an elaborate triangulation of England and Ireland based on 218 measuring stations. Later, in 1791, the Ordnance Survey was founded and was charged with the task of producing a map of Great Britain based on exact astronomic measures. In this way, this agency became the first institution in Europe to collect geodesic measures for the systematic topographic survey of the realm on a continuing basis (Gardiner, 1977; Skelton, 1962).

Scientific Cartography and Territorial Organization in the Nineteenth Century

The rise of the representative state, after the collapse of the absolutist states, transformed the ideological use of cartography to essentially political and social ends. The state bureaucratic structure itself underwent substantial changes directed at making it a more effective agency for administering and managing the natural and human resources of the territory under popular sovereignty. In France, the destruction of the provincial system was replaced in 1789 with the establishment of a departmental administrative division. The explicit purpose of the change was to dismantle the atavistic particularisms that sustained the French monarchy and replace them with the principles of equality and fraternity. For 12 years after the *coup de Termidor*, the production of departmental information was undertaken on a decentralized basis (Bourguet, 1988). This intense activity of survey and statistical collection did not lead immediately to a local thematic cartography. The cartographic knowledge of the territory remained within the context of the geodesic and topographic work done for the construction of a map of France undertaken by Cassini and others since Colbert times. The main enterprise was the finalization of an uncompleted detailed statistical survey of France started by Chaptal between 1801 and 1804. This provided a significant amount of departmentally organized information that grew into a culture almost like a reinvention of the reporting style under the provincial system (Woolf, 1981).

The advent of the Napoleonic empire took the situation almost back to the period of the absolutist monarchs. Territorial statistical information was again a state secret. Two types of activities were, however, undertaken during the Napoleonic period. One was the statistical-cartographic surveys by military engineers; the other the descriptive works of places, countries and regions produced by civilians (more or less referred to as geographers) interested in the resources and potentialities of colonial territories. Both activities were invaluable as a politically relevant means of facilitating the rational control of society and territory and enhancing the technical capacity of government in the management of colonial areas (Adas, 1989). In this sense, imperialism came to be associated with the idea of a legitimate mission of diffusing progress, particularly French civilization, all over the world by conquest, if need be.

In other European countries, the idea of undertaking topographic surveys of national territories also had been taking root since the sixteenth century. In the Holy Roman Empire and under the influence of the German chorographic school, a great inventory of Saxony was undertaken between 1550 and 1600 on a scale of 1:26,000. Bavaria was surveyed between 1554 and 1563 and its maps (considered as the best of its time) comprised 40 sheets on a scale of 1:50,000. Denmark, in the following century, produced a cadastral map of 37 sheets together with the mapping of regional areas. Sweden also finished in 1626 a more modest map made up of 6 sheets. Meanwhile, Russia produced the first map of its imperial areas in 1720 and Austria produced some partial surveys of its territory between 1768 and 1790. In short, a certain correlation was noticeable between the process of political modernization of dynastic powers and cartographic representation of sovereign territories. Related to this development was the emergence of organized institutions with responsibility for the production of such official cartography. Since 1742, the Danish Royal Academy of Science promoted the production of a geodesic survey of the country which resulted with the first triangulation of the kingdom published in 1776. In Sweden, the first triangulation was prepared in 1747 by the Bureau of Land Surveying which in 1805 became the Swedish Field Survey Corp. Russia also in 1739 created through its Academy of Sciences a geographical department of which responsibilities were transferred to the general staff in 1763. The latter undertook the first triangulation of the Vilna territory in 1816 and, 30 years later, the territory of the whole Empire, though at different scales for its European and Asiatic portions. Austria undertook the topographic survey and triangulation of its different regions between 1760 and 1860, a reflection of its low level of political centralization. Finally, after the German Unification in 1871, a state survey was organized which improved and standardized the survey and triangulation undertaken by many German principalities during the eighteenth century.

On the other side of the Atlantic, the revolt and emancipation of the 13 British colonies in 1776 gave rise to the first truly democratic state of the modern period. At first organized on a confederal basis in 1782, the states agreed to become a federal state in 1795 (Sack, 1986). This development gradually led the United States to a different orientation of state modernization and to the kinds of geographical surveys and explorations that came to be required in the circumstances. Surrounded as it was at the time by other British and French colonial dependencies, the eastern coastal zone occupied by the new federal state came to function, especially after the independence wars of the early nineteenth century, as a platform from which to launch its westward expansion as a continental territorial state (Goetzmann, 1986). Thus, unlike French imperialist designs, defended as a civilizing mission by a nation with superior political institutions, national culture and science, the United States did not have similar ideological pretensions. Here, the emphasis was on the consolidation of territory of the initial 13 states and the exploration and occupation of the large, continuous tract of land in the rest of the continent. The need to secure the new territorial acquisitions within the Union encouraged the growth of a decentralized structure of public administration (Duchacek, 1986; Elazar, 1962). Nonetheless, it must be recognized that the situation of the United States was quite unique in many ways. Lacking any contiguous rival state and having resolved the instability posed by the original confederal structure, the country did not have to contend with problems of delimiting and defending colonial territories.

The continental dimensions of exploration and development led to the emergence, quite early, of the idea of freedom in an individualistic and democratic society. This was necessary as large virgin territory constituted a challenge to industrious and enterprising citizens (Toal, 1989). A patriotic fervour became a critical element in this process of territorial expansion. The idea of developing an American science necessary to aid the nation's destiny became characteristic of statements not only of politicians like President Jefferson but also of explorers, such as Lewis and Clark between 1779 and 1830 (Livingstone, 1992). Science, including cartographic knowledge, came to be institutionalized and considerable activities were undertaken for the geodesic and topographic surveys of the territory.

Indeed, the United States was remarkable in the extent to which it emphasized cartographic documentation of its political-administrative divisions and established necessary institutions and offices in the different states for the purpose. The first expansion to the west, from the Appalachians to the Mississippi–Missouri basin in the last decade of the eighteenth century and first decade of the nineteenth century provided the laboratory for the first tracing of linear frontiers. This was capitalized upon in the second period of expansion to the far west and Pacific coast. On both occasions, the territory was treated geometrically, fitting the state districts and counties to the rectangular shape of the country itself (Johnston, 1967). Nonetheless, lacking the urgency of European states for representing and legitimizing the extent of their domains, most of the surveying work in the United States was initially sporadic in character. It did, however, combine exploration with detailed and systematic field survey, description and classification of the content of the areas discovered.

However, by the second half of the nineteenth century, the position began to change. The Treasury Department had established a Coastal Survey in 1807 to update the coastal cartography with trigonometric triangulations (Raiz, 1937). This was confirmed by Congress in 1847 and provided with a stable budget. Similarly, in 1845, the US Census Office was established to develop a regular programme of collecting statistics and representing them cartographically for the whole federation. In 1813, a corps of topographic engineers was created within the army who since 1863, pursued an active programme of hydrological survey, trigonometric triangulation and astronomic determination of the country. The section was formally institutionalized within the war department in 1879. In the same year, a geological survey was established and placed in charge of the first attempts at territorial planning during the expansion towards the far west, as well as in the production of most of the maps required by the war department and the general staff (Brown, 1949). This diversity of institutions emphasizes the impact of a federal state structure on topographic survey, geodesic measurements and cartographic representation of the United States. It also underscores their importance in a country that seeks to link government directly to the political representation of its citizens and to sensitize the administration to the modernizing aspirations and material progress of the people (Raiz, 1937).

A completely different situation was presented by the other south and central American states which became independent a few decades later. These countries formed, with the United States and France, the only representative democratic states of the nineteenth century. The original Spanish colonial territories were transformed into a set of democratic republics that had to institutionalize their inherited administrative jurisdictions into emergent nation states. Like the United States, the new states had to contend with the challenges of effectively occupying their territories through expanding into areas the geographical characteristics of which were inadequately known. The case of Brazil was slightly different because, like the United States, it had become by the nineteenth century a single, territorially integrated, independent state, albeit the only monarchical state on the American continent at this time (Osorio Machado, 1990). The Brazilian Empire (1822–89) and the Republica Velha (1889–1929) had a civilizing ideology of territorial development as the basis for the exploration and occupation of its extensive, virtually virgin interior regions (Lippi Oliveira, 1990; Ortiz, 1985; Robloff de Mattos, 1987; Teixeira Soares, 1972; Zusman, 1993).

By the second half of the nineteenth century, all of these states, in Europe and in North and South America, had laid the foundation for their cartographic institutions. By then, capitalism was at high flood, transforming the social structure of these countries, changing their productive and consumption patterns, altering their labour relations and the relative weight of rural and urban population in their societies (Hobsbawm, 1989). This was the climax of the restructuring of territory that had begun in the late fifteenth century and consequently of the administrative system devoted to its management and control. But their growth accelerated during the nineteenth century until they had built in their own image a global colonial world covering nearly the whole planet (Said, 1993).

Since the 1870s, massive education has slowly crystallized this global image of metropolitan and colonial territories which were then widely represented on maps. The administrative limits imposed by Europe on the rest of the world gradually hardened until they began to function as a new basis of state nationalities, which emerged during the decolonization process of the twentieth century. Geographical societies emerged in the different metropolitan countries as the privileged locus for socializing citizens to the practical use and intellectual importance of geographical knowledge. The educational system developed school curricula which included the study of the natural and human characteristics of state territories and of their colonial extensions (Anderson, 1986; Escolar, Quintero and Reboratti, 1994). This social and educational valuation of geographical knowledge helped further in the institutionalization of cartographic representation as a major function of the state.

Conclusion

At the beginning of the process of modern state building, knowledge about territory was associated with the capacity of the monarch to exert dominion over his jurisdictional and eminent domain. The centralization of political power since the late fifteenth century encouraged the search for various ways of representing the territory over which the sovereign could exercise dominion and power, and anticipate the consequences of that exercise. As a result, exploration was not only a means of establishing and acknowledging the limits of dynastic territory within Europe but also the basis for overseas expansion. In this manner, the geographical characteristics of the real world were established and helped to end the stranglehold of antiquity and the middle ages on the expansion of human knowledge concerning the earth. During the seventeenth and eighteenth centuries, the ability to represent this knowledge cartographically became a factor in the capacity to control and manage territory. The coercive centralization of absolutist states defined the use of information for purposes of control and political manipulation.

Beginning with the French and American revolutions in the late eighteenth century, the territorial basis of national power became well defined and institutionalized. States acquired a high degree of political autonomy in administering the affairs of people within their territorial jurisdiction. This opened the way for a non-personalized idea of power and state entity. Representativeness became the mechanism for delegating sovereign power by diverse peoples within a given territorial entity to an elected body or individual. In this circumstance, the national map became a tool for planning the development of the state. With the consolidation of cartography and the state bureaucratic structure as official institutions, survey and exploration became a practice of recording the locational specification of the diverse characteristics of the national territory. And it was this development that paved the way in the closing decades of the nineteenth century for geography to emerge as an autonomous scientific discipline, a handmaiden to colonial expansion but also the basis for the orderly planning and development of resources of emergent nation states.

References

Adas, M. (1989) *Machines as the Measure of Men* (Ithaca: Cornell University Press).

Albert, W. (1983) The Turnpike Trusts, in D.H. Aldcroft and M.J. Freeman (eds) *Transport in the Industrial Revolution* (Manchester: Manchester University Press).

Alliès, P. (1980) *L'invention du territoire* (Grenoble: Presses Universitaires de Grenoble).

Alpers, S. (1983) L'il de l'histoire: l'effet cartographique dans la peinture hollondaise au siecle, *Actes de la recherche en sciences sociales*, 49:71–101.

Anderson, J. (1986) Nationalism and geography, in J. Anderson (ed.) *The Rise of the Modern State* (Sussex: Wheatsheaf).

Anderson, P. (1979) *Lineages of the Absolutist State* (London: New Left Books).

Baker, J.N.L. (1928) Nathanael Carpenter and English geography in the seventeenth century, *Geographical Journal*, 71:261–71.

Bann, S. (1990) *The Invention of History: Essays on the Representation of the Past* (Manchester: Manchester University Press).

Barber, P. (1987) England II: monarchs, ministers and maps,1550–1625, in B. Buisseret (ed.) *Monarchs, Ministers and Maps: The Emergence of Cartography as a Tool of Government in Early Modern Europe* (Chicago: University of Chicago Press).

Berthon, S. and Robinson, W. (1991) *The Shape of the World: The Mapping and Discovery of the Heart* (London: George Philip).

Boorstin, D.J. (1983) *The Discoverers: A History of Man's Search to Know his World and Himself* (New York: Random House).

Bourguet, M.N. (1988) Dechiffrer la France, in *La statistique departmentale a l'epoque napoléonienne* (Paris: Editions des archives contemporaires).

Bowen, M. (1981) *Empiricism and Geographical Thought: From Francis Bacon to Alexander von Humboldt* (Cambridge: Cambridge University Press).

Broc, N. (1974) *La Geographie des Philosophes: Géographes et Voyageurs Français au XVIII Siècle* (Paris: Ophrys/Association des Publications près les Universités de Strasbourg).

Broc, N. (1980) *La Geographie de la Renaissance* (Paris: CTHS).

Brown, L.A. (1949) *The Story of Maps* (New York: Dover Publications).

Buisseret, D. (1987) Monarchs, ministers and maps in France before the accession of Louis XIV, in B. Buisseret (ed.) *Monarchs, Ministers and Maps: The Emergence of Cartography as a Tool of Government in Early Modern Europe* (Chicago: University of Chicago Press).

Coleman, C. and Starskey, D. (1986) *Revolution Re-assessed: Revisions in the History of Tudor Government and Administration* (Oxford: Oxford University Press).

Cormarch, L.B. (1991) Twisting the lion's tail: theory and practice in the court of Henry, Prince of Wales, in T. Bruce and E. Moran (eds) *Courts, Academies and Societies in Early Modern Europe* (London: Boydell and Brewer).

Cormarck, L.B. (1991) Good fences make good neighbours: geography as self-definition in early modern England, *Isis*, 82:639–61.

De Vrij, M. (1967) *Le Monde sur le papier* (Amsterdam: Theatrum Orbis Terrarum).

Dematteis, G. (1985) *La Metaforce della Terra. La geografia umana tra mito e scienza* (Milano: Filtrinelli).

Dockès, P. (1969) *L'espace dans la pensee economique du XVI au XVIII siecle* (Paris: Flammarion).

Duchacek, I.D. (1986) *The Territorial Dimensions of Politics Within, Among and Across Nations* (Boulder, CO: Westview Press).

Dumbar, G.S. (1985) *The History of Modern Geography. An Annotated Bibliography of Selected Works* (New York: Garland Publishers).

Elazar, D.J. (1962) *The American Partnership: Intergovernmental Coordination in the United States* (Chicago: Chicago University Press).

Elton, G.R. (1953) *The Tudor Revolution in Government: Administrative Changes in the Reign of Henry VIII* (Cambridge: Cambridge University Press).

Escolar, M. (1995) Territorios de representacion y territorios representados, *V Reuniao de antropologia do (merco), sul* (Tramandai-Rs. Brasil).

Escolar, M., Quintero Palacios, S. and Reboratti, C. (1994) Geographical identity and patriotic representation in Argentina, in D. Hooson (ed.) *Geography and National Identity* (Oxford: Blackwell Publishers).

Fordham, H.G. (1929) *Some Notable Surveyors and Map Makers of the Sixteenth, Seventeenth and Eighteenth Centuries and Their Work: A Study in the History of Cartography* (Cambridge: Cambridge University Press).

Friis, H.R. (1965) A brief review of the development and status of the geographical and cartographical activities of the United States government: 1776–1818, *Imago Mundi*, 19:68–80.

Gallois, L. (1909) L'Académie des Sciences et les origines de la carte de Cassini, *Annales de Geographie*, 18, 99:193–204.

Gardiner, R.A. (1977) William Roy, surveyor and antiquary, *Geographical Journal*, 143, 3:439–50.

Godleweska, A. (1994) Napoleon's geographers (1797–1815): imperialists and soldiers of modernity, in A. Godleweska and N. Smith (eds) *Geography and Empire* (Oxford: Blackwell).

Goetzmann, W. (1986) *New Lands, New Men: America and the Second Great Age of Discovery* (New York: Viking).

Gottmann, J. (1944) Vauban and modern geography, *Geographical Review*, 34:120–9.

Harris, G.L. (1963) Mediaeval government and state-craft, *Past and Present*, 24:24–35.

Heawood, E. (1965) *A History of Geographical Discovery in the Seventeenth and Eighteenth Centuries* (New York: Octagon Books).

Heckscher, E.F. (1935) *Mercantilism* (London).

Hobsbawm, E.J. (1989) *The Age of Capital, 1848–1945* (London, Weidenfeld and Nicolson).

Jackson, R.H. (1990)*Quasi-states: Sovereignty, International Relations and the Third World* (Cambridge: Cambridge University Press).

Jacob, C. (1992) *L'empire des cartes: Aproche theoreque de la cartographie a travers l'histoire* (Paris: Albin Michel).

Johnston, H.B. (1967) *Order Upon the Land: The U.S. Rectangular Land*

Survey and the Upper Mississippi Country (New York: Oxford L Press).

Klein, P.W. (1969) Entrepreneurial behaviour and the economic rise and decline of the Netherlands in the 17th and 18th centuries, *Annales Cisalpines d'Histoire Social* I, 1:7–19.

Lippi Oliveira (1990) *A cuestao nacional na Primeira Republica* (São Paul: Brasiliense).

Livingstone, D. (1992) *The Geographical Tradition: Episodes in the History of a Contested Enterprise* (Oxford: Blackwell Publishers).

Marino, J. (1987) Administrative mapping in the Italian states, in B. Buisseret (ed.) *Monarchs, Ministers and Maps: The Emergence of Cartography as a Tool of Government in Early Modern Europe* (Chicago: University of Chicago Press).

Oritz, R. (1985) *Cultura brasileira e identidade nacional* (São Paulo: Brasiliense).

Osorio Machado, L. (1990) Artificio politico en el origen de la unidad territorial de Brasil, in H. Capel and H. Comp (eds) *Los Espacios Acotados Geografia y Dominacion Social* (Barcelona: PPU).

Parker, G. (1987) Maps and ministers: the Spanish Habsburgs, in B. Buisseret (ed.) *Monarchs, Ministers and Maps: The Emergence of Cartography as a Tool of Government in Early Modern Europe* (Chicago: University of Chicago Press).

Pastoureau, M. (1980) Les atlas imprimes en France avant 1700, *Imago Mundi*, 32:45–72.

Raiz, E. (1937) Outline of the history of American cartography, *Isis*, 26, 2:373–91.

Recalde, J.R. (1982) *La construccion de las naciones* (Madrid: Siglo XXI).

Revel, J. (1989) Connaissance du territorie, production du territoire: France XII–XIX siecle, in A. Burguiere and J. Revel (dir.) *Histoire de la France, vol. I – L'espace francais* (Paris: Seuil).

Robloff de Mattos (1987) *O tempo Saquarema* (São Paulo: Hucitec – Ministere du Culture, Instituto Nacional do Livro).

Sack, R.D. (1986) *Human Territoriality: Its Theory and History* (Cambridge: Cambridge University Press).

Said, E.W. (1993) *Culture and Imperialism* (New York: Knopf).

Skelton, R.A. (1962) Landmarks in British cartography, III: the origins of the ordnance surveys in Great Britain, *Geographical Journal*, 128, 3:415–26.

Solé, C. (1976) *Modernizacion: un analisis sociologico* (Barcelona: Peninsula).

Teixeira Soares (1972) *Historia da formacao das fronteiras do Brasil* (Rio de Janeiro: Conselho Federal de Culture).

Texeira de Mota, A. (1976) Some notes on the organization of the hydrographical services in Portugal before the beginnings of the nineteenth century, *Imago Mundi*, 28:51–60.

Tilly, C. (1989) The geography of European state-making and capitalism since 1500, in E.D.Genovese and L. Hochberg (eds) *Geographical Perspectives in History* (Oxford: Blackwell).

Toal, G. (1989) Critical geopolitics: the social construction of space and place in the practice of statecraft (Syracuse: Syracuse University, Ph.D. Dissertation).

Wallerstein, I. (1974) *The Modern World System: I – Capitalist Agriculture and the Origins of the European World Economy in the Sixteenth Century* (New York: Academic Press).

Wallerstein, I. (1980) *The Modern World System II – Mercantilism and the Consolidation of the European World Economy* (New York: Academic Press).

Woolf, S. (1981) Contribution a l'histoire des origines de la statistique: France 1789–1815, in *La statistique en France a l'epoque napoléonienne* (Brussels: Centre G. Jacquemyn).

Zusman, P. (1993) El Instituto Historico y Geografico Brasileiro y la Sociedad de Geografia de Rio de Janeiro. Continuidades y Rupturas en el marco del proyecto geografico imperial, in M. Santos, M.A. Souza and M. Arroyo (eds) *O Novo Mapa do Mundo Vol. III – Natureza e Sociedade Hoje: Uma Leitura geografica* (São Paulo: Hucitec – Anpur).

12 Transnational Corporations and the Nation-State

A maelstrom of debate continues to swirl around the extent to which the geography of the world economy is being transformed from a shallowly integrated international system into a deeply integrated global system. Within the raging currents of that debate is a long-standing dispute over how far the basic institutional driving forces of global economic change and the relationships between them have also been transformed. For some 300 years, from its emergence in the mid-seventeenth century, the nation-state was regarded, no doubt correctly, as the dominant actor in international economic relationships. Historically, the state was the primary regulator of its national economic system. The world economy could realistically be envisaged as a set of interlocking, but independent, national economies, each of which exercised a high degree of autonomy over its own economic affairs. As Hobsbawm (1979:313) aptly observed, the production process 'was primarily organized within national economies or parts of them. International trade . . . developed primarily as an exchange of raw materials and foodstuffs . . . [with] . . . products manufactured and finished in single national economies . . . [Thus], in terms of production, plant, firm and industry, [these] were essentially national phenomena.'

In stark contrast to such nationally bounded production, the world as it approaches the new millennium is, in the view of Robert Reich (1991:3), living through a transformation that will rearrange the politics and economics of the coming century. There will be no *national* products or technologies, no national industries. There will no longer be national economies, at least not in the sense that we have come to understand that concept. Kindleberger (1969:207) puts it even more

This chapter is based on a contribution by Peter Dicken.

succinctly when he asserted that the nation-state is just about through as an economic unit.

The actual extent to which the nation-state has lost its significance in the economic sphere is, in fact, strongly disputed by political scientists. That debate need not concern us here. What is of concern – and is, in fact, the focus of this chapter – is the claim that one of the primary causes of the alleged demise of the nation-state is the rise of another institution: the transnational corporation (TNC). The argument for such a claim is deceptively simple. It runs more or less as follows. First, transnational corporations are giant enterprises with assets/sales which are larger than the gross national product of many nation-states. One study claimed that of the 100 largest economic units in the world today, half are nation-states and the other half TNCs (Benson and Lloyd, 1983:77). Second, the multi-locational nature of TNCs, with their operations spread over a number of countries, renders them insensitive to any individual country's needs. Rather, it enables them rapidly to switch and to reswitch operations from one country to another in response to changing circumstances and, therefore, to reduce the power of national governments to implement their own economic policies. If a TNC does not like what a government – or a labour union – is doing, it can simply move elsewhere (or at least threaten to do so). In the face of such immense power and geographical flexibility, it is argued, the poor, old institution of the nation-state, trapped within its fixed territorial boundaries, has no chance.

Tribesmen at Long Kue overlooking one of five bridges newly constructed over Tutoh river in Sarawak by a Japanese firm to provide access for timbercutters. *Source:* Alberto Venzago/Bruno Manser *Voix de la forêt pluviale: témoignages d'un peuple menacé*. Paris: Georg.

There is, as in the case of most myths, a kernel of truth in this view. TNCs, especially the very large ones, do have considerable flexibility. Insofar as they control parts of individual national economies through their affiliates, they do place some constraints on national economic autonomy. But it is a gross error to extrapolate from this to the view that TNCs are now the controllers of the global economy and that nation-states are mere pawns on the TNC chess-board. The actual position is far more complex than this simple stereotype suggests. It is far more realistic to regard both the TNCs and the nation-states as locked into extremely intricate and dynamic interactions in which there is a high degree of mutual interdependence and bargaining. Gordon's view (1988:61, 64) is that 'it is perhaps most useful to view the relationship between multinationals and governments as both cooperative and competing, both supportive and conflictual. They operate in a fully dialectical relationship, locked into unified but contradictory roles and positions, neither the one nor the other partner clearly or completely able to dominate . . . TNCs are neither all-powerful nor fully equipped to shape a new world economy by themselves.' Similarly, writing from a very different political perspective, Ostry asserts that the international environment of the coming decades will be shaped not by governments or international institutions but by the *interaction* of the two main actors, governments and the global corporations. The basic argument then is that the changing geographical structure of the global economy is the outcome of a complex combination of processes involving both TNCs and nation-states (Dicken, 1992a, 1992b, 1994).

This chapter is organized into two major sections. The first focuses upon transnational corporations themselves. A brief history of their evolution provides the basis for a discussion of their contemporary organization and of the key trends in their current development. The second part focuses upon those aspects of the activities of nation-states which impinge more directly on the operations of TNCs and discusses some of the major interactions between TNCs and nation-states. This section, however, is especially concerned with one particular, and very important, manifestation of TNC–nation-state interaction namely, the emergence of regional economic blocks.

The Diversity and Complexity of Transnational Corporations

The business firm operating across national boundaries and controlling or co-ordinating production and distribution outside its home country is not a recent phenomenon (Dunning, 1993; Jones, 1993, 1994; Wilkins, 1991). Its origins can be traced back to the activities of the early merchant capitalists from the fourteenth century onwards, including such trading companies as the Hanseatic League, the British and Dutch East India Companies, and the Hudson's Bay Company. However, the first really major development of TNC activity occurred during the nineteenth century with the rise of industrial capitalism. The modern TNC emerged, in particular, in the second half of the nineteenth century and, especially, after 1870. The expansion of TNC activity was part and parcel of the dramatic increase in international economic activity which occurred between 1870 and the First World War. Indeed, modern research suggests that the scale and extent of foreign direct investment (FDI) in that period were very much greater than has been assumed hitherto, with British, American, German, French and Dutch firms especially active. It was during that period that the two primary motivations for firms to extend their operations across national boundaries become apparent: to seek new markets and to acquire productive resources. Although the operations of TNCs today are infinitely more complex than those of the nineteenth and early twentieth centuries these two basic motivations still apply. However, they are manifested through increasingly intricate organizational networks, both internal and external to the firm.

Although it is now patent that the importance of TNCs in the pre-1914 period was underestimated, it is equally clear that the really spectacular expansion occurred after 1945 and, especially, from the 1960s onwards (Dicken, 1992a). Not surprisingly, the post-war expansion was led by United States firms building upon the unprecedented strength of their domestic economy, their technological superiority and their huge reserves of investment capital. In 1960, the United States accounted for almost 50 per cent of world foreign direct investment (the British share was 18 per cent and that of Germany and Japan a mere 1.2 per cent and 0.7 per cent respectively). By 1993, however, the geographical composition of this investment had changed dramatically (UNCTAD, 1994). The United States share had declined to 25.4 per cent. Japan had become the second most important source of FDI with 12.4 per cent of the world's total, followed by the United Kingdom (11.6 per cent), Germany (9.2 per cent) and France (8.6 per cent). At the same time, an increasing number of TNCs were emerging from the newly industrializing economies in both Asia and Latin America.

Julius (1990) calculates that FDI growth displayed an even greater upsurge during the 1980s than in the 1960s. According to her, whereas in the 1960s FDI grew at twice the rate of GNP, in the 1980s it has grown more than four times as fast as GNP. After a slow down of FDI growth during the recession of the early 1990s, the upward trend has resumed. Not only has FDI been growing faster than GNP, but it has also been growing at a much faster rate than world exports, particularly since 1985. This figure alone suggests that FDI has become a more significant integrating force in the global economy than the traditional indicator of such integration, namely trade. Indeed, because TNCs are themselves responsible for a large proportion of international trade (much of this as intra-firm transactions), their global significance becomes even more marked.

The United Nations (UNCTAD, 1994) reckons that some 37,000 parent firms controlled over 206,000 foreign affiliates in the early 1990s. But this almost certainly underestimates the true scale of TNC activity. This is because of the immense difficulty involved in identifying the bewildering variety of organizational forms involved in modern production. Simply equating TNC activity with a firm which *owns* overseas operations is to reveal only the tip of a very large iceberg. It is becoming increasingly widely accepted that we have to adopt a much broader and more flexible definition of the transnational corporation if we are to capture the real complexity and diversity of this type of economic institution. Ownership of overseas assets obviously matters; all firms that do so are clearly TNCs. But what about those firms which operate internationally through modalities other than actual ownership? Insofar as they have the capacity to *co-ordinate* international production they, too, should be regarded as TNCs.

There is need to recognize that TNCs, when defined in this broader manner, are extremely diverse in size and characteristics. Although the greatest power and influence is obviously exerted by the truly global corporations, it should be remembered that, numerically, such giant firms form only a small proportion of the total population of TNCs. Most TNCs operate in only a small number of countries outside their home base. But even among the larger TNCs, there is far greater diversity in how they are organized and how they operate than is often believed. One, though by no means the only, reason for such diversity is that all TNCs bear some imprint of the specific characteristics of their country of origin. It has become fashionable in many quarters to depict TNCs as placeless institutions. In fact, TNCs are emphatically *not* placeless. All of them have an identifiable home base, a base that ensures that every TNC is essentially embedded within its domestic environment. Of course, the more extensive a firm's international operations, the more likely it will be to take on additional characteristics derived from the different environments in which it operates. TNCs, like all social institutions, are learning institutions. Yet, it is the influence of the firm's home-country base that remains the dominant influence. Set against what has become a fairly conventional view that TNCs, regardless of their geographical origin, are tending to converge in terms of their organizational characteristics, there is now a growing literature that emphasizes geographical diversity.

Hu (1992), for example, concludes from his analysis of the organization and operations of a substantial sample of TNCs that they are national firms with international operations rather than globally-integrated corporations with no clear geographical base. Stopford and Strange (1991:233) reach similar conclusions when they observe that the firm does psychologically and sociologically belong to its home base. More recently, Ruigrok and van Tulder (1995:159) demonstrated through a very detailed analysis of the 100 largest international firms, what might be regarded as global corporations, that of the largest 100 core firms in the world, not one is truly global, footloose or borderless.

The major reason why even firms with very extensive international operations retain the imprint of their geographical origins is that, like all firms, they are produced through a complex historical process of embedding in which the cognitive, cultural, social, political and economic characteristics of the national home base play a dominant part (Dicken and Thrift, 1992). This does not mean, however, that all TNCs from a particular country are virtually identical. This is certainly not the case. But what can be argued is that there are greater similarities than differences among such firms. In particular, insofar as the nation-state acts as a container of distinctive institutions and practices, it inevitably exerts an influence on the nature of TNCs headquartered within its borders.

The influence of a firm's home base can be seen in the different organizational tendencies as TNCs have evolved over time. Again, this is not to argue for a direct one-to-one relationship between national origin and organizational form. Rather, it is to stress that firms from specific origins have tended to have a predisposition to organize their international operations in distinctive ways. Bartlett and Ghoshal's (1989) study of TNCs illustrates this point quite well. They construct a threefold typology of existing TNC operations – the multinational organization, the international organization and the global organization. Each has rather different characteristics. The 'multinational' organization is characterized by a decentralized federation of activities. The company's world-wide operations are organized as a portfolio of national businesses in which each national unit has a substantial degree of autonomy; each has a predominantly local orientation. Such an organizational form has been common amongst European TNCs. In contrast, Bartlett and Ghoshal's international organization is characterized by far greater formal co-ordination and control by the corporate headquarters over the overseas subsidiaries, a feature common among many current United States TNCs. The third type of TNC organization, that is, the global, is even more centralized and allocates very little autonomy to overseas subsidiaries. Such an organizational form was quite common among Japanese firms, especially during their earlier stages of internationalization.

Each of these forms of TNC organization is still evident in the world economy. But intensifying competitive pressures and accelerating technological change are inevitably stimulating very substantial organizational (and geographical) restructuring. It is too simplistic to suggest that a single organizational form is appearing; diversity will continue to rule. However, there are some clearly observable trends. Perhaps the most widely noted trend is the growing emphasis on *network* forms of organization; less hierarchical and flatter organizational forms, more flexible relationships and more emphasis on co-ordination. The central feature of these emerging organizational forms is their particular emphasis on the rich variety of external relationships within production networks. Some of these relationships reflect the changing links between firms and their suppliers; others reflect the various types of collaborative ventures (strategic alliances) which have become increasingly central to many TNC strategies.

The 1994 *World Investment Report* (UNCTAD, 1994) demonstrates some of the major ways in which the strategies of TNCs are contributing towards a qualitative change in the nature and the degree of international integration within the world economy as a whole. An *integrated international production system* is beginning to emerge, although in a geographically and sectorally uneven manner. The report argues that at the heart of this shift are changes in TNC strategies. It identifies a progression of TNC strategies in each of which the role of TNC affiliates and subsidiaries and their relationships with each other and with outside firms is differently defined. It is necessary to quote from the Report *in extenso* in order to underscore the emerging diversity of strategies:

> *In the past, the foreign production of national firms was typically characterized by a clear division of tasks between parent companies and foreign affiliates. This division was a reflection of the fact that, in most cases (with the natural-resources sector being an obvious exception), foreign affiliates would follow a* stand-alone strategy, *replicating more or less in total the entire value chain of the parent firm (with the exception, typically, of technology and finance which were imported from parent firms), thus performing all tasks necessary for servicing the host country and/or neighbouring markets. In this respect, the initial acceleration of international production after the Second World War was the extension of corporate strategies already visible in the limited FDI flows in manufacturing during the period 1870–1913. However, some of the same pressures driving shallow integration – particularly liberalization and technological progress – steadily altered the way in which international production is being undertaken. The cost-competitiveness of standardized goods, the convergence of consumption patterns and falling transportation and travel costs have expanded the geographical reach of corporate strategy, enabling large oligopolistic firms – in industries such as automobiles, aerospace and electronics – to combine economies of scale with the organization of low-cost suppliers on a worldwide basis. This has led to the adoption of* simple integration strategies, *where affiliates undertake – typically with technology obtained from the parent firm – a limited range of activities in order to supply their parent firms with specific inputs that they are in a more competitive position*

to produce. Such strategies have given rise to new forms of cross-border linkages (such as subcontracting relations) and allow for greater two-way flows of information, technology and value-added activities between parent firms and affiliates. A feature of both stand-alone and simple integration strategies is that production within a TNC remains quite fragmented and cross-border internalization of economic activity is rather limited. However, enabled by the liberalization of the framework for international economic transactions, the spread of information technology, and driven by competition, TNCs have begun to redefine the ways in which they manage and organize their worldwide productive assets. More specifically, as part of complex integration strategies, *TNCs are turning their geographically dispersed affiliates and fragmented production systems into regionally or globally integrated production and distribution networks.* (UNCTAD, 1994:137–8)

The development of more complex internal organizational networks within TNCs has been accompanied by an upsurge in the number and significance of strategic collaborations with other firms, often firms in direct competition. Such strategic alliances themselves represent complex business networks and are a reflection of the kinds of changes occurring in the global economy, notably intensification of competition, acceleration of technological change, and increased costs of developing, producing and marketing new products. Strategic alliances represent one method of risk and cost sharing, especially in those industries where such considerations are most evident (for example, automobiles, biotechnology, information technology, electronics). Overall, it would appear that TNCs are restructuring their activities in ways that involve:

1 reorganizing the co-ordination of their activities in a complex realignment of internalized and externalized network relationships. In some cases, this also involves allocating greater autonomy to at least some of their affiliates;
2 reorganizing the *geography* of their production chains internationally and, in some cases, globally; and
3 transforming their relationships with their supplier firms in ways which differentiate between different tiers of suppliers.

Important as these tendencies are, they are just *tendencies* and not universals. There still remains very substantial variation between firms and between (and even within) sectors in the nature and extent of globally-integrated strategies. One source of such variation has already been noted namely the nationality of TNCs. The most important point to emphasize is the existence of a spectrum of organizational forms, a diversity of developmental trajectories in which consciously planned global operations exist side by side with firms that have internationalized in an unplanned, often adventitious, way. Across this spectrum, complex restructuring is occurring at all geographical scales, from the global to the local, as TNCs have to make strategic decisions regarding the organizational co-ordination and geographical configuration of production-chain functions. Proposals to centralize or decentralize decision-making powers, or to cluster or disperse some or all of the firm's functions in particular ways are a continuing concern for corporate decision-makers as they struggle with the fundamental geographical tension facing all international firms, namely whether to strive to globalize completely or whether to remain responsive to the forces of local differentiation.

Transnational Corporations, Nation-States and Regional Economic Blocs

The international production system is primarily articulated and co-ordinated by transnational corporations. As the previous section demonstrated, although there is still a great deal of geographical variation in how and where production takes place, the general trend is towards far more complex *networks* of production in which firms of different kinds are co-ordinated. The networks themselves operate at different geographical scales, from the local to the global. In so doing, they integrate *places* into a more intricate economic and social structure. The scale and extent of TNC activity inevitably poses problems for *states*. It has often been pointed out, for example, that some TNCs are as large as, or even larger than, some nation-states. By extension, it is then argued that TNCs are displacing states as the key economic units in the world economy. There is some truth in this. The global reach of the major TNCs does, indeed pose a threat to the autonomy of nation-states simply because such firms, in effect, incorporate *parts* of the economy of the nation-state into their own domain. It is also true that nation-states have become increasingly active in competing against each other to attract international investment projects to locate within their own territories. Such competitive bidding occurs at all geographical scales: global, regional, national and local. As a result, TNCs do have the capability to play off one state against another in order to win the best deal.

However, the relationship between TNCs and states is far more complex than this. The bargaining power does not inevitably lie with the TNC in all cases although it is certainly true, as Stopford and Strange (1991:215) observe, that governments *as a group* have indeed lost bargaining power to the multinationals. Nonetheless, they also point out that intensifying competition among states seems to have been a more important force for weakening their bargaining power than have the changes in global competition among firms. In fact, while recognizing the undoubted power of the TNC to shape and reshape the geography of international production, it can be argued that the changing structure of the global economy is the outcome of a complex combination of processes involving both TNCs and nation-states (Gordon, 1988). All states attempt to regulate the economic transactions occurring across, and within, their borders. They operate regulatory structures which determine the extent to which firms may have *access* to markets or resources contained within state territory. They operate regulatory structures which define the *rules of operation* for firms operating within their particular jurisdiction. The real issue, then, has to do with the effectiveness of such regulatory mechanisms. Here the problem is not just the extent to which TNCs have the power to circumvent state policies but also the extent to which states in general implement competitive policies among themselves which undermine the effectiveness of each individual state. This is the inevitable outcome of a world in which all states have become overtly *competition states.*

Within the last few decades a new mercantilism has emerged in which the leading industrial countries have become strategically competitive in a more overt way. In some respects, indeed, states have taken on some of the characteristics of business firms as they strive to develop strategies to achieve competitive advantage (Dicken, 1994). It is no accident that one of the world's leading writers on the competitive strategies of business firms has turned his attention to the competitive advantage of nations (Porter, 1990). Specifically, states compete to enhance their international trading position and to capture as large a share as possible of the gains from trade. They compete to attract productive investment and to build up their national production-base which, in turn, enhances their competitive position. During the past ten to fifteen years, many states have engaged in a flurry of *deregulation*, especially in certain sectors, notably financial services and telecommunications, in a further attempt to enhance their share of value-adding activities.

Nation-states, like firms, therefore, *compete* to increase their material advantage *vis-à-vis* other nation-states. Like firms, which increasingly engage in strategic alliances with their competitors, nation-states also do *collaborate*. They too form economic alliances with other states. As in the case of firms, interstate collaboration can range from the simple bilateral arrangement over a single issue to the complex collaborative networks of a supranational economic bloc.

An increasingly widely accepted view of the world economy is that it is crystallizing around three major regional blocs: North America, Europe and East Asia. Certainly, it is true that these three major regions – what Ohmae (1985) called the global triad – contain the major share of the world's economic activity. To a considerable extent, the formation of such geographical concentrations of economic activity can be explained simply in terms of the basic geographical phenomenon of proximity. Firms benefit from being close to their major markets as well as being close to their suppliers. Spatial agglomeration is a concept that can be applied at different geographical scales, not only at the micro-scale. A spatial division of labour within such a large-scale agglomeration becomes increasingly feasible and attractive to business firms, especially TNCs. In terms of the long-established principles of cumulative causation, growth begets further growth.

Thus, one of the forces leading to regional economic integration in the world economy is simply geographical (Thomsen, 1994). It is important to make this basic statement of the significance of economic-geographical processes in the formation of regional economic blocs because, too often, they are explained solely in political terms. Too often also, the term regional economic bloc is used in a highly undiscriminating way to imply a homogeneity of form. But, as Cable and Henderson (1994:1) point out:

> *In some cases, regional integration is taking place spontaneously, through market forces, impelled by the dictates of economic geography. In others, formal structures are being created in the shape of free trade areas, customs unions, common markets and various types of preferential association. These approaches – which differ considerably in*

their motivations and effects – have unfortunately attracted the all-purpose designation of trade blocs. This term carries connotations both of confrontation and uniformity of style which may be quite wrong.

Whether regional blocs are spontaneously created through the economic geography of market-driven forces (as in the Asia–Pacific) or are the deliberate creation of political agreement, their fundamental basis is that of trade. Short of complete political union, we can identify four types of politically-negotiated regional economic arrangements of increasing degrees of integration:

1 The *free trade area* in which trade restrictions between member states are removed by agreement but where member states retain their individual trade policies towards non-member states.
2 The *customs union* in which member states operate both a free trade arrangement with each other but also establish a common external trade policy (tariff and non-tariff barriers) towards non-members.
3 The *common market* in which not only are trade barriers between member states removed and a common external trade policy adopted but also the free movement of factors of production (capital, labour, etc.) between member states is permitted.
4 The *economic union* is the highest form of regional economic integration short of full-scale political union. In an economic union, not only are internal trade barriers removed, a common external tariff operated and free factor movements permitted, but also broader economic policies are harmonized and subject to supranational control.

The number of regional trading arrangements has grown dramatically during the past half century. As data collected by the World Trade Organization show, some 109 regional trading arrangements were notified to GATT between 1948 and 1994. Although many of these were initially established (at least in embryonic form) between 30 and 40 years ago there was a very substantial development of such arrangements during the late 1980s and early 1990s (Gibb and Michalak, 1994). In the 1980s, membership of the European Community expanded further from nine to twelve. In 1991, an agreement was reached with all but one of the members of the European Free Trade Association (EFTA) to form a European

Economic Area (EEA). Towards the end of the 1980s, the political processes aimed at the completion of a single European Community Market by 1992 reached their climax. In 1989, the United States and Canada implemented the Canada–United States Free Trade Agreement. In 1994, the North America Free Trade Agreement came into being whereby Mexico joined Canada and the United States in the first example of a developing country becoming fully integrated with highly developed countries.

In the first few months of 1995, the regionalizing trend accelerated further. In Latin America, the MERCOSUR customs union between Argentina, Brazil, Paraguay and Uruguay came into being; further development of the Andean Pact (involving Bolivia, Colombia, Ecuador, Peru and Venezuela) occurred; and a free trade area involving Mexico, Colombia and Venezuela came into force. In Europe, the European Union (successor to the European Community) expanded its membership from 12 to 15 with the incorporation of Austria, Finland and Sweden; free trade agreements were implemented between the European Union (EU) and the three Baltic republics of Estonia, Latvia and Lithuania. In the Pacific Basin, where, apart from ASEAN, there is no formal regional trade bloc, the 19 states of the Asia–Pacific Economic Forum agreed to remove all regional trade barriers by 2020. Other, more tentative regional trading arrangements are at least on the agenda for discussion. The United States has ambitions for a free trade area encompassing the whole of the Americas (from Anchorage to Tierra del Fuego), while it has even been proposed that there might be a Transatlantic Free Trade Agreement – a TAFTA – between NAFTA and the EU.

The vast majority of the regional economic groupings fall into the first two categories of the classification shown above (the free trade area and the customs union). There is a small number of common-market arrangements but only one group – the European Union – has come close to being a true economic union. In fact, not only is there enormous variation in the scale, nature and effectiveness of these regional trade groupings but there is also, in some cases, a considerable overlap of membership of different groups, especially in Latin America. Such diversity must be borne in mind when considering the likely geographical effects of regional integration both internally (on member states and communities) and externally (on the rest of the world economy).

Regional trading blocs are essentially *discriminatory* in nature. Most have a strongly defensive character; they represent an attempt to gain advantages of size in trade by creating large markets for their producers and protecting them, at least in part, from outside competition. Consequently, the most important of the regional blocs, particularly the European Union and NAFTA, have a very considerable influence on patterns of world trade. The classic analysis of the trade effects of regional blocs (specifically of customs unions) identifies two opposing outcomes: *trade creation* and *trade diversion.* According to Smith (1994:18):

> *When trade barriers are reduced between partner countries, trade between them will increase; trade diversion refers to the replacement of trade with other countries by trade with partners, while trade creation refers to trade replacing home production or associated with increased consumption. There can also be external trade creation if the integration process leads to a reduction in trade barriers with the rest of the world.*

Regional trading blocs also have a major influence on the investment decisions of transnational corporations. A United Nations study (UNCTC, 1990) examined both the general influence of regional economic integration on TNC strategies and also the specific effect of the completion of the Single European Market on the investment decisions of European and non-European firms. The UNCTC study demonstrated that the effects of regional integration on direct investment could, like that on trade, also be conceptualized in terms of creation and diversion.

Empirical observation suggests that the introduction of regional economic integration results in an increase in inward direct investment. Three kinds of such investment-creating strategy can be identified. First, *defensive export-substituting investment* is characteristic of firms which formerly served the regional market through trade but which, with the introduction of regional trade barriers, need to replace that form of market access by investing in productive facilities inside the market. In the UNCTC's view, this kind of upsurge in inward investment – effective tariff jumping – will be short term in its effect and will taper off after the firms have completed their entry. The two other types of investment-creating investment are more dynamic and longer term in their effects. *Rationalized foreign direct investment* is the response of firms to the scale economy effects of regional integration whereby the larger integrated market stimulates the rationalization of previously dispersed production oriented towards individual national markets into larger, more efficient regional market operations. *Offensive export-substituting investment* differs from defensive export-substituting investment insofar as it refers to investment made by firms in anticipation of, rather than in defence of, market share in the integrated region (UNCTC, 1990:2).

As in the case of trade creation, the notion of investment creation in response to the formation of a regional bloc refers to the aggregate geographical scale of the regional bloc as a whole. In contrast, investment diversion occurs at the level of individual member states within the regional bloc; the removal of internal trade (and other) barriers may lead firms to engage in *reorganization investment* whereby they 'realign their organizational structures and value-adding activities to reflect a regional rather than a strictly national market . . . As firms reorganize their value-adding activities in the region, this may in turn attract still more inflows of FDI, if the associated rationalization of the region's industries and services allows for new economies of scale, technical progress and the associated higher returns to capital' (UNCTC, 1990:2). Whether or not such further investment increase occurs, the primary result of reorganization investment is an increase in cross-border investment activity within the region. This, by definition, diverts investment from some locations in favour of others.

One of the fears generated by the growth of regional trading arrangements is that each bloc will become inward looking and will, therefore, divert trade from non-member countries. Such a fear has been articulated very strongly, for example, in the case of the Single European Market process and is expressed in the popular term Fortress Europe. Whether or not such fears are exaggerated, there is no doubt that concern about subsequent exclusion from, or restricted access to, the European Union had a major influence on the investment decisions of transnational corporations from outside Europe. In the run-up to the target date of December 1992, there was a major upsurge in new investments in Europe by North American, Japanese and other Asian firms in particular. Some of the investment was in the form of new 'greenfield' operations but much of it was through the acquisition of European firms, as well as through the forging of strategic alliances between European and non-European firms.

The European Union provides the most developed example of such processes although, as Thomsen (1994) argues, it was to a large extent the economic pressures created by major firms that accelerated the political process of integration. Many of the leading firms had begun to reorganize their European operations on a regional, rather than on a national, basis long before the single-market process began. A good example is the American automobile manufacturer, Ford, which established a European-wide organization as early as 1967 (Dicken, 1992a). However, it is certainly the case that Europe has seen a veritable tidal wave of massive organizational and geographical restructuring by both European and non-European firms. Most, if not all, European industries are becoming increasingly *regionalized* as firms strive to create organizational networks oriented to the European market as a whole rather than to individual national markets.

Such regionally-extensive reorganization of production contributes to the redirection and intensification of intraregional trade and investment. It also leads to shifting patterns of uneven development between different parts of the region. Some areas undoubtedly benefit; others undoubtedly do not. Internal regional differentiation in levels of economic and social well-being are the inevitable outcome of the processes of regional integration. These problems are intensified where there are major economic differentials between the member states of a regional bloc. Of course, in the kind of economic union exemplified by the European Union, there are redistributive mechanisms in the form of the European Social Fund and the European Regional Development Fund. These are expected to help alleviate the extremes of regional poverty and economic decline. But in the less politically developed regional blocs, including NAFTA, such redistributive mechanisms do not exist. It is not surprising, for example, that the creation of the North American Free Trade Agreement led to immense concern in both the older urban-industrial and poor rural areas of the United States over the possible flight of investment to the far lower labour-cost areas of Mexico.

A final consideration in this brief discussion of the regionalization of the global economy concerns the extent to which this process is likely to continue and even intensify. Lawrence (1993:63–4) argues that at least in the near future – a decade, if not longer – increased regional integration is inevitable. The critical question, though, is whether the regional arrangements will become building blocks in a more integrated global system or stumbling blocks that cause the system to 'fragment'. There are real concerns amongst outsiders about the likely future openness of regional blocs. Such concerns include fear about increased trade diversion; about possible increasing introvertedness as the regional integration process proceeds; about the possible dominance in policy making by the more protectionist member-states which might lead to new external barriers to trade and investment.

Not surprisingly, prognoses about the global implications of regional economic integration are strongly polarized. The critical issue will be the extent to which such blocs are open to external interaction. The optimistic view, as expressed by Lawrence (1993:48), is that open regional blocs can actually promote and facilitate external liberalization, that is, trade with parties outside blocs. He points to the history of the European Community in which 'with the noteworthy exception of agriculture . . . increased regional integration among the original six members of the EC was associated with extensive participation in multilateral traffic reductions . . . The European experience also demonstrated that excluded countries may have stronger incentives to liberalize in a system with emerging regional arrangements' (Lawrence, 1993:49). But not everybody takes such a sanguine view. In many parts of the world, there is a genuine fear that the regional integration of some states will lead to the exclusion and further peripheralization of others.

Conclusion

The primary aim of this chapter has been to clarify and correct some of the misconceptions and stereotypes which litter a good deal of the literature on the contemporary global economy. There is no doubt that in the process of economic transformation that has been occurring over recent decades, and especially since the early 1960s, the role of the nation-state has changed. It is no longer accurate to conceive of the world economy as being made up simply of nation-states in interaction with one another and being the primary regulators of the world economic system. The degrees of freedom available to individual states have undoubtedly been reduced. However, to attribute such an alleged decline in national autonomy and sovereignty simply to the rise of the transnational corporation is invalid reasoning. Whilst it is certainly true that TNCs have the capacity to operate across national boundaries they cannot do so without reference to the actions and policies of the nation-states within the territories of which they locate their operations.

Similarly, to conceive of TNCs as a homogeneous population of giant firms all pursuing somewhat similar objectives in similar ways is equally inadequate. TNCs are enormously varied in their size, their structures and their strategies. Although it is certainly possible to identify some general empirical tendencies in their modes of operation, it is also true to say that diversity, rather than uniformity, is the most significant feature. Some of that diversity arises from the particular relationships between a TNC and its home country. Since TNCs are socially and culturally produced organizations, we should not be surprised to find that, at least in certain respects, TNCs from one country may differ somewhat from TNCs from another country.

In fact, to understand contemporary change in the global economy, we need to focus upon the complex interactions between both TNCs and nation-states as they each pursue their specific objectives. In a very real sense, nation-states need firms but firms also need nation-states. The relationship, as observed in the introductory section, is at once both cooperative and competitive, both supportive and conflictual. The emergence of regionally integrated economic blocs which have become an increasingly pervasive feature of the modern global economy, reveals something of the essence of TNC–nation-state interaction. Regional economic integration would probably have occurred at the level of firms as an almost inevitable outcome of the geographical phenomenon of proximity even without political initiatives. In fact, once set in motion by whatever initiating force, one reinforces the other. TNCs are anxious to develop and maintain access to large markets; nation-states are anxious to protect themselves against the ravages of global economic competition. The regional bloc goes some way towards meeting both sets of goals.

References

Barlett, C.A. and Ghoshal, S. (1989) *Managing Across Borders: The Transnational Solution* (Boston: Harvard Business School Press).

Benson, I. and Lloyd, J. (1983) *New Technology and Industrial Change* (London: Kogan Page).

Cable V. and Henderson, D. (eds) (1994) *Trade Blocs? The Future of Regional Integration* (London: Royal Institute of International Affairs).

Dicken, P. (1992a) *Global Shift: The Internationalization of Economic Activity* (London: Paul Chapman Publishing).

Dicken, P. (1992b) International production in a volatile regulatory environment: the influence of national regulatory policies on the spatial strategies of transnational corporations, *Geoforum*, 23:303–16.

Dicken, P. (1994) Global–local tensions: firms and states in the global space-economy, *Economic Geography*, 70:101–28.

Dicken, P. and Thrift, N. (1992) The organization of production and the production of organization: why business enterprises matter in the study of geographical industrialization, *Transactions of the Institute of British Geographers*, 17:279–91.

Dunning, J.H. (1993) *Multinational Enterprises and the Global Economy* (Reading, MA: Addison-Wesley).

Gibb, R. and Michalak, W. (eds) (1994) *Continental Trading Blocs: The Growth of Regionalism in the World Economy* (Chichester: John Wiley).

Gordon, D.M. (1988) The global economy: new edifice or crumbling foundations?, *New Left Review*, 168:24–64.

Hobsbawm, E.J. (1979) The development of the world economy, *Cambridge Journal of Economics*, 3:305–18.

Hu, Y.S. (1992) Global firms are national firms with international operations, *California Management Review*, 34:107–26.

Jones, G. (ed.) (1993) *Transnational Corporations: A Historical Perspective* (Aldershot: Avebury).

Jones, G. (ed.) (1994) *The Making of Global Enterprise* (London: Frank Cass).

Julius, DeAnne (1990) *Global Companies and Public Policy: The Growing Challenge of Foreign Direct Investment* (London: Pinter).

Kindleberger, C.P. (1969) *American Business Abroad* (New Haven: Yale University Press).

Lawrence, R.Z. (1993) Futures for the world trading system and their implications for developing countries, in M. Agosin and D. Tussie (eds) *Trade and Growth: New Dilemmas for Trade Policy* (London: Macmillan).

Ohmae, K. (1985) *Triad Power: The Coming Shape of Global Competition* (New York: The Free Press).

Porter, M.E. (1990) *The Competitive Advantage of Nations* (London: Macmillan).

Reich, R.B. (1991) *The Work of Nations: Preparing Ourselves for 21st Century Capitalism* (New York: Knopf).

Ruigrok, W. and van Tulder, R. (1995) *The Logic of International Restructuring* (London: Routledge).

Smith, A. (1994) The principles and practice of regional economic integration, in V. Cable and D. Henderson (eds) *Trade Blocs? The Future of Regional Integration* (London: Royal Institute of International Affairs).

Stopford, J.M. and Strange, S. (1991) *Rival States, Rival Firms: Competition for World Market Shares* (Cambridge: Cambridge University Press).

Thomsen, S. (1994) Regional integration and multinational production, in V. Cable and D. Henderson (eds) *Trade Blocs? The Future of Regional Integration* (London: Royal Institute of International Affairs).

UNCTAD (1994) *World Investment Report 1994: Employment and the Workplace* (New York: United Nations).

UNCTC (1990) Regional economic integration and transnational corporations in the 1990s: Europe 1992, North America, and developing countries, *UNCTC Current Studies, Series A*, 15.

Wilkins, M. (ed.) (1991) *The Growth of Multinationals* (Aldershot: Avebury).

13 Geography in the Age of the Mega-Cities

In the latter half of the twentieth century the world has witnessed the rapid growth of human population and its attendant concentration in cities, particularly in the developing world. Both of these processes are now well known. Between 1950 and 1990, the world's population increased from 2513 million to 5289 million, more than a 100 per cent increase. Of the latter total, the urban population accounted for 2389 million. During the same four decades, the urban population in developed countries increased from 448 million (some 54 per cent of the total population) to 875 million or 73 per cent of the total. In developing countries, the urban population soared from 286 million (17 per cent) to 1514 million (37 per cent). It is thus clear that over the period under review, the bulk of the increase in urban population has been derived from developing countries. By 1990, although the level of urbanization in developing countries was a little less than half of that in developed countries, the total urban population in this group of countries was almost twice as large as that in developed countries.

What is even more important is the fact that these trends are likely to persist. The forecast is that cities in developing countries are likely to continue to grow faster than those in developed countries. The fastest growing urban centres are all located in developing countries, many growing at an annual rate of over 4 per cent in the period 1980–90 (table 13.1). By the year 2000, more than one person in two in the world will be living in a city and shortly before 2010, countries in the developing world will cross the urban–rural divide; that is, more than 50 per cent of their population will live in urban areas (United Nations, 1991).

Against the background of a rapidly urbanizing developing world,

This chapter is based on a contribution by Yue-man Yeung.

Table 13.1 The world's mega-cities, 1950–2000

Mega-city	*Population (in million)* 1950	1980	1990	2000	*Average annual rates of growth* 1980–90 (%)		*Percentage of country's total population* 1990
Africa						3.9	
Cairo	2.4	6.9	9.0	11.8	2.6		37.0
Lagos	0.3	4.4	7.7	12.9	5.6		20.2
Latin America						2.8	
Buenos Aires	5.0	9.9	11.5	12.9	1.5		41.3
Lima	1.0	4.4	6.2	8.2	3.4		41.3
Mexico City	3.1	14.5	20.2	25.6	3.3		31.4
Rio de Janeiro	2.9	8.8	10.7	12.5	2.0		9.5
São Paulo	2.4	12.1	17.4	22.1	3.6		15.4
Asia						3.3	
Bangalore	0.8	2.8	5.0	8.2	5.7		2.2
Bangkok	1.4	4.7	7.2	10.3	4.1		56.8
Beijing	3.9	9.0	10.8	14.0	1.8		2.8
Bombay	2.9	8.1	11.2	15.4	3.3		4.9
Calcutta	4.4	9.0	11.8	15.7	2.7		5.1
Dhaka	0.4	3.3	6.6	12.2	7.0		35.0
Delhi	1.4	5.6	8.8	13.2	4.6		3.8
Istanbul	1.1	4.4	6.7	9.5	4.1		19.4
Jakarta	2.0	6.0	9.3	13.7	4.4		16.4
Karachi	1.0	4.9	7.7	11.7	4.4		19.6
Manila	1.5	6.0	8.5	11.8	3.5		31.9
Seoul	1.1	8.3	11.0	12.7	2.8		35.7
Shanghai	5.3	11.7	13.4	17.0	1.3		3.5
Tehran	1.0	5.1	6.8	8.5	2.9		21.9
Tianjin	2.4	7.3	9.4	12.7	2.5		2.5
More developed regions						0.7	
Los Angeles	4.0	9.5	11.9	13.9	2.2		6.4
Moscow	4.8	8.2	8.8	9.0	0.8		4.7
New York	12.3	15.6	16.2	16.8	0.4		8.7
Osaka	3.8	8.3	8.5	8.6	0.2		9.0
Paris	5.4	8.5	8.5	8.6	–0.03		20.4
Tokyo	6.7	16.9	18.1	19.0	0.7		19.1

Source: United Nations, *World Urbanization Prospects 1990* (New York, 1991).

Street in the Ginza district, Tokyo, at twilight.
Source: J. Aaronson/ Aspen/Cosmos.

the salience of very large cities is striking. Between 1950 and 1990, the number of cities with over a million inhabitants tripled. In 1950, there were only 78 such cities; in 1990, the number exceeded 276. By 2010, the number is projected to grow to 511. Just in Asia alone the number of cities with a million plus inhabitants during this period rose from 24 to 115. In 1990, 33 per cent of the world's urban population resided in 'million plus' cities and 10 per cent lived in very large urban agglomerations of at least 8 million inhabitants, otherwise referred to by the United Nations as 'mega-cities' (Dogan and Kasarda, 1988; United Nations, 1991). There is little doubt that these very large cities are destined to dominate the economic and social life of their nation-states throughout the world.

A recent United Nations study indicated that there are 28 mega-cities in the world at present, with the vast majority of them located in developing countries. Mexico City, São Paulo and Tokyo top the list of mega-cities (United Nations, 1991). The dominance of certain mega-cities on the social and economic life of their nation-states can hardly be disputed, especially where up to 30 to 40 per cent of the national urban population are concentrated in such mega-cities. Bangkok represents the extreme case of having 56.8 per cent of Thailand's population living in that single city.

The prominence and impact of 'million plus' and mega-cities on national life has spurred a recent flurry of studies on these cities (Dogan and Kasarda, 1988), mega-cities (Brennan and Richardson,

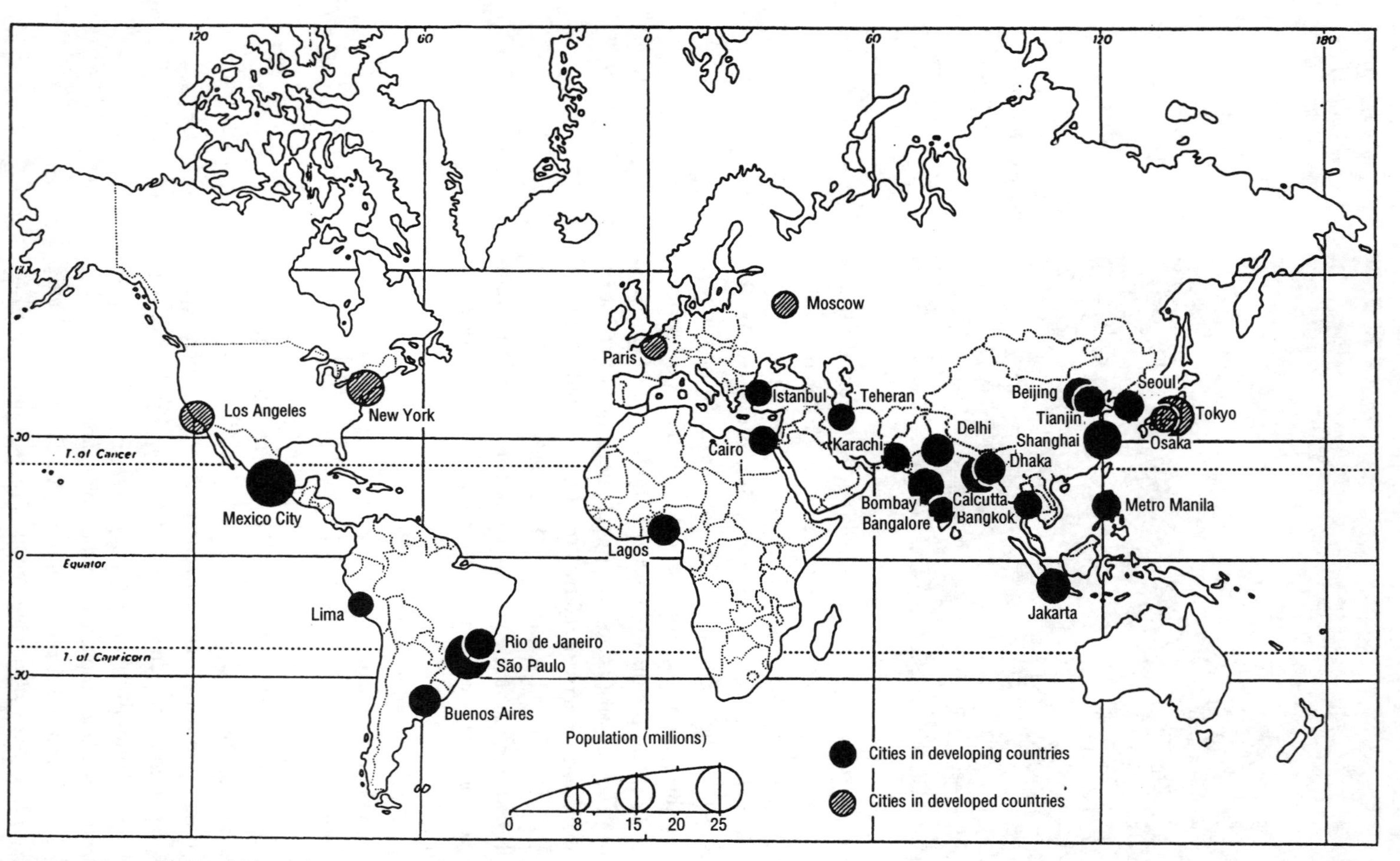

Figure 13.1 Mega-cities of the world *c.*2000

1989), world cities (Hall, 1984; Friedmann, 1986) and global cities (King, 1990; Sassen, 1991). While mega-cities are defined primarily by their demographic weight, world cities and global cities are particularized because of the vital functions they perform in the global economy. The list of mega-cities indicated in this chapter and of world cities as proposed by Friedmann are different, with only 11 cities being on both lists. Notwithstanding, the chapter will compare both groups of cities, noting the difference in their characteristics, orientations and challenges. The chapter will focus on the rise of both types of cities and examine their *raison d'être*, and the problems and opportunities they face. It will conclude with a prospective statement on their role in the twenty-first century.

The Rise of Mega-Cities

By the closing years of the twentieth century, the world has become increasingly urban. Even the developing world shows a prominent and pronounced concentration of population in mega-cities (figure 13.1). These mega-cities account for 12.1 per cent of the world's urban population in 1990, or 5.5 per cent of the world's total population. By the year 2000, the corresponding percentages will be 12.7 and 5.9 respectively. The importance of such population concentrations pale into relative insignificance when put against the economic dominance of some of the mega-cities on their national economic landscape. For example, Tokyo accounted for 27 per cent of Japan's retail and wholesale sales, 60 per cent of invested capital and 50 per cent of university students; Seoul has 78 per cent of headquarters of all business firms in South Korea, 90 per cent of large business enterprises, and 65 per cent of all loans and deposits in South Korea (Yeung, 1988:162).

Mega-cities have grown rapidly. With the exception of Dhaka, Calcutta and Bombay, they have grown faster in the first 25 years after 1950 than in the following 25 years. This deceleration of growth may be explained in many ways. There are the national demographic trends which have slowed down in many countries, improved data-collection methods, specific policies to slow down big-city growth, and a less favourable growth environment in many developing countries since the oil crisis of 1973 (Richardson, 1993). In absolute terms, however, during the four decades 1950–90, mega-cities in developing countries have experienced explosive growth. Many

Table 13.2 Living standards in the world's mega-cities

Mega-city	*Public safety*	*Food costs*	*Housing standards*	*Education*	*Traffic flow*	*Urban Living standard score*
Cairo	56.4	47	94	53	12.4	36
Lagos	–	58	50	31	17.4	19
Buenos Aires	7.6	40	86	51	29.8	55
Lima	–	70	82	55	3.7	33
Mexico City	27.6	41	94	62	8.0	38
Rio de Janeiro	36.6	26	92	55	18.6	51
São Paulo	26.0	50	100	67	15.0	50
Bangalore	2.8	62	67	60	16.0	37
Bangkok	7.6	36	76	71	13.0	42
Beijing	2.5	52	89	97	25.7	55
Bombay	3.2	57	85	49	10.4	35
Calcutta	1.1	60	57	49	13.3	34
Dhaka	2.4	63	73	37	21.4	32
Delhi	4.1	40	66	49	14.0	36
Istanbul	3.5	60	89	67	11.2	42
Jakarta	5.3	45	85	77	16.3	40
Karachi	5.7	43	75	65	17.6	36
Manila	30.5	38	91	67	7.2	43
Seoul	1.2	34	100	90	13.8	58
Shanghai	2.5	55	95	94	15.3	56
Tehran	–	–	84	58	7.5	39
Tianjin	2.5	52	82	71	20.2	51
Los Angeles	12.4	12	94	90	19.0	69
Moscow	7.0	33	100	100	31.5	64
New York	12.8	16	99	95	8.7	70
Osaka	1.7	18	98	97	22.4	81
Paris	2.4	21	99	99	8.5	72
Tokyo	1.4	18	100	97	28.0	81

Notes:
Public safety: murders per 100,000 people
Food costs: percent income spent on food
Housing standards: percent homes with water/electricity
Education: percent children in secondary school
Traffic flow: miles per hour in rush hour
Urban living standard score, ranking from 100 to 0: 100–75: very good; 74–60: good; 59–45, fair; 44 and below, poor
Source: 'Cities: Life in the World's 100 Largest Metropolitan Areas', spread sheet. Population Crisis Committee, Washington, DC, 1991.

urban agglomerations such as Lagos, Lima, Mexico City, São Paulo, Bangalore, Bangkok and so on more than quadrupled their populations within this period (table 13.2).

Many reasons have been advanced for the extraordinary growth of mega-cities in the developing world. Population dynamics relating to the conjuncture of declining death rates and rising birth rates, coupled with large-scale and persistent rural–urban migration have been a powerful determinant of sustained urbanization in developing countries. China is perhaps the only exception. Until the recent relaxation with the adoption of a relatively more open policy in 1978, the country successfully implemented a policy of controlling urban growth and holding down population growth in the rural area through a programme of household registration and food rationing. Mega-cities in all of these countries, notwithstanding their heterogeneous economic and social situations, have been confronted with horrendous problems of having to create jobs, provide basic services, and keep body and soul together for their teeming masses.

Another reason for the extraordinary rate of growth of mega-cities in developing countries is economic. Gilbert and Gugler (1981) perceived urbanization in Third World countries as dependent development, resulting from the capitalist penetration of the economies of these countries by industrially advanced economies. As a consequence, dependent urbanization has led to patterns of inequality and poverty, arising from the fact that, while cities in these Third World countries form part of the world economy, their inhabitants do not have equal access to the world's resources. In like vein, Armstrong and McGee (1985) conceptualized cities in Asia and Latin America as theatres of capital accumulation. They viewed cities in these regions in a world political economy framework, with their institutional structures providing the vehicles for capital expansion and accumulation via the processes of production, circulation and consumption. In both studies, Third World cities are situated in the context of a new international division of labour with transnational corporations (TNCs) playing a central role in the global economy.

There are other reasons that have been advanced to explain the rapid growth of mega-cities. The cultural role of these cities, particularly when they are the capital cities of newly independent countries, the concentration of political power in such centres, and the attraction of facilities for better education and other services, among others, figure greatly in the enormous pull of mega-cities for their national population, generating a sustained impetus of growth. Large cities continue to grow because of economies of scale. However, beyond a certain threshold, negative externalities set in, such as polluted air and water, snarled traffic, festering slum and squatter settlements, criminality, and acute difficulties in maintaining law and order. In fact, the increasing negative externalities of some mega-cities in developing countries has encouraged a spontaneous process of deconcentration or 'polarization reversal'. Such trends have been reported from mega-cities in Latin America: São Paulo, Mexico City and Buenos Aires (Gilbert, 1990).

Indeed, the growing prominence of these problems in Third World cities became so overpowering that, beginning in the early 1970s, the World Bank led a concerted institutional and international attack on urban poverty in the developing world. It evolved special programmes of assistance, initially in the housing and regional development sector, later developing a more integrated approach towards urban poverty alleviation. The 1990s have been viewed as a decade of opportunity in which the World Bank will support appropriate national and city-wide policies, promote institutional developments to improve urban productivity and protect the urban environment (World Bank, 1991). Similarly, the United Nations Development Programme (UNDP) is implementing a programmatic agenda for the 1990s with considerable emphasis on people-centred development (UNDP, 1991).

As the 1990s unfolds, there has been a noticeable increase in international efforts and resources being mobilized for the support of research, networking and information dissemination concerning urban problems in developing countries. Between 1991 and 1993, the Ford Foundation supported an ambitious project on urban research in the Third World with the participation of 12 research teams mustered from Latin America, Asia and Africa. The findings of the first phase of the research have been published in four volumes (Stren, 1994). The second phase began in 1994 and is anticipated to be completed in 1996. Another initiative has been launched concerned with actually creating an awareness among policy makers and planners on urban problems and solutions in developing countries. This is the ten-year Urban Management Programme, which has been making a strong impact in all developing regions of the world. The programme, supported by the UNDP and bilateral donors, involves the professional cooperation of many leading institutions, notably the World Bank and the United Nations Centre for Human Settlements (UNCHS), and has actively built up regional and country focal points and expertise since 1992 (Clarke, 1991).

World Cities and the New Global Economy

In the new global environment that has fostered the growth of mega-cities in developing countries, a new class of cities – world cities or global cities – is emerging and coming to the fore. Some of them may be demographic giants, in which case they are also mega-cities; but others are world cities on the basis of their economic importance especially in terms of the functions they perform in the new global economy. World cities are control or command centres in the network of similar cities around the world, often finding themselves as headquarters of TNCs, as major financial centres, transport and telecommunication hubs, and tourist meccas. These are cities of novel importance and it is necessary to examine the factors that have promoted their growth.

Briefly, since the early 1980s three factors have affected the global economy in fundamental ways, giving due emphasis to comparative advantage in economic production and open competition. The first is the collapse of commodity and oil prices in the early 1980s. This, along with the long-run decline of the material input in production in industrially advanced economies, has resulted in the decreased importance of primary resources in economic production. Industries have tended to shift from labour-intensive to knowledge-intensive modes. Second, along with the decline in the role of material resources in production has been the rise of capital, in particular, transnational capital, as an engine of national economic growth. Foreign direct investment (FDI) has globally become a new indicator of economic growth. In this regard, the Plaza Accord of 1985, with the resultant dramatic appreciation of the yen against the US dollar, has had a far-reaching effect on the pattern of flow of FDI. Japanese investment has actively sought foreign destinations, with the United States, Europe and Pacific Asian countries benefiting the most from this Japanese export of capital. Between 1985 and 1987, Japanese FDI increased world-wide almost three times to US$33.4 billion (Nakakita, 1988). Third, the last two trends have been facilitated by technological changes that have been rapid, revolutionary and widespread since the 1980s. The range and speed of innovations in micro-electronics and communications have been breath-taking. A new era of telecommunications, information technology and widened global opportunities is generally believed to have arrived. These mega-trends have transformed the essence of the global economy. National boundaries have become a less inhibiting factor in many aspects of life, especially in manufacturing production. Consequently, urban life, economic production, finance, banking and services have become globalized. A borderless domain has become characteristic of the new global economy in which world cities have risen to play pivotal roles.

The international economic order that had governed much of the post-Second World War period was thrown into disarray in the early 1970s. A new global economy has since been created in which a new international division of labour (NIDL) has favoured the newly industrializing countries as new centres of productive activities and, consequently, as theatres of major capital accumulation. The forces driving the NIDL essentially stem from four sources. First, a strong desire among firms to utilize less expensive sources of labour and more profitable situations for production; second, response by many foreign firms to the growing bargaining power of certain developing countries in stimulating and developing industrialization; third, growing international competition forcing firms out of developed countries to search for new investment opportunities; and finally the role of well-organized labour and government regulations in developed countries inducing firms to look overseas for a less contentious labour market.

These forces, as part of NIDL, reshaped the spatial hierarchy of economic activities and the relationship and relative importance of industrially advanced and newly industrializing economies. The new global economy is characterized by the international spread of productive facilities, the international spread of corporate-related services and the rise of a system of international capital markets (Cohen, 1981). As Dicken (1992) has highlighted, economic activity is becoming not only more internationalized but increasingly globalized. The relatively simple trade flows of past decades have become far more complex. Economic production has become highly fragmented and complex, involving many production processes and geographical locations on a global scale in ways that slice through national boundaries. Robert Reich (1991) illustrates vividly the nature of this global shift:

Consider some examples. Precision ice hockey equipment is designed in Sweden, financed in Canada, and assembled in Cleveland and Denmark for distribution in North America and Europe, out of alloys whose molecular structure was researched and patented in Delaware and fabricated in Japan . . . A microprocessor is designed in California and financed in America and West Germany, containing dynamic random-access memories fabricated in South Korea.

As a result of these kinds of fine-tuning production processes, there have been changes in the international competitiveness of a number of industries and cities throughout the world. Indeed, a new class of cities – world cities – has emerged or is re-emerging from old forms to drive this new global economy. The greatest impact has been felt in the urban centres of developed nations and the larger cities of developing countries (Cohen, 1981:303).

To be more explicit, Sassen (1991) identifies four key functions for world cities namely, as command posts in the organization of the global economy, as key locations for finance and specialized services replacing manufacturing as the leading economic sector, as sites of production and innovation in leading industries, and as markets for the products and innovations produced. These multiple functions of world cities enable them better to articulate the global economy through interactions with other world cities, regionally and globally, within an effective and functionally efficient network. The emerging global economy requires the unrestricted flow of capital, the internationalization of the markets for commodities and labour, as well as for information, raw materials, management and organization (Castells, 1992:5).

The new global economy has had an impact on cities in developed and developing nations in different ways. In industrially advanced economies, previously thriving cities in manufacturing production such as Detroit, Chicago, New York, Manchester and Osaka lost many factory jobs and had to restructure their economy. Some cities rose to meet these new challenges effectively, notably Los Angeles, Miami, Berlin and Vienna. Cities bordering on Eastern Europe such as Vienna and Berlin assumed new roles or recaptured old ones as international business centres for the central European region (Sassen, 1994). However, the dominant financial centres of the world remained by far New York, Tokyo and London. In 1994, in terms of capitalization, their stock markets were the largest in the world in that order. New York has more *Fortune* 500 corporate headquarters than any other city, and yet it is struggling to maintain its leading position globally and nationally. London is similar to New York in its absolute population decline, absolute losses in manufacturing employment and slow overall job growth (Markusen and Gwiasda, 1994). London's claim to leadership in world finance has been greatly strengthened by the deregulation of financial markets in the United Kingdom during the 1980s. In fact, the general climate in the world is for less regulation, more diversification and more competition. This has also benefited some Asian markets, notably Hong Kong, Bangkok and Taipei. Compared to New York and London, Tokyo's dominance on the Japanese economy is unrivalled, serving as a centre for finance innovation and international trade. Japan accounts for nearly half of the total value of the global top-1000 companies. Within the country, some 84.6 per cent of total foreign companies numbering some 1251 have their headquarters in Tokyo (Fujita, 1991).

Sassen (1994) emphasizes that under the new global economy, a limited number of cities are emerging as the transnational location for investment for firms, for the production of services and financial investments, and for various international markets. By the 1980s, such cities already included New York, London, Tokyo, Paris, Frankfurt, Amsterdam, Zurich, Los Angeles, Hong Kong, Singapore, São Paulo and Sydney. These cities have acquired new roles and importance within the emergent global economy. They have become the focal points for the processes of economic globalization, international migration, and new producer services and finance.

In an examination of the control and co-ordination of the international exchange of capital and commodities by business intermediaries (financiers, wholesalers, corporate head offices), Meyer (1991) concluded that the reactions of these bodies to competition – their ability to alter transaction costs, to differentiate and de-differentiate among control markets and to appeal to force – all offer insights into their long-run capabilities to influence change in the world system of cities. In an earlier but similar study, Meyer (1986) explored the relations between international financial metropolises and South American cities and revealed that the world system of cities is organized independently of national boundaries and regional borders.

As world cities are internationally oriented, they are subject increasingly to the forces of international migration of population. For example, Europe's world cities, such as Paris and Frankfurt, are at the centre of major transport-network and international migration flows. Berlin and Vienna have been reported to be recapturing their past role and are favoured as destination points for vast regional migration streams from central and eastern Europe (Sassen, 1994). Tokyo has similarly experienced, for the first time, legal and illegal immigration from neighbouring Asian countries including those from South Asia. Immigration has become a new factor in Japanese urban life, providing needed labour for menial and other jobs, occasionally involving the shady underworld of crime. Even in Hong Kong, as a result of its greater international orientation in recent years, the rise of foreign populations has been very rapid. Between 1980 and 1992, the number of Filipinos, Thais and Canadians in the population has more than tripled (Yeung, 1995).

In order for world cities to discharge their functions well, they need to be engaged in the process that can best be described as world-city formation. They need a strong physical and social infrastructure. For example, office buildings of distinct architectural design convey images of power and prestige; major international airports, super-fast trains, and information highways sustain the global reach of a world city. The social networks are nourished by a large variety of cultural and entertainment facilities (Dieleman and Hamnett, 1994:358). For illustrative purpose, the Los Angeles metropolitan area expanded its land area by over eight times from a base of 4070 sq m in 1985. Many new and massive airports in Asia, such as Osaka, Seoul and Hong Kong are being or have been constructed on reclaimed land in response to their accession to world-city status. Expansion of land area into the ocean by reclamation and by building intensively and vertically constitute two major methods for Asian global cities to create new space to meet their functional needs.

So far, discussion in this section has centred on world cities in an emergent global economy. Some of these world cities, by achieving certain population size, are also mega-cities. However, for the most part, mega-cities are designated by population size and as such, they may or may not qualify as world cities. Calcutta, Beijing, Shanghai and Lagos have huge population concentrations, but they are not as yet world cities. Their economic and social orientations remain by and large national and regional. Indeed, many mega-cities in the world are carried forward in their growth and evolution by the forces that are generated within their national and regional space-economies. For many, particularly those in developing countries, they are beset by a different set of problems that are more immediate and intractable. It is on these that the next section will focus.

Challenges Confronting Mega-Cities in Developing Countries

As mega-cities have been shown to be increasingly a phenomenon of the developing world, a profile of mega-cities, at present, is largely one of cities in this part of the world. Indeed, the population policy section of the population division of the United Nations has been conducting a mega-city research project involving 20 such cities in developing countries. To date, reports on 13 of these cities have been published (Bangkok, Bombay, Cairo, Calcutta, Dhaka, Delhi, Jakarta, Karachi, Madras, Metro Manila, Mexico City, São Paulo and Seoul). Some of the highlights of these studies may be summarized as follows (Brennan, 1994; Brennan and Richardson, 1989).

There is concern about the economic efficiency of mega-cities. Richardson (1993) posed this question and marshalled data to show that although big cities have some advantages that could make them attractive to certain types of industry, the empirical evidence is inconclusive. The spatial concentrations of manufacturing in mega-cities, he maintained, may reflect less inherent productivity advantages than the intended and unintended effects of government policies. Hamer (1994) went even further and questioned the relevance of size as a factor in the economic impact of Third World mega-cities. He alleged that their size *per se* may not be the issue. Instead, it is management at both the regional and local levels. He argued that these mega-cities may be no more than the product of wrong-headed national urbanization policies promoted by physical planners with vision of optimal geography and very little sense of economics (Hamer, 1994:175). In any event, mega-cities in developing countries perform national and/or international economic functions, such as banking, insurance and finance, depending on their relative connectivity within the global economy. Manufacturing employment features prominently in Asian mega-cities ranging from 23 per cent in Manila and Dhaka to 30 per cent in Bangkok, Madras, Delhi and Karachi, and to a high 35–40 per cent in Bombay, Calcutta and Seoul. The informal sector provides a sizeable share of the total employment, reaching as high as 37.5 per cent in Manila and 65 per cent in Dhaka.

Mega-cities in developing countries recognize the need to adopt spatial strategies that would allow them to adjust to their growing size and functions. An almost universal planning goal is to evolve a policentric structure by promoting the growth of sub-centres. Calcutta and Madras have pursued ambitious, but largely ineffective, decentralization strategies. Similarly, Cairo and Jakarta have attempted to reorientate their axis of growth, changing it from the existing north–south to an east–west direction, with only limited success. Also, Dhaka adopted a northern-development corridor strategy in 1981, but development since then has departed from this goal. Only Seoul appeared to have achieved a measure of growth control and directional strategy by adopting physical measures, such as new towns, industrial estates and green belts, complemented by social and tax policies (Yeung, 1986). One major reason for many mega-cities in the Third World not being able to effect spatial restructuring is the lack of a mechanism to enforce land-use controls. This has resulted in a substantial difference between actual development and prescribed plans. In Manila, Bangkok and Dhaka, it has been reported that no effective measures existed to influence or control land development. Consequently, the price of land has increased rapidly at rates much faster than the consumer price index and is susceptible to manipulation and speculation. The central issue for these mega-cities, therefore, is to learn how, when and where to intervene to promote efficient policentric structures critical to their growth rather than inhibiting their formation.

In terms of urban services, the United Nations mega-city study generated rich comparative perspectives. Housing appears to be a problem in all mega-cities. Population growth is always ahead of housing supply. The efforts of both the public and the private sectors in the housing field are usually not well co-ordinated. A physical manifestation of inadequate housing in Third World mega-cities is the considerable proportion of their population living in slums and squatter settlements. For example, in Manila, the proportion is 32 per cent. It is 37 per cent in Karachi, 40 per cent in Mexico City and as much as 50 per cent in Bogota. In Cairo, the City of the Dead (the necropolis) is home to hundreds of thousands of inhabitants. Half a million people live and sleep on the streets of Calcutta. Water is also a severe problem in mega-cities, especially in Bombay, Madras,

Karachi, Dhaka and Jakarta. Beijing and Tianjin, not yet covered in the study, suffer from the same fate. Bombay rations water to between two and eight hours per day; Karachi supplies piped water to only 40 per cent of its population, again on a rationed basis of a few hours daily. About half of Jakarta's population obtain their water from vendors at as high as 13 times the cost of piped water.

With respect to those basic services that directly affect the environment, most mega-cities suffer from ancient sewerage facilities, poor solid-waste disposal systems, and polluted air and water. The percentage of metropolitan populations connected to a sewage system amounts to only 11 per cent in Metro Manila, 18 per cent in Dhaka, 20 per cent in Karachi, 80 per cent in Mexico City and 86 per cent in Seoul. The proportion of solid waste collected is 25 per cent in Jakarta, 33 per cent in Karachi, 55 per cent in Calcutta, 70 to 80 per cent in Madras, Metro Manila, Mexico City and Bangkok, and almost 100 per cent in Seoul. It should be noted that a recent research project has examined systematically, and in a retrospective and prospective manner, policy innovations in the delivery of basic infrastructure services, in Asia, Africa and Latin America (Menendex, 1991; Yeung, 1991).

Going beyond the United Nations Megacity Study, table 13.2 presents in another more systematic and comprehensive manner a comparison of mega-cities from the standpoints of a number of key indicators. On any of the five selected indicators, mega-cities in developing countries with very few exceptions scored less favourably than those in developed countries. Thus, in the summarized urban living-standard score, only mega-cities in developed countries are in the range of good and very good. All mega-cities in developing countries are, without exception, rated fair to poor. The difference in living standards between the two groups of cities cannot be more stark.

From another viewpoint, the general pathology of modern urban life has also affected mega-cities wherever they are located. Laquian (1994:201) aptly describes the situation:

Drug-related bombings and assassinations plague Bogota, while gangs battle each other for turf in Miami, Boston and Toronto. Riots break out in Berlin, Amsterdam and London as squatters are evicted from dilapidated buildings or as ethnic conflicts erupt between Asians and Caucasians. A survey has revealed that more than two-thirds of New Yorkers feel unsafe to go out at night; thousands of families have been fleeing the inner city for the safety of the suburbs.

Law and order has become a problem in most mega-cities and random terrorist acts can afflict any city in the world. The nerve-gas poisoning of the subway system in Tokyo and the massive bombing of a government building in Oklahoma City in 1995 are chilling reminders of the vulnerability of urban living.

The main challenge of mega-cities to policy makers in developing countries in the 1990s and beyond is thus: how, against the background of growing poverty, basic urban services can be provided and economic growth promoted? The dimension of urban poverty in the Third World is awesome. Some 329 million urban dwellers, representing 27.7 per cent of the total urban population in developing countries, lived below the poverty line in 1988. Mathur (1994) reminds us that the incidence of poverty in mega-cities is quantitatively huge, and qualitatively complex and different. The two, taken together, present the biggest and most formidable urban agenda for such cities now and in the future. Poverty-alleviation programmes, with the benefit of experience to date, must be well conceived and targeted to the poor in a general environment of healthy economic growth.

If the foregoing survey of the mega-cities of the world has not yielded too many bright spots, a brief sketch of some world cities will. Indeed, because of their specific position, which enables them better to articulate the new global economy with its emphasis on technology, knowledge, finance and telecommunications, traditional urban agglomerations such as New York, London, Tokyo and Paris have found new roles and functions. They have successfully transformed themselves to meet new challenges and capture fresh opportunities. They have positively responded by transforming their built environment, undergoing spatial, economic and social restructuring, and cooperating more closely between the public and private sectors in pursuing growth and development (Beauregard, 1991; Machimura, 1992). However, it has also been found that the new development has resulted in growing occupational, income and social polarization (Sassen, 1991), although this thesis is yet to be tested in other global cities beyond those studied. For instance, Hamnett (1994) found, at least for cities in the Netherlands, that the major role of the Dutch welfare system has been to ameliorate social inequality.

Other relatively new world cities have witnessed recent spurts of rapid growth and beckoning vistas ahead. Los Angeles, for example, has undergone a major transformation to become an important financial, producer service, and manufacturing centre on the west coast of North America, rivalling New York in almost every way (Beauregard, 1991). Frankfurt has found new prosperity in its ability to combine local politics with global economics. World-city formation in Frankfurt has taken into account international capital as well as the interests of the local inhabitants (Kell, 1992). Similarly, Miami recently blossomed in its transformation from a resort town to a Caribbean entrepôt propelled by Cuban immigration and its growth in international business transactions, especially with Latin America whose economies have become increasingly open to foreign trade and investment (Sassen and Portes, 1993). Even a Dutch group of cities, notably Amsterdam, the Hague and Rotterdam, is positioning itself collectively as Randstad Holland in order to strengthen its position to be a key player in the new global economy. Individually, these cities would be too small to make any significant impact on world trade and development (Shachar, 1994). The renewed and reinvigorated roles Berlin and Vienna have carved for themselves in view of the rapidly changing situation in eastern and central Europe have been referred to earlier. Finally, there are many cities in the developing world that have already benefited from the recent global emphasis on transnationalization of capital, production and services. Global cities along the Western Pacific Rim such as Seoul, Taipei, Hong Kong, Manila, Bangkok, Kuala Lumpur, Singapore and Jakarta have done particularly well (Lo and Yeung, 1996). Johannesburg, in a reborn, multiracial South Africa, should be poised for accelerated growth in the global economy as probably the only world city in Africa. As trade liberalization and economic development progress further in Latin America, the world cities of Mexico City, Caracas, Rio de Janeiro, São Paulo and Buenos Aires should discover new functions and positions in the global economy.

Conclusion: Future Outlook

As the twenty-first century approaches, mega-cities are becoming a dominant force in directing global processes and reshaping the global economy. In this respect, they are reconfiguring the nature of global geography. One key factor in this reconfiguration is their continued growth in the years ahead. Over time, the relative growth of mega-cities is however expected to be moderated. From an annual growth rate of 3.9 per cent for all mega-cities in the period 1950–60, the rate dropped to 3.3 per cent in the next decade. From the decade 1970–80 and on to the year 2000, the average annual growth rate has stabilized at around 2.5 per cent. There are still exceptions to this stabilization. Mega-cities such as Lagos, Bangkok and Dhaka are some of the fastest growing agglomerations in the world today with an estimated annual growth rate of 5.1, 5.0 and 6.0 per cent respectively, between 1990 and 2000 (United Nations, 1991). In the age of mega-cities, their growth is propelled by global forces over which neither cities nor states have full control. These forces, which underlie industrialization and urbanization world-wide, and are undermining traditional settlement patterns, are becoming increasingly global in nature (Knight, 1993:38).

With the increasing globalization of the world's economy, the pattern of development of mega-cities is bound to influence considerably the subject matter of geography. What matters to geography is where, how and when human beings have made a success of their relationship with the natural environment, and can lead a good and sustainable life. To the traditional geographical concern with location, commodity flows, geopolitics and international trade will be added a heightened interest in human and cultural resources, information flows, knowledge infrastructures, technologies and amenities (Knight, 1989:328–9). These are interwoven into the new pattern of global geography with its emphasis on networks and processes that are as intriguing and complex to disentangle. The geography of national-states will need to be reconfigured in the age of mega-cities and the world cities. It is, of course, not too early to begin to explore the geographical implications and the social changes entailed in the globalization process.

For mega-cities themselves to be able to meet the challenges of the new age, Knight (1993) argues that their development will have to be increasingly intentionally planned. Knowledge-based development provides the possibility of the city shaping its own development and regaining control over its own destiny. Singapore, for example, is often quoted as the classical case of a world city that has, since its independence in 1965, planned and purposefully directed its progressive development over time. By the 1980s, it strove to build a modern industrial economy based on science, technology, skills and knowledge. In the twentieth century, its aspiration has been to become a fully developed world city.

It is in this context that it has been suggested that, particularly in developing countries, the challenges of mega-cities would become central to the whole issue of nation building. As mega-cities strive to position themselves to be competitive in the emerging global economy, they would constrain the policies and programmes of their national government towards national integrative goals. Indeed, Knight (1989:326, 332) puts the situation succinctly when he observed:

> *Given the nature and power of the global forces that are now shaping them, all cities must redefine their role in the context of the expanding global society. Global cities . . . will not be determined by locational or geopolitical considerations but by their capacity to accommodate change and provide continuity and order in a turbulent environment . . . With the advent of the global economy, nation building is becoming synonymous with city building.*

References

Armstrong, W. and McDee, T.G. (1985) *Theatres of Accumulation: Studies in Asian and Latin American Urbanization* (London: Methuen).

Beauregard, R.A. (1991) Capital restructuring and the new built environment in global cities: New York and Los Angeles, *International Journal of Urban and Regional Research*, 15, 1:90–105.

Brennan, E. (1994) Mega-city management and innovation strategies: regional views, pp. 233–55 in R.J. Fuchs et al. (eds) *Mega-City Growth and the Future* (Tokyo: United Nations University Press).

Brennan, E. and Richardson, H.W. (1989) Asian megacity characteristics, problems and policies, *International Regional Science Review*, 12, 2:117–29.

Castells, M. (1992) *European Cities, the Information Society and the Global Society* (Amsterdam: Amsterdam University, Centre for Metropolitan Research).

Clarke, G. (1991) Urban management in developing countries, *Cities*, 8, 2:93–107.

Cohen, R.B. (1981) The new international division of labour: multinational corporations and the urban hierarchy, pp. 287–315 in M. Dear and A.J. Scott (eds) *Urbanization and Urban Planning in Capitalist Society* (London: Methuen).

Dicken, P. (1992) *Global Shift: The Internationalization of Economic Activity*, 2nd edition (London: Paul Chapman).

Dieleman, F.M. and Hamnett, C. (1994) Globalization, regulation and the urban System, *Urban Studies*, 31,3:357–64.

Dogan, M. and Kasarda, J.D. (eds) (1988) *The Metropolis Era: A World of Giant Cities*, Vol. I (Newbury Park: Page).

Friedmann, J. (1986) The world cities hypothesis, *Development and Change*, 17:69–83

Fujita, K., (1991), A world city and flexible specialization: restructuring of the Tokyo Metropolis, *International Journal of Urban and Regional Research*, 15, 2:219–84.

Gilbert, A. (1990) Urbanization at the periphery: reflections on changing dynamics of housing and employment in Latin American cities, pp. 73–124 in D.W. Drakakis-Smith (ed.) *Economic Growth and Urbanization in Developing Areas* (London: Routledge).

Gilbert, A. and Gugler, J. (1981) *Cities, Poverty and Development: Urbanization in the Third World* (London: Oxford University Press).

Hall, P. (1984) *The World Cities*, 3rd edition (London: Weidenfield and Nicolson).

Hamer, A.M. (1994) Economic impacts of Third World mega-cities: is size the issue?, pp. 172–91 in R.J. Fuchs et al. (eds) *Mega-City Growth and the Future* (Tokyo: United Nations University Press).

Hamnett, C. (1994) Social polarisation in global cities: theory and evidence, *Urban Studies*, 31, 3:401–24.

Kell, R. (1992) Frankfurt: global city – local politics, *Comparative Urban and Community Research*, 4:39–69.

King, A.D.(1990) *Global Cities: Post Imperialism and the Internationalization of London* (London: Routledge).

Knight, R.W. (1989) City building in a global society, pp. 326–34 in R.W. Knight and G. Gappert (eds) *Cities in a Global Society* (Newbury Park: Sage).

Knight, R.W. (1993) Sustainable development – sustainable cities, *International Social Science Journal*, 135:35–54.

Lacquian, A. (1994) Social and welfare impacts of mega-city development, pp. 192–214 in R.J.Fuchs et al. (eds) *Mega-City Growth and the Future* (Tokyo: United Nations University Press).

Lo, F.C. and Yeung, Y.M. (eds) (1996) *Emerging World Cities in Pacific Asia* (Tokyo: United Nations University Press).

Machimura, A. (1992) The urban restructuring process in Tokyo in the 1980s: transforming Tokyo into a world city, *International Journal of*

Urban and Regional Research, 16, 1:114–28.

Markusen, A. and Gwiasda (1994) Multipolarity and layering of functions in world cities: New York City's struggle to stay on top, *International Journal of Urban and Regional Research*, 18, 2:167–93.

Mathur, O.P. (1994) The dual challenge of poverty and mega-cities: an assessment of issues and strategies, pp. 349–89 in R.J. Fuchs et al. (eds) *Mega-City Growth and the Future* (Tokyo: United Nations University Press).

Menendex, A. (1991) *Access to Basic Infrastructure by the Urban Poor* (Washington, D.C.: World Bank, Economic Development Institute, Policy Seminar Report, No. 28).

Meyer, D.R. (1986) The world system of cities: relations between international financial metropolises and South American cities, *Social Forces*, 64, 3:553–81.

Meyer, D.R., (1991) Change in the world system of metropolises: the role of business intermediaries, *Urban Geography*, 12, 5:393–416.

Nakakita, T. (1988) The globalization of Japanese firms and its influence on Japan's trade with developing countries, *Developing Economies*, 26, 4:306–22.

Reich, R.B. (1991) *The Work of Nations: Preparing Ourselves for 21st Century Capitalism* (New York: Knopf).

Richardson, H.W. (1993) Efficiency and welfare in LDC mega-cities, pp. 32–57 in J.D. Kasarda and A.M. Parnell (eds) *Third World Cities* (Newbury Park: Sage).

Sassen, S. (1991) *The Global City: New York, London, Tokyo* (Princeton: Princeton University Press).

Sassen, S. (1994) The urban complex in a world economy, *International Social Science Journal*, 139:43–62.

Sassen, S. and Portes, A. (1993) Miami: a new global city? *Contemporary Sociology*, 22, 4:471–80.

Shachar, A. (1994) Randstad Holland: a world city?, *Urban Studies*, 31, 3:381–400.

Stren R. (ed.) (1994) *Urban Research in the Developing World, Vol. 1: Asia; Vol. 2: Africa; Vol. 3: Latin America; Vol. 4: Perspectives on the City* (Toronto: University of Toronto, Centre for Urban and Community Studies).

United Nations (1991) *World Urbanization Prospects 1990* (New York: United Nations).

United Nations Development Programme (UNDP) (1991) *Cities, People and Poverty* (New York: UNDP).

World Bank (1991) *Urban Policy and Economic Development: An Agenda for the 1990s* (Washington, D.C.: World Bank).

Yeung, Y.M. (1986) Controlling metropolitan growth in eastern Asia, *Geographical Review*, 76, 2:125–37.

Yeung, Y.M. (1988) Great cities of eastern Asia, pp. 155–86 in M. Dogan and J.D. Kasarda (eds) *The Metropolis Era: A World of Giant Cities*, vol. 1 (Newbury Park: Sage).

Yeung, Y.M. (1991) Past approaches and emerging challenges, pp. 27–81 in Asian Development Bank (ed.) *The Urban Poor and Basic Infrastructure Services in Asia and the Pacific*, vol. 1 (Bangkok: Asian Development Bank, Seminar Report).

Yeung, Y.M. (1995) Hong Kong's hub functions, paper presented at the Conference of Planning Hong Kong for the 21st Century held in Hong Kong, 12–13 April.

Epilogue

It is perhaps a simple truism to state that 'all geography is historical geography'. This is no more than a recognition of the inexorable passage of time and the changes that form the content of this passage. What we have done in this volume is to capture to a certain extent the richness of that content. Chapter 1, indeed, presented us with a broad review of changing geographical perspectives over the last century. These perspectives of geography were influenced greatly by the reality of conditions and of knowledge at each particular era of human development. But, more than anything else, those perspectives reflected an appreciation of the capacity of the human agency in transforming the character of the earth's surface. When that capacity was minimal and of little consequence, environmental processes assumed such importance as to be regarded as deterministic of cultural and societal development. As that capacity increased, especially following on the age of enlightenment and the industrial revolution initiated in Europe, it led to an overweening confidence that every problem of the physical environment can be easily resolved by the technological and engineering creativity of the human mind.

Until the Second World War, such resolution of environmental challenges was the product of chance inventions or innovations. One of the major changes deriving from the prosecution of that war was the systematization of the process of technological development. Programmes of collaborative research were initiated to tackle specific problems. Of course, in the process some other incidental discoveries were made. The consequence of all this has been an almost incredible acceleration in the pace of technological progress, enhancing the human capacity to alter the environment on a clearly unprecedented scale. Such accelerated progress particularly in the fields of electronics and materials have increased the range of goods and services available to human beings, increased their economic well-being, led to the tremendous enhancement of life chances of individuals and assured a prolonged life expectancy for the majority of humankind. Development in bio-technology is also opening up the prospect of a third agricultural revolution, and offering the vastly increased size of the human population the possibilities of an abundant and diversified supply of food.

More than this, such ordered scientific and technological research and development have made possible the initial attempt to land human beings on the moon and usher in the era of space exploration. These events have had the unprecedented implication that towards the end of the twentieth century there is a reality to the appreciation of the position of the earth as part of not only the solar system but also a universe the limits of which are infinite. No one at present knows enough about what the continued space probe may reveal about other planets and other worlds. What is clear is that the technological advances which are bringing about the expansion in our knowledge about the universe are bound to lead in the twenty-first century to a geographical awareness vastly different from our present perception.

The enhanced technological capacity has had quite varied consequences for organizational development. At the private-sector level, it has encouraged the growth of transnational business conglomerates, some of which control resources far and beyond those available to many nation-states. The activities of these conglomerates are of such magnitude, and their influence of such amplitude, that the concept of 'globalization' as a means of capturing the measure of their span is increasingly gaining in currency and actuality. These conglomerates have been instrumental in virtually obliterating the constraints of distances and territorial boundaries and in diffusing and disseminating new standards of living and somewhat uniform life-styles around the world. Their activities have fostered the concentration of populations in mega-cities in different countries of the world and have linked these cities up in a wired network that makes them operate as a global system on their own terms. Moreover, the tremendous financial resources available to these conglomerates are forcing on nation-states an agonizing reappraisal of the role of their public and quasi-public agencies *vis-à-vis* private enterprises in the provision of public utilities and amenities which used to be regarded as undisputed responsibilities of governments. Privatization is thus becoming a feature of service provision in most countries. Its implications for the continued relevance of the nation-state as a major feature of societal organization are part of the challenges that are bound to confront the geography of the twenty-first century.

At present, one of the most important functions of the nation-state is to minimize sub-national regional inequalities affecting the life chances of its citizens. This it has done customarily through generally progressive taxation and non-discriminatory investment in infrastructural facilities and social services, especially education and health care. Nonetheless, in spite of the variety of policies, programmes and projects developed to reduce social and geographical inequalities, the most notable feature of virtually every country remains the persistence of these internal differences in regional opportunities and economic performance. Compared to the outreach and impact of transnational corporations, the capacity of nation-states effectively to address the problems of economic well-being and social welfare of their citizens continues to prove inadequate, even amongst the most successful. One of the reasons adduced for the increasing incapacitation of the nation-state is the widening mismatch between the technological scale of productive enterprises and the size of the market necessary to support them economically and profitably. The result has been the search for a higher, supranational level of integration of groups of nation-states. Although there remains serious concern about national sovereignty and the preservation of differing cultural heritages, the emergence of multinational, regional unions could represent the beginning of 'globalization' on a socio-political plane.

These emergent changes in the configuration of human society at the end of the twentieth century have tremendous implications for the relationship between humans and the natural environment. The acceleration in the rate of change has meant that it does not take very long before deleterious consequences to the environment as a result of mindless, productive processes are noticed and effective remedial actions proferred or implemented. The alarm signals, with regard to the environmental impact of rapid population growth, of air, water and land pollution, of global warming and the depletion of the ozone layer have all been followed by swift, if not as yet totally successful, actions. There has been a heightened awareness on a global scale of the interrelations between environmental quality and human economic activities and a deepening appreciation that the earth has to be treated as one single home for the whole of the human race. This has provoked attempts to understand further the linkages amongst the earth's processes, leading to a global concern for the preservation of such disparate features of the earth's surface as the tropical rainforests and the polar icy wilderness.

Three developments at the end of the twentieth century – the technological and economic reach of private-sector transnational corporations, the reduced importance of the nation-state, and the greater understanding of the interrelationship between human beings and their environment – do underscore the potential for a new and more potent role for international organizations, such as the United Nations. In just over half of a century, the United Nations has proved its value in having been able to prevent global wars whilst, in spite of some blunderings, it has also been able to oversee the amicable resolution of numerous regional conflicts. It has overseen the global provision of social services in the field of education, health care, recreation and shelter. It has tried to assist with enhancing the productivity of labour and the development of trade and industry. Through associated agencies such as the World Bank and the International Monetary Fund, it has tried to monitor the performance and assist nation-states in their task of development and economic management. More than this, it promotes an appreciation everywhere of a global responsibility for caring for the environment and seeks paths to resource development that are environmentally sustainable.

All this has been achieved largely on the basis of suasion and international sanctions. Yet, there is need to monitor and keep in line transnational corporations as they stride across the globe like colossi, confronting nation-states the capabilities of which to make them comply with their laws and policies, or show concern for the need to keep their environment in good health, are becoming relatively weaker. This growing asymmetry in the capabilities of nation-states to control effectively the operations of transnational corporations even within their own territories is bound sooner or later to provoke the search for an international organization with a mandate more potent and compelling than that of the present United Nations.

The prospect of geography in the twenty-first century, thus, has tremendous fascination because of the numerous uncertainties that characterize the relationship between human beings and their environment. The continued urbanization of the global population in a limited number of mega-cities should result in an unusual pattern of spatial distribution and interaction, both within and between national entities. The increasing capacity for bio-technological production of food could lead to such massive output from limited areas as to create new hazards for the environment or represent the evolution of a more sustainable system of feeding the earth's large population. Large areas of the earth's surface may, thus, be enabled to return to its natural state and be enjoyed essentially for recreational and tourist resource-value. The size of the human population may reach its optimal replacement level as a result of advanced reproductive technology and practice. And the political configuration of the world could come to comprise large, regionally integrated entities within which a new concept of human rights and national sovereignty would need to be defined.

Whatever the outcome, the dramatic changes which the world has witnessed within the last half of the twentieth century are likely to accelerate further as we progress into the first half of the twenty-first century. Dramatic changes are the essence of future geography. The challenge to humankind will be not only how to cope with such rapid changes but also how to ensure that they are of a benign and beneficial nature, especially with regard to their impact on the environment and the prospect for sustainable development. One important way of achieving such a goal is through the informed socialization of present and future generations of human societies in every country of the world. The discipline of geography provides a valuable means of promoting such a socialization process. It seeks to inculcate a good understanding, not only of the constantly changing relations between human beings and their environment, but also of the imperative of developing a thoughtful and caring attitude to that relationship. To the extent that the present volume on geography seeks to contribute to such an understanding, the United Nations Educational, Scientific and Cultural Organization must be complimented for their far-sighted initiative in these closing years of the twentieth century.

Select Annotated Bibliography

The following bibliography is intended to serve as a non-exhaustive guide to professional geographers and to students.

Selected Books

Alexander, D., 1993. *Natural Disasters* (New York: Chapman Hall)
A comprehensive text on natural hazards that emphasizes physical processes with secondary considerations given to human responses.

Alliès, P., 1980. *L'invention du territoire* (Grenoble: Presses Universitaires de Grenoble)
A historical and theoretical study of the institutional, administrative and ideological processes involved in the social construction of the territory with special emphasis on the French case.

Alonso, W., 1964. *Location and Land-Use: Towards a General Theory of Land Rent* (Cambridge, MA: Harvard University Press)
The first text to elaborate a theory of urban land use, based on competition for rent, somewhat akin to the von Thunen scheme for agricultural land use.

Anon. 1987. *Expert Systems: Principles of Operation and Examples* (Moscow: Radio i svyaz), in Russian
A high-level book containing theoretical and empirical analyses of expert systems with applications to a number of substantive scientific areas.

Aronoff, S., 1991. *Geographical Information Systems: A Management Perspective* (WDL Publications)
Provides a comprehensive coverage of the fundamentals of the GIS technology, applications, and management issues that arise in implementing and operating these systems.

Barnes, T. and Duncan, J. (eds), 1992. *Writing Worlds: Discourse, Text and Metaphor in the Representation of Landscape* (London: Routledge)
Essays displaying the great variety of ways that the geographic perspective has been informed by recent trends in literary history.

Berry, J.K., 1993. *Beyond Mapping: Concepts, Algorithms, and Issues in GIS* (GIS World)

Explores the concepts of GIS technology and discusses the issues involved as GIS moves from the researcher to the general user.

Berthon, S. and Robinson, M., 1991. *The Shape of the World: The Mapping and Discovery of the Heart* (London: George Philip).

A historiographic approach that sheds light on the process of cartographic figuration and inventory, survey and scientific classification of the world through the parallel advance of discoveries, and cartographic representation techniques.

Blackie, P., Cannon, T., Davis, I. and Wisner, B., 1994. *At Risk: Natural Hazards, People's Vulnerability and Disasters* (London: Routledge)

Utilizing a political economic perspective, this book examines the human factor in disasters with special reference to developing world contexts.

Blackie, P. and Brookfield, H.C., 1987. *Land Degradation and Society* (London: Methuen)

A classic work on the social causes and consequences of land degradation. Case studies are used to discern different patterns of land use and land degradation in selected world regions.

Blumenau, D.I., 1989. *Information and Information Service* (Leningrad: Nauka Publ.), in Russian

A theoretical analysis of information.

Braudel, F., 1988, 1990. *The Identity of France, Vols.I and II* (New York: Harper and Row)

The last work of a great historian whose thinking was saturated with geographical ideas and awareness. Originally published (1986) as *L'Identité de la France.*

Broc, N., 1974. *La Géographie des Philosophes. Géographes et voyageurs français au XVIII siècle* (Paris: Ophrys/Association des Publications près les Universités de Strasbourg)

The most exhaustive analysis of French overseas exploration during the eighteenth century, highlighting its impact on the development of science in that country, as well as the promotion of empirical research and textual representation of the world and the state's geography.

Broc, N., 1980. *La Géographie de la Renaissance* (Paris: C.T.H.S.), 2nd edition – 1986

A wide and erudite study of the different practices and knowledge of regional and planetary geography during the Renaissance.

Budyko, M.I., 1974. *Climate Change* (Leningrad: Gidrometeoizdat) in Russian

Presents basic features of present-day climate and past climates, and discusses factors influencing climate change. A semi-empirical theory of thermal regime of the atmosphere is proposed and used to explain current climatic changes and the mechanism of the quaternary glaciation's development. Probable climatic changes in the future are appraised and the probability of an emerging environmental crisis due to changing climatic conditions is discussed.

Budyko, M.I., 1980. *Climate in the Past and in the Future* (Leningrad: Gidrometeoizdat), in Russian
Considers the regularities of natural and human-induced climatic changes and discusses the physical mechanism of climatic changes in the geological past and in the current epoch. Proposes a theory to facilitate calculation of climatic change and uses this to project future climate. Discusses climatic conditions of the next few decades.

Budyko, M.I., 1984. *Evolution of the Biosphere* (Leningrad: Gidrometeoizdat), in Russian
Provides information on the structure of the biosphere and the interrelations of its basic components. Results of the impact of human activity on the biosphere are evaluated with a view to examining their effects on climate changes.

Budyko, M.I., Ronov, A.B. and Yanshin, A.L., *Istoria atmosfery (History of the Atmosphere)* (Leningrad: Gidrometeoizdat)
Uses data of carbon deposits in the sedimentary mantle of the earth to calculate the amount of oxygen and carbon-dioxide in the atmosphere of the earth's geological past. Based on this, it attempts to evaluate the effect of the varying chemical composition of the atmosphere on the evolution of the biosphere.

Buisseret, B. (ed.) 1987. *Monarchs, Ministers and Maps: The Emergence of Cartography as a Tool of Government in Early Modern Europe* (Chicago: University of Chicago Press)
A collection of articles discussing the institutional relations between the centralization of monarchical power, the modernization of state apparatuses and the development of cartographic representation in France, Spain, England, Poland, Austria and Italy during the late Renaissance period.

Burton, I., Kates, R.W. and White, G.F., 1993. *The Environment as Hazard*, 2nd edition (New York: Guilford Press)
An update of the classic monograph on human adjustments to hazards written by the leading hazards geographers in the field.

Buttimer, A., 1974. *Values in Geography* (Washington, D.C.: Association of American Geography, Commission on College Geography, Resource Paper 24)
The first exploration of the role of values in geography, connecting the social relevance movement with the development of a humanistic approach.

Button, K., 1994. *Transport Policy – Ways into Europe's Future: Strategies and Options for the Future of Europe, Basic Findings* (Gütersloh: Bertelsmann Publishers)

This work focuses on the problems of transportation in the framework of recent global transformations. Emphasizes the potential role Europe can play in the structuring of the international terrain in which transportation makes an important contribution.

Capel, H., 1981.*Filosofía y ciencia en la geografía humana contemporánea* (Philosophy and Science in Contemporary Human Geography, (Barcelona: Barcanova)

A comparative study of the process of institutionalization of contemporary human geography, including a historical analysis of the context and cognitive conditions in the discipline's development.

Cassettari, S., 1993. *Introduction to Integrated Geo-Information Management* (London: Chapman & Hall)

Introduces the idea of integrated information systems, the importance of management in the successful implementation of such systems and the key role of spatial data.

Christaller, W., 1966. *Central Places in Southern Germany* (translated by C.W. Baskin) (Englewood Cliffs, NJ.: Prentice-Hall)

The classic exposition of the central place theory, originally published in 1933, setting out the basis of a regular pattern of service centres and market areas.

Cloke, P., Philo,C. and Sadler, D., 1991. *Approaching Human Geography: An Introduction to Contemporary Theoretical Debates* (London: Paul Chapman)

An introduction to the broad scope of contemporary approaches in human geography, including postmodernism.

Cosgrove, D. and Daniels, S., 1988. *The Iconography of Landscape* (Cambridge: Cambridge University Press)

Essays on the emergence of meaning of landscapes approached as texts for interpretation.

Cressie, N., 1993. *Statistics for Spatial Data* (Chichester: John Wiley)

Designed for scientists and engineers with solid quantitative skills. Spatial data are used throughout the book to illustrate spatial theory and methods.

Crosby, A.W., 1986. *Ecological Imperialism: The Biological Expansion of Europe, 900–1900.* (Cambridge: Cambridge University Press)

An empirical study of the ecological consequences, often unintended, of European overseas exploration, and how it changed many New World landscapes irrevocably.

Cutter, S.L., 1993. *Living with Risk: The Geography of Technological Hazards* (London: Edward Arnold)

Reviews the social science contributions to risk and hazards research and provides geographical case studies of hazardous waste, chemical risks and nuclear power issues.

Cutter, S.L. (ed.), 1994. *Environmental Risks and Hazards* (Englewood Cliffs: Prentice-Hall)

An edited volume that contains some of the seminal articles in the risk and hazards field.

Dematteis, G., 1985.*La Metafore della Terra: La geografia umana tra mito e scienza* (The Heart Metaphor: Human Geography between Myth and Science), (Milano: Feltrinelli)

Discusses the social origin of geographical thought and institutions, and considers the relation between the epistemology of the discipline and the social perception of collective identification and territorial representation.

Dezert, B. and Wackermann, G., 1991. *La nouvelle organisation internationale des échanges* (Paris: SEDES)

Stresses the role of basic global transformation with regard to new requirements in the area of exchange. It underscores the various factors and mechanisms at work in the current organization of the world as concerns the flow of both people and goods.

Dicken, P., 1992. *Global Shift: The Internationalization of Economic Activity* (London: Paul Chapman)

First and most comprehensive book by a geographer dealing with the processes of contemporary global economic change during the past four decades, with emphasis on the role of transnational corporations.

Dogan, M. and Kasarda, J.D. (eds), 1988. *The Metropolis Era: A World of Giant Cities, Vol. I* (Newbury Park: Sage)

A comprehensive exposition of the phenomenon of large world cities, focusing on regional patterns and individual cities.

Dollfus, O., 1994. *L'espace monde* (Paris: Economica)

This work has an original outlook on the new configuration of relations between societies and space in the light of globalization. Presents the world as an ensemble of interdependent ties and of tension that can be understood only through a global approach.

Doyal, L. and Gough, I., 1991. *A Theory of Human Need* (London: Macmillan)

An exploration of human need by a philosopher and a political scientist, with great geographical relevance, especially with respect to development indicators.

Eyles, J. and Smith, D.M. (eds), 1988. *Qualitative Methods in Human Geography* (Cambridge: Polity Press)

The first book-length exposition of the application of qualitative methods to geographical problems, demonstrated through case studies.

Fotheringham, S. and Rogerson, P. (eds), 1994 *Spatial Analysis and GIS* (Taylor & Francis)

Focuses on the relative lack of research into the integration of spatial analysis and GIS, and the potential benefits in developing such an integration.

French, R.A. and Hamilton, F.E.I. (eds), 1979. *The Socialist City: Spatial Structure and Urban Policy* (Chichester: John Wiley)
The first text to draw attention to the possibility of a distinctively socialist city in eastern Europe and the USSR.

Friedmann, J., 1992. *Empowerment: The Politics of Alternative Development* (Oxford: Blackwell Publishers)
A text with an explicitly moral dimension, focusing on development strategies designed to meet the basic needs of poor people in the underdeveloped world.

Fuchs, R.J. et al., 1994. *Megacity Growth and the Future* (Tokyo: United Nations University Press)
Collection of papers presented at a mega-city conference in 1990, covering a wide spectrum of issues relevant to the role of these large cities in the future.

GIS World Inc., 1994. *GIS World Sourcebook 1995* (GIS World Inc.)
The sixth edition presents an expert forum on the technological and methodological issues in GIS.

GIS World Inc., 1994. *InterCarto: GIS for Environmental Studies and Mapping* (Conference Proceedings, GIS World Inc.)
Proceedings of the first-ever international GIS Conference in Russia held at Moscow, 23–25 May 1994. Includes more than 30 papers highlighting new developments in the field.

Glacken, C., 1967. *Traces on the Rhodian Shore* (Berkeley: University of California Press)
A classic account of the development of human–environment relationships from earliest times to the end of the eighteenth century.

Glantz, M.H. (ed.), 1987. *Drought and Hunger in Africa: Denying Famine a Future* (New York: Cambridge University Press)
Examines the impact of drought on Africa and drought as a social process. There is particular focus on the differential impact of drought at the local and regional levels, as well as on subgroups of people.

Godleweska, A. and Smith, N. (eds), 1994. *Geography and Empire* (Oxford: Blackwell Publishers)
Collection of articles analysing the relations between geography and the development of European and American imperialism during the nineteenth and twentieth centuries.

Goodchild, M.F., Parks, B.O. and Steyaert, L.T. (eds), 1993. *Environmental Modelling with GIS* (Oxford: Oxford University Press)
Brings together a collection of invited interdisciplinary perspectives on the topic of environmental modelling with GIS.

Goudie, A., 1993. *The Human Impact* (Oxford: Blackwell Publishers)

A standard and highly successful contemporary textbook on anthropogenic modification of climate, biota and landforms.

Gould, P., 1992. *The Slow Plague: A Geography of the AIDS Pandemic* (Oxford: Blackwell Publishers)

An account written for the educated layperson, of the geographic diffusion and intensification of the AIDS pandemic world-wide.

Gould, P. and White, R., 1974. *Mental Maps* (Harmondsworth: Penguin)

A study of the geography of perception: the mental images people form of places, which are inevitably shaped by a selective channelling of information.

Guiomar, J-Y., 1974. *L'idéologie nationale: Nation, Représentation, Proprieté* (Paris: Champ Libre)

A detailed analysis of the genesis of national ideology during the French Revolution. Considers the main ideas of a nation of patriots, the doctrine of popular sovereignty and the idea of the territorial nation. Emphasizes the relations between the system of political representation and territorial delimitation of citizenship.

Haas, J.E., Kates, R.W. and Bowden, M. (eds), 1977. *Reconstruction Following Disaster* (Cambridge, MA: MIT Press)

A survey of disaster reconstruction issues and processes using case studies of the San Francisco, Anchorage and Managua earthquakes, and the flood disaster in Rapid City.

Hägerstrand, T., 1968. *Innovation Diffusion as a Spatial Process* (Chicago: University of Chicago Press)

The first major exposition of the process of spatial diffusion of innovation.

Haggett, P., 1965. *Locational Analysis in Human Geography* (London: Edward Arnold)

Possibly the most influential work in the early years of the re-orientation of human geography from its traditional ideographic approach towards a more analytical and quantitative approach stressing the regular geometry of spatial form.

Haines-Young, R., Green, D.R. and Cousins, S.H. (eds), 1994. *Landscape Ecology and GIS* (Taylor & Francis)

Focuses on the way in which a range of processes interact, and provides a framework to facilitate the understanding of human impact on the environment and the development of suitable management strategies.

Hall, P. (ed.), 1966. *Von Thunen's Isolated State* (translated by C.M. Wartenberg) (Oxford: Pergamon)

The classic nineteenth century study of agricultural land use, which predicted a pattern of concentric zones about a central city on an isotropic plain on the basis of competition for land rent.

Hall, P., 1984. *The World Cities*, 3rd edition (London: Weidenfeld and Nicolson)

A study of selected world-city profiles. Useful for general background reference.

Harrington, J.W. and Warf, B., 1995. *Industrial Location: Principles and Practice* (Berlin: Reimer)

The upheavals that are inherent in post-Taylorism have modified the traditional principles of industrial location. The author sets out to delineate the new theories and practices of industrial siting, particularly in industrialized countries.

Harris, C., Matthews, G., Gentilcore, R., Kerr, D., Holdsworth, D. and Laskin, S., 1987–92. *Historical Atlas of Canada, Volumes I–III* (Toronto: University of Toronto Press)

Possibly the greatest historical atlas of all time, breaking new ground in scholarly exposition and graphic presentation. Published in both French and English editions.

Harvey, D., 1973. *Social Justice and the City* (London: Edward Arnold)

The first major exploration of social justice in geography, spanning the author's transition from a liberal position to Marxism.

Harvey, D., 1982. *The Limits of Capital* (Oxford: Blackwell)

A close examination of Marx's *Das Kapital*, its theoretical strengths and limitations, and implications for spatial and geographical analysis.

Harvey, D., 1985. *Consciousness and the Urban Experience* (Baltimore: Johns Hopkins University Press)

A Marxist account of the restructuring of nineteenth-century cities by emerging capitalism.

Hewitt, K. (ed.), 1983. *Interpretations of Calamity* (Winchester, MA: Allen and Unwin)

A critique of the human-ecological paradigm in hazards research. The original articles support claims that hazards are socially constructed where people and their activities often aggravate the impact of hazards.

Hooson, D. (ed.), 1994. *Geography and National Identity* (Oxford: Blackwell Publishers)

A collection of articles analysing relations between geography and the development of nationalism in the nineteenth and twentieth centuries, emphasizing the role of long-established imperial identities, long-submerged identities and the newly-emerging identities.

Huxhold, W.E., 1991. *An Introduction to Urban Geographic Information Systems* (Oxford: Oxford University Press)

Describes the fundamental value of urban geographic information, leads the reader through the definitions of an urban GIS and discusses its various applications.

Jackson, P., 1989. *Maps of Meaning: An Introduction to Cultural Geography* (London: Unwin Hyman)

The first text to outline the 'new' cultural geography, in which culture is recognized as an arena in which meanings are contested.

Jacob, C., 1992. *L'empire des cartes: Approche théorique de la cartographie à travers l'histoire* (Paris: Albin Michel)

A theoretical approach to the cartographic historiography, developing the different levels of semiology and iconography present in the graphic representation, and the determinations of the social and cultural context of the continued relation between pictography and cartography.

Johnson, B.B. and Covello, V.T. (eds), 1987. *The Social and Cultural Construction of Risk: Technology, Risk, Society* (Dordrecht: D. Reidel)

An edited collection that examines how social and cultural factors influence the selection of and response to risks.

Johnston, R.J., 1991. *Geography and Geographers: Anglo-American Human Geography since 1945,* 4th edition (London: Edward Arnold)

A comprehensive review of the development of the field, incorporating summaries of major contributions.

Johnston, R.J., Gregory, D. and Smith, D.M. (eds), 1994. *The Dictionary of Human Geography*, 3rd edition (Oxford: Blackwell Publishers)

A comprehensive review of the content of contemporary human geography, from brief definitions to extended essays, with reference to relevant literature.

Johnston, R.J., Taylor, P.J. and Watts, M. (eds), 1995. *Geographies of Global Change: Remapping the World in the Late Twentieth Century* (Oxford: Blackwell Publishers)

A collection of individually authored chapters which includes a section on 'geoeconomic change'.

Jones, J. and Casetti, E., 1992. *Applications of the Expansion Method* (London: Routledge)

Essays and applications of a formal methodology allowing both space and time co-ordinates to be explicitly incorporated into geographical analysis.

Kansky, K.J., 1963. *The Structure of Transportation Networks* (Chicago: University of Chicago. Department of Geography, Research Paper No. 84)

Transportation systems are now part of vast global networks the performance of which is of great importance in national and international competition. This complex organization has changed the very conditions of exchange and the way in which they operate. The author shows how transportation plays an important role in the global economy, though it is often at the service of shippers.

Kasarda, J.D. and Pannell, A.M., 1993. *Third World Cities: Problems, Policies and Prospects*, (Newbury Park: Sage)

Collection of essays that addresses problems of mega-cities in developing countries from the perspective of development, migration and other key socio-economic issues.

Kates, R.W. and Burton, I. (eds), 1986. *Geography, Resources, and Environment. Vol. II. Themes from the Work of Gilbert F. White* (Chicago: University of Chicago Press)

A festschrift honouring the work of Gilbert F. White with thematic updates on how the field has progressed.

Kates, R.W., Hohenemser, C. and Kasperson, J.X., 1985. *Perilous Progress: Managing the Hazards of Technology* (Boulder, CO: Westview Press)

A broad-based review of the range of technological hazards facing modern society.

King, A.D., 1990. *Global Cities: Post Imperialism and the Internationalization of London* (London: Routledge)

Detailed treatise on the evolution of London from imperial times to the present.

Kirby, A. (ed.), 1990. *Nothing to Fear: Risks and Hazards in American Society* (Tuscon: University of Arizona Press)

Examines the ways in which the public discourse on risk and hazards is generated and how management systems arise.

Knox, P., 1995 *Urban Social Geography: An Introduction* (London: Longman)

A contemporary review of a growing field.

Knox, P. and Agnew, J., 1994. *The Geography of the World Economy: An Introduction to Economic Geography* (London: Edward Arnold)

A useful introductory text that takes a historico-geographical approach to the evolution of the world economy and emphasizes the processes of uneven development.

Kondratyev, K. Ya., 1992. *Globalnyi Klimat (Global Climate)* (St. Petersburg: Nauka)

Analyses interaction between the atmosphere and the oceans in climate formation and discusses the effects of human activities and volcanic eruptions on climate change.

Koshkariov, A.V. and Tikunov, V.S., 1993. *Geoinformatics* (Moscow: Kartgeocentr-Geodezizdat), in Russian

Discusses the theory, methodology and applications of GIS technology based on international and Russian investigations.

Kotlyakov, V.M., 1994. *Mir Snega i L'da* (*The World of Snow and Ice*) (Moscow: Nauka)

Considers the development and importance of snow-and-ice phenomena and processes for human life. Places special emphasis on the variability of snow and ice on the earth in geological and historical past and the implications for the future of the environment.

Krimsky, S. and Golding, D. (eds), 1992. *Social Theories of Risk* (Westport, CT: Praeger)

Original articles discuss the origins of social science perspectives on risk and detail the need to broaden the debate beyond the technical issues to include the role of social choices in risk management and decision making.

Laurini, R. and Thompson, D., 1992. *Fundamentals of Spatial Information Systems* (New York: Academic Press)
Covers a variety of situations in the spatial domain, including semantics, mapping and geographical problems.

Livingstone, D., 1992. *The Geographical Tradition: Episodes in the History of a Contested Enterprise* (Oxford: Blackwell Publishers)
An account of the development of geographic thought as it was structured by its changing times.

Lo, F-C and Yeung, Y.M. (eds), 1996. *The World Cities of Pacific Asia* (Tokyo: United Nations University Press)
A United Nations University sponsored study on the adjustment of the urban system of Pacific Asia, including its global cities, to the new global economy. Most countries and world cities along the Western Pacific Rim are covered.

Lösch, A., 1954. *The Economics of Location* (New Haven, CT: Yale University Press)
A spatial extension of general equilibrium theory in economics, providing a comprehensive theoretical account of the formation of economic landscape, with special reference to central place systems. A work of rare elegance and originality.

Lowenthal, D., 1985. *The Past is a Foreign Country* (Cambridge: Cambridge University Press)
A major study of ways in which we know and recover the past with insights derived from landscape and architectural perspectives.

Macmillan, B. (ed.), 1989. *Remodelling Geography* (Oxford:Blackwell)
Essays along a broad spectrum of geographical analysis employing formal and informal modelling approaches.

Maguire, D.J., Goodchild, M.F. and Rhind, D.W. (eds), 1991. *Geographic Information Systems: Principles and Applications* (London: Longmans)
Discusses the need for awareness of the potential of GIS and accurately documents principles vital for continued progress in research in the field.

Martin, D., 1991. *Geographic Information Systems and their Socio-economic Applications* (London: Routledge)
Provides a concise, non-technical introduction to the field of geographic information systems.

Massey, D., 1984. *Spatial Divisions of Labour: Social Structures and the Geography of Production* (London: Macmillan)
An account of the way social relations and other local conditions can influence economic development and redevelopment in the form of successive layers of investment.

Massey, D., 1994. *Space, Place and Gender* (Cambridge: Polity Press)

Essays rethinking basic issues of space and place sensitive to gender differences and their social constructions.

Medyckyj-Scott, D. and Hearnshaw, H.M. (eds), 1993. *Human Factors in Geographic Information Systems* (Belhaven Press)

A collection of new contributions by leading experts in the fields of human–computer interaction and geographical information systems. Covers all the human and organizational factors that relate to the users of computer systems handling spatial data.

Meinig, D., 1986–95. *The Shaping of America. Vols I–III* (New Haven: Yale University Press)

A grand sweeping account of the settlement of the United States by Europeans from a distinctly geographical perspective.

Mitchell, B. and Draper, D., 1982. *Relevance and Ethics in Geography* (London: Longman)

An early account of the role of social relevance and ethical considerations in human geography research.

Monin, A.S. and Shishkov, Yu.A., 1979. *Istoria Klimata (History of Climate)* (Leningrad: Gidrometeoizdat)

Sets out the history of the earth's climate from the currently accepted stand of the general history of the planet. Uses palaeoclimatic reconstructions to present climate history from pre-Cambrian to the present.

Monmonier, M., 1993. *Mapping It Out: Expository Cartography for the Humanities and Social Sciences* (Chicago: University of Chicago Press)

A concise and practical book. Provides an introduction to the fundamental principles of graphic logic and design from the basics of scale to the complex mapping of movement or change. Of considerable value to writers and researchers in deciding when maps are most useful and what formats work best in a wide range of subject areas from literary criticism to sociology.

Montgomery, G.E. and Schuch, H.C., 1993. *GIS Data Conversion Handbook* (GIS World)

Draws on the extensive experiences of the authors to provide a comprehensive discussion of the multi-faceted and multi-disciplinary conversion process.

Olsson, G., 1991. *Lines of Power/Limits of Language* (Minneapolis: University of Minnesota Press)

An important, though idiosyncratic, series of essays pointing to power relations in all accounts while rejoicing in language's ambiguities.

Palm. R.I., 1990. *Natural Hazards: An Integrative Framework for Research and Planning* (Baltimore: Johns Hopkins University Press)

Offers a theoretical framework for assessing hazards and applies this to a study of earthquake hazards in California.

Pred. A., 1967. *Behaviour and Location: Foundations for a Geographic and*

Dynamic Location Theory (Lund: Lund Studies in Geography, Series B. 27)
An early attempt at the theoretical exposition of a behavioural approach to location decision making, introducing the behavioural matrix identifying information available and ability to use it.

Pred, A., 1990. *Lost Words and Lost Worlds* (Cambridge: Cambridge University Press)
An extraordinarily sensitive account relating changes in the nineteenth century world of Stockholm to parallel and reflective changes in language.

Price, M.F. and Heywood, D.I. (eds), 1994. *Mountain Environments and Geographical Information Systems* (Taylor & Francis)
Evaluates and discusses GIS use for research and management in mountain environments by looking at experiences from five continents.

Roberts, N., 1989. *The Holocene. An Environmental History* (Oxford: Blackwell)
The story of the transformation of nature over the last ten thousand years due to climatic change and human impact, and its subsequent reconstruction.

Rose, G., 1993 *Feminism and Geography: The Limits of Geographical Knowledge* (Cambridge: Polity Press)
The first text to explore the possible implications of a dominantly masculine kind of knowledge in geography.

Sassen, S., 1991. *The Global City: New York, London, Tokyo* (Princeton: Princeton University Press)
A pioneering study of three of the most important global cities based on a thematic approach, drawing theoretical implications from solid empirical data.

Sauer, C.O., 1965. *Land and Life* (Berkeley: University of California Press)
A selection from the writings of the great mid-twentieth century cultural geographer, Carl Otwin Sauer, edited by John Leighly.

Scholten, H.J. and Stillwell, J.C.H. (eds), 1990. *Geographic Information Systems for Urban and Regional Planning* (Kluwer Academic Publishers)
Contains contributions from a diverse group of international experts demonstrating the progress achieved in developing and applying GIS to urban and regional planning.

Simmons, I.G., 1989. *Changing the Face of the Earth: Culture, Environment, History* (Oxford: Blackwell)
A scholarly and comprehensive account of how the human relationship with the natural world has evolved through time.

Smith, D.M., 1977. *Human Geography: A Welfare Approach* (London: Edward Arnold)
An attempt to restructure human geography around the concept of welfare, with an emphasis on the description and explanation of spatial patterns of human well-being.

Smith, D.M., 1981. *Industrial Location: An Economic Geographical Analysis*, 2nd edition (New York: John Wiley)
A review of the main contributions to industrial location theory and an elaboration of extensions to the variable cost theory, associated with Alfred Weber, incorporating some elements of market area analysis.

Smith, D.M., 1994. *Geography and Social Justice* (Oxford: Blackwell Publishers)
A review of alternative theories of social justice and of links between geography and moral philosophy, illustrated by case studies which include the American city, eastern Europe and South Africa.

Smith, K., 1992. *Environmental Hazards: Assessing Risk and Reducing Disaster* (London: Routledge)
A good introductory text in the hazards field combining both physical and social science perspectives.

Smith, N., 1990. *Uneven Development: Nature, Capital and the Production of Space* (Oxford: Blackwell)
A broad approach to the study of uneven development, grounded in the process of capital accumulation.

Socolow, R., Andrew, C., Burkoutt, F. and Thomas, V. (eds), 1994. *Industrial Ecology and Global Change* (Cambridge: Cambridge University Press)
A collection of papers from a workshop addressing the role of industrial activity in producing global changes.

Star, J. and Estes, J., 1990. *Geographical Information Systems: An Introduction* (Englewood Cliffs: Prentice-Hall)
Emphasizes data integration as a philosophical basis for modern GIS and the role of the GIS as an appropriate tool for integrating the technologies used in gathering, analyzing and assessing spatial data.

Stren, R. (ed.), 1994. *Urban Research in the Developing World*. 4 vols (Toronto: University of Toronto, Centre for Urban and Community Studies)
Arising from a Ford Foundation sponsored study, these four volumes discuss thematic issues and investigate the peculiarities of urban development in the three continents of Africa, Asia and Latin America.

Symanski, R., 1988. *The Immoral Landscape: Female Prostitution in Western Societies* (Toronto: Butterworths)
A geographical analysis initially greeted with great antagonism that has re-emerged as a highly respected and pioneering study in sociology.

Thomas, W.L. (ed.), 1956. *Man's Role in Changing the Face of the Earth* (Chicago: University of Chicago Press)
For several decades this massive edited text, resulting from a Wenner-Gren symposium in Princeton, was the standard reference for the human transformation of the earth's surface.

Tikunov, V.S., 1985. *Modelling in Socio-economic Cartography* (Moscow: MGU), in Russian

Based on the author's own research into the development of modelling of the thematic contents of socio-economic maps.

Tomlin, C.D., 1990. *Geographic Information Systems and Cartographic Modelling* (Hemel Hempstead: Prentice-Hall)

An introduction to cartographic modelling. Emphasizes environmental decision-making applications and develops a high-level computer language presenting cartographic data, processing conventions, capabilities and techniques.

Turner, B.L., II, Clark, W.C., Kates, R.W., Richards, J.F., Matthews, J.T. and Meyer, W.B. (eds), 1990. *The Earth as Transformed by Human Action* (Cambridge: Cambridge University Press)

An up-to-date version of Thomas (1956) following a symposium at Clark University in 1987, with both thematic and regional chapters. Chronicles the global and regional changes in the biosphere over the last 300 years. A classic in the field of nature–society interactions.

Wackermann, G., 1994. *Loisir et tourisme: une internationalisation de l'espace* (Paris: SEDES)

Analyses the transformation of the local, regional and nationally-oriented recreational economy to one that is internationalized and globalized. Also shows the extent to which each branch of activity has become integrated into the general economy and its financial machinery, which demands efficiency and profitability.

Wackermann, G. 1995. *De l'espace national à la mondialisation de l'espace* (Paris: Ellipses)

Discusses the mechanisms which have driven local, regional and national socio-spatial ensembles to internationalize, and then globalize. Insists on the necessity of continuing the post-globalization trend that is already underway, which is based on taking account of regional and national identities.

Wackermann, G. 1995. *Le transport de marchandises dans l'Europe de demain* (Paris: Le Cherche-Midi)

Modal disequilibrium and new possibilities for the redistribution of the flow of merchandise call into question the traditional structuring of land, sea and air transportation. Pluri- and intermodal transportation, as well as combined transportation systems, would seem to contribute to putting some order into the chaotic state of circulation if there exists a political will that favours these developments and integrates external costs with internal ones.

Watts, M., 1983. *Silent Violence: Food, Famine, and the Peasantry in Northern Nigeria* (Berkeley: University of California Press)

An exhaustive case study on how farmers perceive their hazardous environment. Written from the standpoint of a political ecologist, the book is a seminal piece on Hausa peasants and their adjustments to hazards.

Weber, A., 1929. *Alfred Weber's Theory of the Location of Industries* (translated by C.J. Friedrich) (Chicago: University of Chicago Press) Reprinted 1971 (New York: Russell and Russell)

The starting point for industrial location theory, originally published in 1909, analysing the least cost location based on transportation, labour and agglomeration economies.

Western, J., 1981. *Outcast Cape Town* (London: Allen & Unwin)

A major case study in the tradition of humanistic geography, showing how apartheid in South Africa assigned place to the so-called coloured population of Cape Town.

White, G.F. (ed.), 1974. *Natural Hazards: Local, National, Global* (New York: Oxford University Press)

Empirical case studies of natural hazard problems in 23 countries and the choice of adjustments available to local managers.

Wijkman, A. and Timberlake, L., 1984. *Natural Hazards: Acts of God or Acts of Man* (London: Earthscan)

An analysis of natural disasters in developing countries.

Wittfogel, K.A., 1957 *Oriental Despotism: A Comparative Study of Total Power* (New Haven: Yale University Press)

A definitive study of the 'hydraulic civilizations' of China, and a comparison with those in other irrigated river valleys such as the Nile and the Indus.

Zhukov, V.T., Serbenyuk, S.N. and Tikunov, V.S., 1980. *Mathematical-Cartographic Modelling in Geography* (Moscow: Mysl.), in Russian

Discusses the unification of mathematical and cartographic models within the system of map compilation, and their use for designing or analyzing the thematic content of maps.

Selected Journals in Geography

In virtually all countries of the world, there are numerous journals and other serial publications in which articles and papers of geographical interest are published. The following is a highly selective list of learned journals that are primarily concerned with publishing geographical articles of academic interest. Serial geographical publications for purely popular consumption have not been included in the list. However, specialized publications that are of a multi-disciplinary nature and in which the geographical contributions have been noteworthy (e.g. Journal of Regional Science) have been included.

Actes de la Recherche en Sciences Sociales (France)
Annals of the Association of American Geographers (USA)
Annals of Tourism Research (USA)
Antipode: A Radical Journal of Geography (USA/UK)
Area (UK)

Business Geographics (USA)
Canadian Geographer
Disasters
Economic Geography
Environment
Environment & Planning (UK)
Geocritica
Geoforum
Geografia i Prirodnye Resursy (Geography and Natural Resources) (Russia)
Geografiska Annaler (Sweden)
Geographical Analysis (USA)
Geographical Journal
Geographical Review
Geographical Systems (Switzerland)
Geography Research Forum (Israel)
GIS Asia/Pacific (USA)
GIS Europe (Holland)
GIS Obozrenie (Russia)
GIS World (USA)
Imago Mundi
Industrial and Environmental Crisis Quarterly
International Journal of Geographical Information Systems (UK)
International Journal of Mass Emergencies and Disasters
International Journal of Urban and Regional Research
Isis
Izvestia Rossiiskoi Akademii Nauk, seria geograficheskaya (Proceedings of the Russian Academy of Sciences, geography series)
Izvestia Russkogo Geograficheskogo Obshchestva (Proceedings of the Russian Geographical Society)
Journal of Contingencies and Crisis Management
Journal of Geography in Higher Education
L'espace Geographique
Leisure Studies: Journal of the Leisure Studies Association (UK)
Mappemonde
Natural Hazards
Political Geography
Professional Geographer (USA)
Progress in Human Geography
Regional Studies
Review of International Political Economy
Risk Abstracts
Risk Analysis

Society and Space
South African Geographical Journal
Tijdschrift voor Economische en Sociale Geografie (Holland)
Transactions of the Institute of British Geographers (UK)
Urban Geography
Vestnik Moskowskogo Universiteta, ser. geogr. (Russia)

The Author

Akin L. Mabogunje is the Chairman of the Development Policy Centre, Ibadan, Nigeria and former Professor of Geography at the University of Ibadan. Between 1972 and 1980 he was Vice-President of the International Geographical Union and became President of the Union between 1980 and 1984. He is a recipient of the Nigerian National Order of Merit, the highest honour for academic and intellectual achievements in the country. His research interests include migration, urban and regional planning and, more recently, geography and public policy. His publications include *Urbanization in Nigeria* (1968), *Regional Planning and National Development in Africa* (1977), and *The Development Process: A Spatial Perspective* (1980, 1989) which is about to go into its third edition.

Contributors

Susan L. Cutter is Professor and Chair of the Department of Geography at the University of South Carolina, Columbia. She has been conducting hazards research for more than a decade, focusing primarily on technological risks as well as the differential burdens of risks and hazards on society. Dr. Cutter also directs the Geography Department's Hazard Research Laboratory.

Peter Dicken is Professor of Geography at the University of Manchester and has held visiting appointments in the United States, Canada, Australia and Asia. His research interests are: global economic change, the spatial behaviour of transnational corporations and economic change in East and South-East Asia. He has written numerous papers and books on these topics, including *Global Shift: The Internationalization of Economic Activity* (1992). He has acted as a consultant to UNCTAD.

Marcelo Escolar is Director of the Institute of Geography in the Faculty of Philosophy and Letters at the University of Buenos Aires. His principal interests are the social history of geography, political and cultural geography and the epistemology and sociology of geographical understanding. His most recent publication is *Crítica do Discurso Geográfico* (1996).

Reginald G. Golledge is Professor of Geography at the University of California Santa Barbara, Santa Barbara, California. He is interested in human spatial behaviour, knowledge acquisition, and cognition, as well as cognitive mapping, disaggregate transport modelling and geography and disability. His most recent co-authored book was *Behavior and Environment.* Forthcoming co-authored books include *Spatial Decision Making and Choice Behavior* and *Spatial and Temporal Reasoning in Geographic Information Systems.*

Peter Gould is Evan Pugh Professor of Geography at Penn State

University, Pennsylvania. He is the author and editor of seventeen books, a consistent research theme of which has been processes of spatial diffusion. One of his most recent books, *The Slow Plague*, examines, in a style accessible to a lay audience, the spread of AIDS. He has served as a consultant in many countries of Africa, Europe and North and South America.

Vladimir Kotlyakov is Director of the Institute of Geography, Russian Academy of Sciences, Moscow. He is also First Vice-President of the International Geographical Union and member of the Earth Council. He has written many books and articles on general geography and glaciology. His most recent publication is *World of Snow and Ice* (1994).

Andrew Leyshon is Reader in Geography at the University of Bristol. Together with Nigel Thrift he has been investigating geographies of money and finance for over a decade. Their most recent publication is *Money/Space: Geographies of Monetary Transformation* (1996, in press).

Neil Roberts is Reader in Physical Geography at Loughborough University. His principal research interests are in Holocene environmental change in Mediterranean and tropical ecosystems; sustainable development of environmental resources, including soil, water and vegetation management; and lake-watershed ecosystems and palaeolimnology. His most recent book publications include *The Changing Global Environment* (1994), and jointly with R. A. Butlin, *Ecological Relations in Historical Times: Human Impact and Adaptation* (1995).

Milton Santos is Professor of Geography at the University of São Paulo. He has published a number of books and articles on the theory and epistemology of geography, as well as on the urbanization of the third world. The last book he published was *Técnica, Espaço, Tempo, Globalização e meio téchnico cientifico informacional* (1994).

David M. Smith has research interests extending from industrial location theory, regional analysis and urban development to spatial inequality, social justice and moral issues in geography. He has worked extensively in the United States, Eastern Europe and South Africa. His most recent book is *Geography and Social Justice* (1994). He is Professor of Geography at Queen Mary and Westfield College, University of London.

Nigel Thrift is Professor of Geography at the University of Bristol. His research interests include the social and cultural determinants of the international financial system, geographies of financial exclusion, social and cultural theory and countries of the Pacific Basin. His most recent publications include *Money, Power and Space* (1994), and jointly with Andrew Leyshon, *Money/Space: Geographies of Monetary Transformation* (1996, in press).

Vladimir Sergeevitch Tikunov is Professor of Cartography and Geoinformatics at the Lomonosov University, Moscow. His scientific interests include geographical information systems (GIS), cartography and socio-economic geography. His most recent monograph is *Geoinformatics* (1993), in collaboration with A.V. Koshkarev. He is a member of the GIS Commission of the International Geographical Union, and a corresponding member of the Russian Academy of Natural Sciences.

Gabriel Wackermann is Professor of Land-Use Planning and Urban Development at the Institute of Geography of the Sorbonne, and Director of the Laboratoire de recherches internationales en transports et échanges (Paris). He is the author of many publications, dealing in particular with land-use planning, transport, tourism and international trade. His most recent book is *De l'espace national à la mondialisation*, 1995. His *Géopolitique et dynamique* will be published in 1996.

Yue-man Yeung is Professor of Geography and Director of the Hong Kong Institute of Asia-Pacific Studies at the Chinese University of Hong Kong. He has worked in Singapore and Canada as well as in Hong Kong. His latest edited volumes are: *Pacific Asia in the 21st Century* (1993), and with David Chu, *Guangdong* (1994). Another edited volume, with F. C. Lo, *Emerging World Cities in Pacific Asia*, appeared in 1995.

Index

Note: Page references in italics refer to figures or tables.